11th International Conference on Advanced Manufacturing Technologies (ICAMaT 2020)

Held online due to COVID-19

IOP Conference Series: Materials Science and Engineering
Volume 1018

Bucharest, Romania
29 - 30 October 2020

ISBN: 978-1-7138-3322-2
ISSN: 1757-8981

Printed from e-media with permission by:

Curran Associates, Inc.
57 Morehouse Lane
Red Hook, NY 12571

Some format issues inherent in the e-media version may also appear in this print version.

Institute of Physics
Dirac House, Temple Back
Bristol BS1 6BE UK

Phone: 44 1 17 929 7481
Fax: 44 1 17 920 0979

techtracking@iop.org

Additional copies of this publication are available from:

Curran Associates, Inc.
57 Morehouse Lane
Red Hook, NY 12571 USA
Phone: 845-758-0400
Fax: 845-758-2633
Email: curran@proceedings.com
Web: www.proceedings.com

TABLE OF CONTENTS

PRODUCT DEVELOPMENT

PROCESSES, TOOLS AND EQUIPMENT

IOP Conf. Series: Materials Science and Engineering **1018**(2021) 011001 doi:10.1088/1757-899X/1018/1/011001

The 11th **International Conference of Advanced Manufacturing Technologies, ICAMaT 2020**, was organized by the Manufacturing Engineering Department (TCM) from University POLITEHNICA of Bucharest, Romania, Faculty of Industrial Engineering and Robotics, on 29th of October 2020, with the remarkable participation of experts from Algeria, Austria, Republic of Moldova, Russia, Serbia and Ukraine. The organizers were happy to welcome participants from 21 academic and research institutions - which include 15 universities, four research centers and two companies.

The first ICAMaT Conference was held in 1999 under the acronym CITAF (derived from the abbreviation in Romanian) at the initiative of the Manufacturing Engineering Department (TCM) and had a national character. The following Conference sessions were held almost every two years and attracted increasingly more specialists.

Since 2005, with the clear prospect of Romania's accession to the European Union, the conference takes on an international character and changes its name to ICAMaT - International Conference on Advanced Manufacturing Technologies. Ever since this Conference session there has been a continuous increase in the number of participants from home and abroad, as well as in the number of published papers.

The Conference successfully brought together engineers, researchers and professors, specialists in various fields, such as Product Development in areas such as medicine, transports, with a focus on railway vehicles, oil and gas and other engineering fields; Materials; Cutting and Welding Processes, Cutting Tools and Forming Technologies; Manufacturing Technologies; Technological Equipment and Automation; Management, Marketing, Logistics, Learning Factories and Industry 4.0.

44 papers were initially entered in the conference and, from the 36 selected to be reviewed, 30 of them were accepted and presented. The Editors are thankful to the authors for their valuable contributions and to the reviewers for their timely reviews and recommendations as improvement guidance for selected papers. We also thank the members of the Organizing Committee, for the consistent and substantial effort of organizing this conference.

Due to the 2020 pandemic, the ICAMaT 2020 conference format had to be changed into a virtual one, so the conference was organized completely online. It was a decision taken not only for maintaining the desired annual frequency of the event, but also because of looking at this way as to a challenge for both the organizers and the presenting authors.

The organizers used the Microsoft Teams platform belonging to the University POLITEHNICA of Bucharest, where a Teams group, named ICAMaT 2020, was created. A link to the group was sent to all participants using the EasyChair mailing list and was also displayed on the conference website. Specific channels were created for the general announcements, for testing the connectivity and for the three conference sections. The links to the sections' channels were also mentioned in the conference program which was uploaded on the conference website. Invitations for joining the team, using the Microsoft Teams specific tools, were sent to all participants several days before the conference's start. A total of 107 participants are registered in the conference Teams group.

IOP Conf. Series: Materials Science and Engineering **1018**(2021) 011001 doi:10.1088/1757-899X/1018/1/011001

The conference took place on October 29th, 2020, starting with the organizers' short welcome speeches, attended by 34 participants. The organizers and the sections' chairs connected either from their homes or from their university offices in Bucharest, Romania. Participants connected from other Romanian cities and from Algeria, Austria, Republic of Moldova, Russia, Serbia, Ukraine.

The papers were presented using the three sections' channels. It was recommended for each paper to be presented in a maximum of 15 minutes, but it was up to the sections' chairs to allow for longer presentations. Questions and comments followed the presentation of each paper.

All the presentations and discussions were recorded and are available on the Teams platform for all participants. While in Section 1 (25 attendees, more than three hours) there were two recordings with a total of 160 minutes for nine papers, separate recordings were made for the other two sections: 30 attendees, more than four hours, 194 recorded minutes for 11 papers in Section 2 and 15 attendees, two and a half hours, 136 recorded minutes for six papers in Section 3.

Authors of two papers from Section 2 and two papers from Section 3 were not able to participate and to present in due time. The authors were asked to record and send papers' presentation and the recordings were uploaded on the respective conference channels.

There were no major technical problems during the conference, the participants were able to attend and understand the presentations and the discussions took place without altering the dialogue due to communication problems. It can be considered that both the authors and the other attendees succeeded to gather useful information from their participation.

Special thanks to all the participants, to the 24 reviewers who brought their essential contribution to the papers' quality and to the sections' chairs who succeeded to create a friendly but rigorous atmosphere during the papers' presentations.

Manuela - Roxana DIJMĂRESCU, Tom SAVU, Ioan - Cristian TARBĂ - Editors

assoc.prof. Gabriela Adriana PLAIASU, Univ. of Pitesti, Romania

prof. Hanan SALEET, Applied Science Private University, Amman, Jordan

prof. Tom SAVU, Univ. POLITEHNICA of Bucharest, Romania

assoc.prof. Jeremie SCHUTZ, Université de Lorraine - Metz, France

prof. Elena SCUTELNICU, Univ. "Dunarea de Jos" of Galati, Romania

prof. Hongwu SONG, Institute of Metal Research, Chinese Academy of Science, Shenyang, China

assoc.prof. Paulina SPANU, Univ. POLITEHNICA of Bucharest, Romania

prof. Nicolae-Doru STANESCU, Univ. of Pitesti, Romania

lect. Ioan - Cristian TARBA, Univ. POLITEHNICA of Bucharest, Romania

lect. Petre Gheorghe TIRIPLICA, Univ. POLITEHNICA of Bucharest, Romania

Reviewers:

assoc.prof. Bogdan ABAZA, Univ. POLITEHNICA of Bucharest, Romania

assoc.prof. Ovidiu ALUPEI, Univ. POLITEHNICA of Bucharest, Romania

prof Alexandra BANU, Univ. POLITEHNICA of Bucharest, Romania

assoc.prof. Diana Irinel BĂILĂ, Univ. POLITEHNICA of Bucharest, Romania

lect. Nadia BELU, Univ. of Pitesti, Romania

assoc.prof. Madalin CATANA, Univ. POLITEHNICA of Bucharest, Romania

dr. Saša ĆUKOVIĆ, University of Kragujevac, Serbia

lect. Manuela - Roxana DIJMARESCU, Univ. POLITEHNICA of Bucharest, Romania

lect. Maria - Cristina DIJMARESCU, Univ. POLITEHNICA of Bucharest, Romania

prof. Andrei DUMITRESCU, Univ. POLITEHNICA of Bucharest, Romania

lect. Ana Cornelia GAVRILUTA, Univ. of Pitesti, Romani

prof. Marian GHEORGHE, Univ. POLITEHNICA of Bucharest, Romania

assoc.prof. Ionut GHIONEA, Univ. POLITEHNICA of Bucharest, Romania prof. Nicolae IONESCU, Univ. POLITEHNICA of Bucharest, Romania

prof. Monica IORDACHE, Univ. of Pitesti, Romania

assoc.prof. Lucian LAZARESCU, Technical Univ. of Cluj-Napoca, Romania

prof. Eduard Laurențiu NIȚU, Univ. of Pitesti, Romania

assoc.prof. Razvan-Ioan PACURAR, Technical Univ. of Cluj-Napoca, Romania

assoc.prof. Gabriela Adriana PLAIASU, Univ. of Pitesti, Romania

prof. Tom SAVU, Univ. POLITEHNICA of Bucharest, Romania

prof. Elena SCUTELNICU, Univ. "Dunarea de Jos" of Galati, Romania

prof. Nicolae-Doru STANESCU, Univ. of Pitesti, Romania

lect. Ioan - Cristian TARBA, Univ. POLITEHNICA of Bucharest, Romania

lect. Petre Gheorghe TIRIPLICA, Univ. POLITEHNICA of Bucharest, Romania

IOP Conf. Series: Materials Science and Engineering **1018** (2021) 011002 doi:10.1088/1757-899X/1018/1/011002

Peer review declaration

All papers published in this volume of *IOP Conference Series: Materials Science and Engineering* have been peer reviewed through processes administered by the Editors. Reviews were conducted by expert referees to the professional and scientific standards expected of a proceedings journal published by IOP Publishing.

- **Type of peer review:** **Single-blind**
- **Conference submission management system:** **EasyChair** https://easychair.org/my/conference?conf=icamat20200#
- **Number of submissions received:** **44**
- **Number of submissions sent for review:** **36**
- **Number of submissions accepted:** **30**
- **Acceptance Rate (Number of Submissions Accepted / Number of Submissions Received X 100):** **68.2 %**
- **Average number of reviews per paper:** **2**
- **Total number of reviewers involved:** **24**
- **Any additional info on review process:**
- **Contact person for queries:** **Tom SAVU** (tom.savu@upb.ro)

Effects analysis of vertical track irregularities on bogie vibration - method based on bogie modelling and wheelsets accelerations measurement

M Dumitriu[1], D Fologea[2], I C Cruceanu[2]

[1] Department of Railway Vehicles, University Politehnica of Bucharest, 313 Splaiul Independenței, Bucharest, Romania
[2] Doctoral School of Transports, University Politehnica of Bucharest, 313 Splaiul Independenței, Bucharest, Romania

E-mail: madalinadumitriu@yahoo.com

Abstract. The paper analyzes the effect of the vertical track irregularities upon the vibrations in a two-axle bogie. The method in use implies the bogie modelling and the measurement of the vibrations in wheelsets generated by the track vertical irregularities. The behaviour of vibrations in the bogie is evaluated via the results derived from applications of numerical simulation developed on the basis of the bogie model, where the inputs are the accelerations measured in the two wheelsets. The vibration features in the bogie are pointed out at according to the spectra of accelerations in the bogie frame, to the standard deviation values of accelerations and to the maximum values of the accelerations calculated against the two wheelsets.

1. Introduction

The main cause of the vibrations of railway vehicles is found in the vehicle-track interaction [1]. Due to imperfections construction, exploitation, change in the infrastructure due to the action of the environment factors or soil movements, the track features as a whole, on the one hand, a number of deviations from the ideal geometrical shape and, on the other hand, defects of the rolling surfaces of the rails [2]. All the above, along with the constructive discontinuities of the rail, constitute major reasons for the vehicle vibrations [3, 4]. At the wheel, the defects in the rolling surface such as eccentricity, ovality, poligonal profile, corrugation, or flattening are also causes for the railway vehicle vibrations [5 - 10].

Running on a track with irregularities will generate vibrations of the wheelsets [11, 12], which are transmitted to the bogies and the carbody [13, 14], so that the dynamic response of the entire vehicle is affected by the track irregularities [15 - 19].

The issues of vibrations in the railway vehicles can be approached both theoretically and experimentally. The theoretical approaches are based on numerical simulations, developed on a vehicle-track system model [19, 20]. Since the railway vehicle is a complex oscillating system, with specific vibration characteristics, it is rather difficult to produce a model of a high degree of reliability and trust that would accurately feature all the factors influencing the vibrations behaviour in the vehicle. On the other hand, the more complex the model, the closer the results are to reality, but it will be complicated to draw some general conclusions regarding the vibration characteristics of the vehicle.

In general, the models used to study the vibrations in a railway vehicle are derived from a simplified representation of the vehicle-track system, which considers the most important factors affecting the behaviour of vibrations in vehicle.

The track irregularities are primary inputs in the software applications of simulating the behaviour of vibrations in the railway vehicles. There are two distinct approaches to include the track irregularities in the numerical models of the vehicle-track system. One is to consider the track irregularities as absolute values, derived from measurements with special vehicles. The other one uses a statistical representation of the track irregularities. This representation has to contain information on the wavelengths, as well as the amplitudes of the track irregularities. A complete description of the track geometry can be given by the power spectral density of the measured track irregularities. The power spectral density can be used either directly as an input for power spectra analysis or they can be retransformed into track irregularities as function of distance [22].

The paper examines the effect of the vertical track irregularities upon the vibrations in a two-axle bogie, by using the results from numerical simulations. The software applications of numerical simulation are developed on basis of a bogie model taking for the rigid modes of vibration in vertical plan - bounce and pitch, where the excitations generated by the vertical track irregularities are transmitted to the bogie through the two wheelsets. The primary inputs in the software applications of the behaviour of vibrations in the bogie are the accelerations measured in the axle boxes in a Minden-Deutz bogie during circulation at a constant velocity on a track section in alignment and crosslevel. The characteristics of the behaviour of vibrations in the bogie are examined based on the spectra of the accelerations in the bogie frame in two points located against the two wheelsets. Similarly, the values of the standard deviation of accelerations (STD accelerations) and of the maximum accelerations in the bogie frame in the two mentioned points are analyzed.

2. Mechanical bogie model and motion equations

To study the effect of the vertical track irregularities upon the behaviour of vibrations in the two-axle bogie, is considered a mechanical model shown in figure 1. The bogie model consists of 3 rigid bodies to help with modelling the bogie and the two wheelsets connected via Kelvin-Voigt type systems that model the suspension in each wheelset. The elastic element of the suspension corresponding to an wheelset has the constant $2k$, while the damping element has the constant $2c$. The vertical displacements of the wheelsets imposed by the vertical track irregularities are noted with $w_{1,2}$.

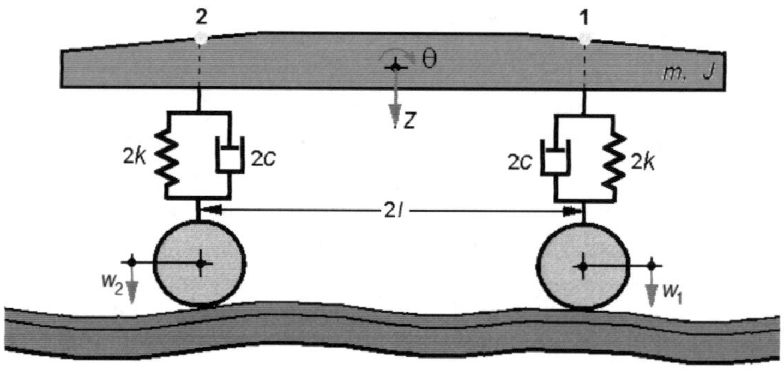

Figure 1. The mechanical model of the bogie.

The modes of vibration of the bogie in a vertical plan – bounce z and pitch θ are being taken into account. The bogie parameters are as such: m – mass, $2l$ - wheelbase, J – moment of inertia. The values of these parameters, as well as the values of the elastic and damping constants are featured in table 1.

IOP Conf. Series: Materials Science and Engineering **1018** (2021) 012001 doi:10.1088/1757-899X/1018/1/012001

Table 1. The parameters of the numerical bogie model.

Bogie mass	$m = 2700$ kg
Wheelbase	$2l = 2.5$ m
Inertia moment	$J = 1.73 \cdot 10^3$ kg·m^2
Elastic constant of the suspension corresponding to a wheel	$k = 0.616$ MN/m
Damping constant of the suspension corresponding to a wheel	$c = 9.05$ kNs/m

The bogie motion equations are as below:

$$m\ddot{z} = F_1 + F_2, \tag{1}$$

$$J\ddot{\theta} = l(F_1 - F_2), \tag{2}$$

where F_1 and F_2 stand for the forces due to the suspension,

$$F_{1,2} = -2(\dot{z} \pm l\dot{\theta} - \dot{w}_{1,2}) - 2k(z \pm l\theta - w_{1,2}). \tag{3}$$

Upon replacing the equations (3) in (1) and (2), the motion equations of the bogie are as follows:

$$m\ddot{z} + 4c\dot{z} + 4kz = 2c(\dot{w}_1 + \dot{w}_2) + 2k(w_1 + w_2), \tag{4}$$

$$J\ddot{\theta} + 4cl^2\dot{\theta} + 4kl^2\theta = 2cl(\dot{w}_1 - \dot{w}_2) + 2kl(w_1 - w_2). \tag{5}$$

3. Calculation of the bogie accelerations

While taking into account the permanent harmonic behaviour of vibrations and the complex quantities associated to real ones, the harmonic functions describing the motions in vertical plan of the wheelsets and bogie are shown below

$$\overline{w}_{1,2}(t) = \overline{w}_{1,2}e^{i\omega t}, \tag{6}$$

$$\overline{z}(t) = \overline{Z}e^{i\omega t}, \quad \overline{\theta}(t) = \overline{\Theta}e^{i\omega t}, \tag{7}$$

where ω is the angular frequency of the excitation, $\overline{w}_{1,2}$ represents the amplitudes of the vertical displacements of wheelsets, while $\overline{Z}$ and $\overline{\Theta}$ stand for the bounce and pitch motions of the bogie. The last motions are as such

$$\overline{Z} = Ze^{i\varphi_z}, \quad \overline{\Theta} = \theta e^{i\varphi_\theta}, \tag{8}$$

where φ_z and φ_θ are the phase shifts.

The accelerations for the wheelsets are as below

$$\overline{a}_{w_{1,2}}(t) = -\omega^2\overline{w}_{1,2}e^{i\omega t} = \overline{a}_{w_{1,2}}e^{i\omega t}, \tag{9}$$

where $\vec{a}_{w_{1,2}}$ represents the amplitudes of the accelerations of the wheelsets.

Upon introducing the equations (6) and (7) in the bogie motion equations, they become

$$(-\omega^2 m + 4i\omega c + 4k)\overline{Z} = 2(i\omega c + k)(\overline{w}_1 + \overline{w}_2), \tag{10}$$

$$(-\omega^2 J + 4i\omega c l^2 + 4kl^2)\overline{\Theta} = 2(i\omega c l + kl)(\overline{w}_1 - \overline{w}_2)\,. \tag{11}$$

The equations (10) and (11) help with determining the amplitudes of the accelerations in the bounce and pitch motions of the bogie,

$$\overline{a}_z = -\omega^2 \overline{Z} = \frac{2(i\omega c + k)}{-\omega^2 m + 4i\omega c + 4k}\overline{a}_w^+\,, \tag{12}$$

$$\overline{a}_\theta = -\omega^2 l\overline{\Theta} = \frac{2l^2(i\omega c + k)}{-\omega^2 J + 4i\omega c l^2 + 4kl^2}\overline{a}_w^-\,, \tag{13}$$

with the following notations

$$\overline{a}_w^+ = \overline{a}_{w_1} + \overline{a}_{w_2}\,,\quad \overline{a}_w^- = \overline{a}_{w_1} - \overline{a}_{w_2}\,. \tag{14}$$

The amplitudes of the bogie frame accelerations against the two wheelsets, namely in the points 1 and 2 (see figure 1), are calculated from the equations

$$\overline{a}_{1,2} = \overline{H}^+\overline{a}_w^+ \pm \overline{H}^-\overline{a}_w^-\,, \tag{15}$$

where, $$\overline{H}^+ = \frac{2(i\omega c + k)}{-\omega^2 m + 4i\omega c + 4k}\,,\quad \overline{H}^- = \frac{2l^2(i\omega c + k)}{-\omega^2 J + 4i\omega c l^2 + 4kl^2}\,. \tag{16}$$

4. Measurement of the wheelsets accelerations

The method proposed to analyze the effect of the track irregularities upon the bogie vibrations involves the measurement of the accelerations in the wheelsets. They represent the inputs in the applications of numerical simulation of the bogie vibration behaviour.

The accelerations of the wheelsets featured in the paper herein were measured in the Minden-Deutz bogie of a passenger train for long and average lead traffic, during running on a current track, on a section in alignment and crosslevel. The maximum constructive speed of the vehicle is 140 km/h.

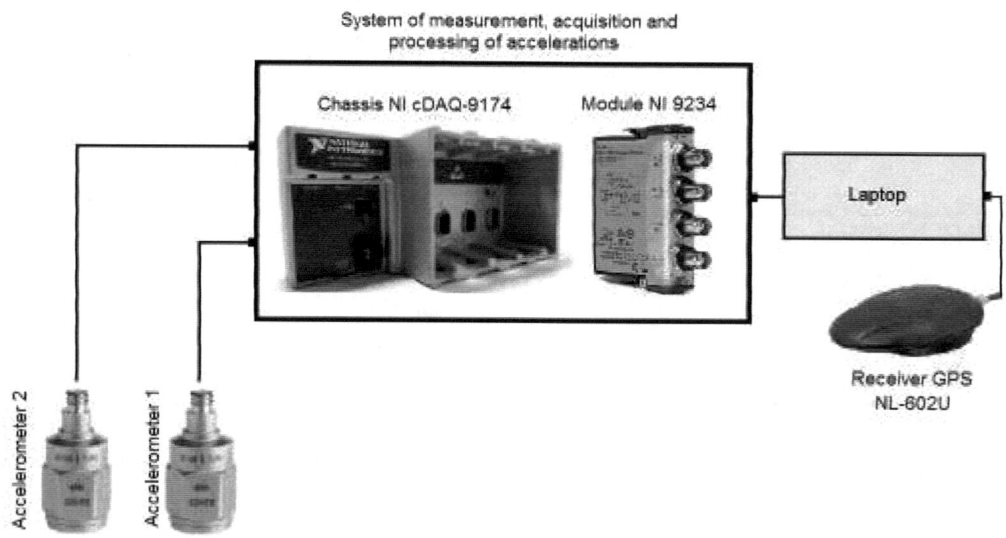

Figure 2. Scheme of the experimental measurement chain.

To perform the experimental determinations, a measurement chain was formed in which they were integrated the components of the measurement, acquisition and processing system – namely two 4514

IOP Conf. Series: Materials Science and Engineering **1018** (2021) 012001 doi:10.1088/1757-899X/1018/1/012001

Brüel & Kjær piezoelectric accelerometers and the set of the NI cDAQ-9174 chassis for data acquisition and the NI 9234 module for acquisition and synthesizing the data from accelerometers, and the NL-602U type GPS receiver for monitoring and recording the vehicle velocity (figure 2).

The accelerometers were installed on a side of the bogie, with one accelerometer on each axle box. The position of the two accelerometers is shown in figure 3.

Figure 3. Mounting the accelerometers on the axle boxes.

Recordings of the wheelsets accelerations at constant speed 137 km/h for sequences of 20 seconds at the sampling frequency of 2048 Hz were made. For instance, figure 4 shows the accelerations recorded at the two wheelsets for a measurement sequence.

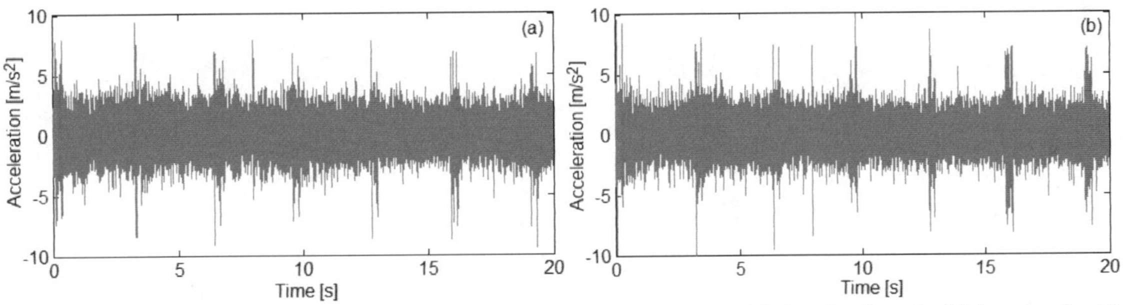

Figure 4. Accelerations recorded during a measurement sequence: (a) in wheelset 1; (b) in wheelset 2.

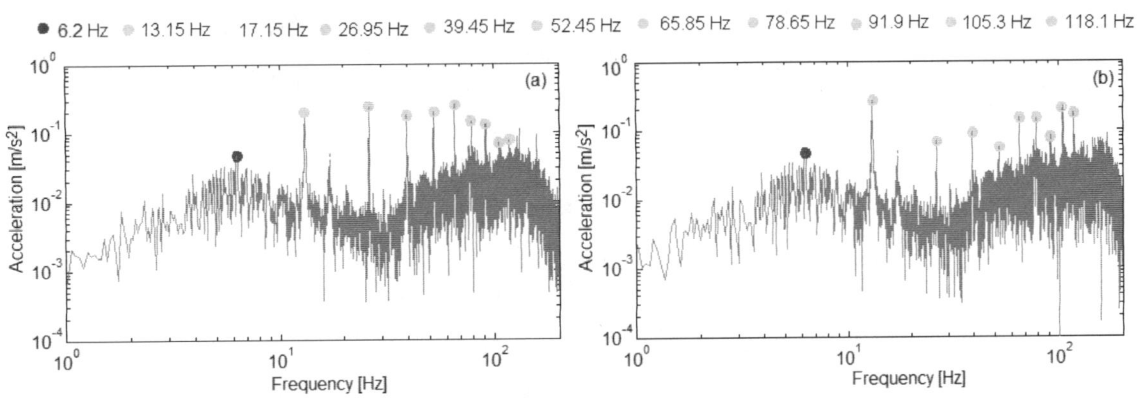

Figure 5. Spectra of the wheelsets accelerations recorded during a measurement sequence: (a) in wheelset 1; (b) in wheelset 2.

Figure 5 exhibits the spectra of the accelerations measured in the two wheelsets for a measurement sequence, in the 1-200 Hz frequency range. The spectra show more peaks between 6.2 and 150 Hz, where the frequency peak at 6.4 Hz corresponds to the resonance frequency of the bogie bounce. For this interval, several local more visible peaks can be seen, in an arithmetic progression. The first peak corresponds with the frequency of 13.15 Hz, while the other peaks have the following frequencies 26.25, 39.45, 52.45, 65.85, 78.65, 91.9, 105.3 and 118.1 Hz. The first peak comes from the wheel eccentricity, the second from its ovality and from the third to the ninth degree from the undulatory wear of the wheel rolling surface. Another peak is seen at 17.15 Hz, which matches the wavelength of 2.2 m of the irregularities of the rail rolling surface, meaning a long-wave rail undulatory wear.

5. Analysis of the vibration behaviour of the bogie

This section deals with the results derived from the numerical simulations in regards to the behaviour of vibrations in the bogie. The numerical simulation applications are developed on basis of the model in section 2, where the inputs are the wheelsets accelerations measured at velocity 137 km/h for 20 measurement sequences. To this purpose, the spectral components of the wheelsets accelerations contained within the frequency range corresponding to the bogie vibration rigid modes in a vertical plan will be considered, namely 0 – 10 Hz (see figure 6). The spectra of the wheelsets accelerations in figure 6 highlight the peaks corresponding to the resonance frequency of the bogie bounce (at 6.2 Hz) and the resonance frequency of the bogie pitch (at 8.9 Hz).

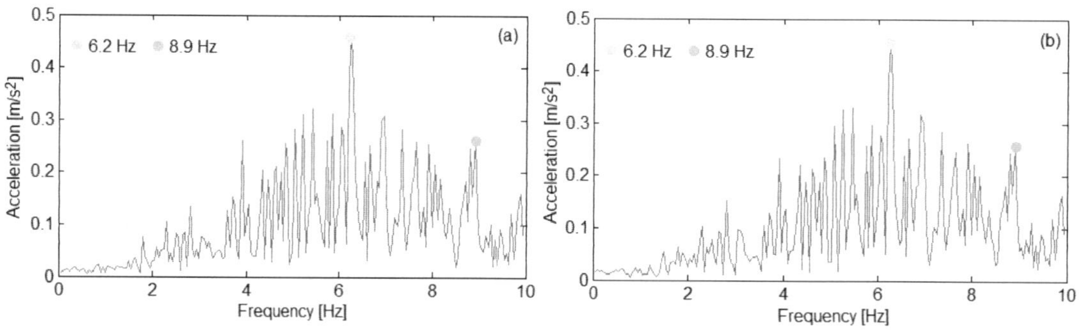

Figure 6. Spectra of the wheelsets accelerations in the frequency range 0 - 10 Hz recorded during a measurement sequence: (a) in wheelset 1; (b) in wheelset 2.

Figure 7 features the values for the STD accelerations of the two wheelsets for the 20 measurement sequences, calculated in the frequency interval of 0 – 10 Hz. The spreading values of the STD accelerations between 0.73 m/s^2 and 1.35 m/s^2 is due to the amplitude variability of the vertical track irregularities alongside. The maximum values of the amplitudes in the wheelsets accelerations are also scattered, due to the amplitude variability of the track defects alongside in the interval between 0.18 m/s^2 and 0.47 m/s^2 (see figure 8).

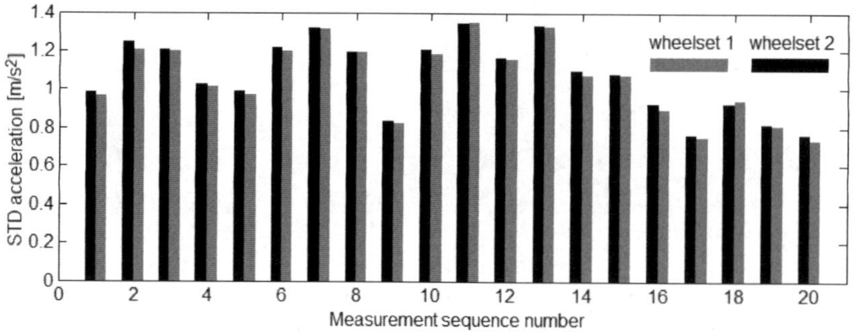

Figure 7. STD wheelsets accelerations for 20 measurement sequences.

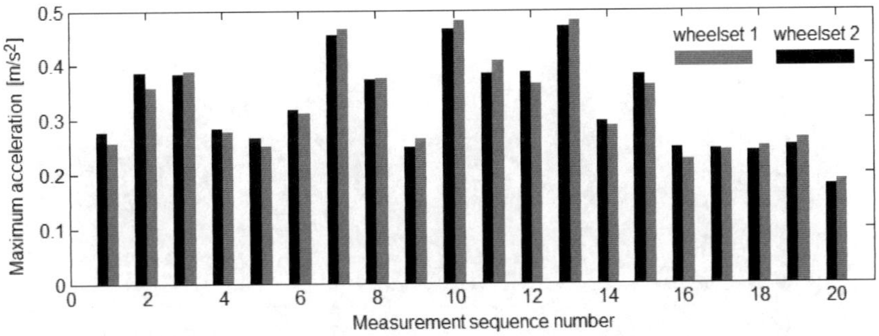

Figure 8. Maximum amplitude of the wheelsets accelerations for 20 measurement sequences.

Figure 9 shows the spectra of the accelerations of the bogie frame against the two wheelsets, namely in the points 1 and 2, within the frequency range of 0 to 10 Hz. It stands out the peak corresponding to the resonance frequency for the bogie bounce (at 6.2 Hz) and the peak for the resonance frequency in the bogie pitch (at 8.9 Hz). The amplitudes of the accelerations against these peaks are noticed to be higher than the amplitudes of the accelerations measured in the wheelsets (see figure 6). For instance, the acceleration in the bogie frame in point 1 is 0.48 m/s^2, at 6.2 Hz frequency, while the acceleration measured in the front wheelset is 0.46 m/s^2; for 8.9 Hz frequency, the acceleration in the bogie frame in point 1 is 0.53 m/s^2, whereas the acceleration measured in the front wheelset is 0.26 m/s^2. This can be explained by the resonant behaviour of the bogie.

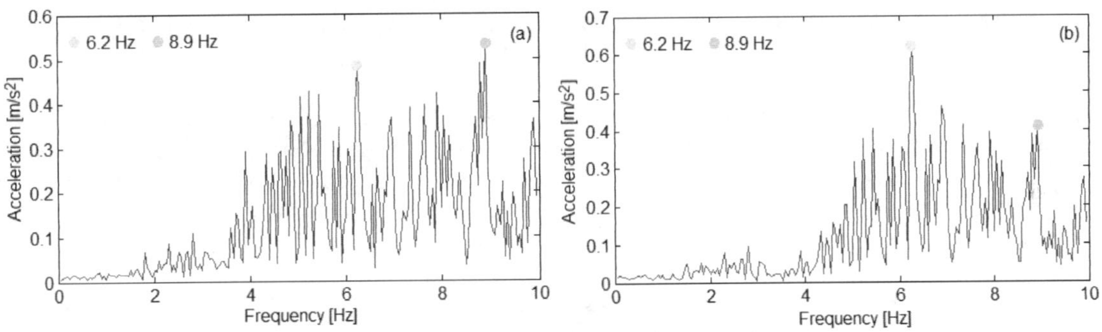

Figure 9. Spectra of the bogie frame accelerations corresponding to a wheelsets accelerations during a measurement sequence: (a) in point 1; (b) in point 2.

The variability of the amplitude in the vertical track irregularities is also reflected in the value of the accelerations in the bogie frame, calculated in points 1 and 2. This is highlighted both on the basis of the STD accelerations values (see figure 10) and on the basis of the maximum values of the bogie frame accelerations (see figure 11). The values of STD accelerations calculated in point 1 range from 1.18 m/s^2 to 1.83 m/s^2, while in point 2 these values are scattered between 1.07 m/s^2 and 1.70 m/s^2. The maximum values of the accelerations in point 1 in the bogie frame are included in the interval of 0.35 m/s^2 and 0.83 m/s^2, and the values in point 2 are between 0.29 m/s^2 and 0.64 m/s^2.

The vibration regime of the bogie frame is not symmetrical in the two points located against the two wheelsets. It is noted that STD accelerations in point 1 are higher than STD accelerations in point 2 and that the differences range from 1.44% to 20%. As for the values of the maximum accelerations in the bogie frame, the maximum acceleration is generally higher in point 1. The differences between the maximum accelerations in the two points in the bogie frame range from 1.12% to 29.56%.

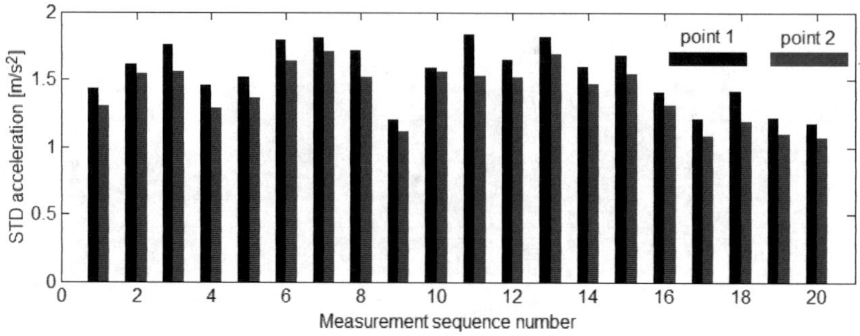

Figure 10. STD bogie frame accelerations in points 1 and 2.

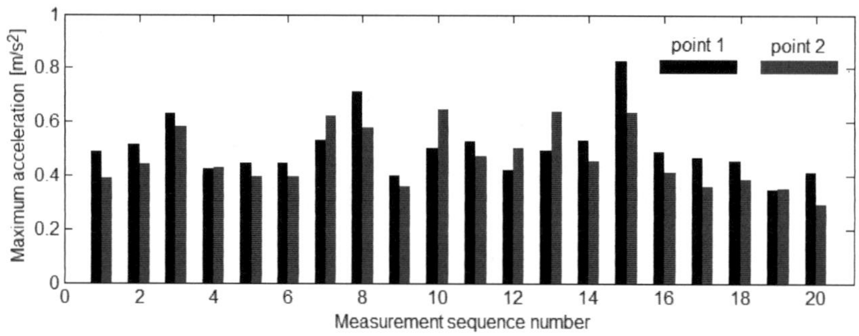

Figure 11. Maximum amplitude of the bogie frame accelerations in points 1 and 2.

6. Conclusions

The analysis in the paper herein has focused on the effect of the vertical track irregularities, transmitted through the wheelsets, upon the vibrations in a bogie for passenger railway vehicles. The results under study were derived from numerical simulations developed on basis of a bogie model that considers the rigid vibration modes - bounce and pitch. The inputs in the software applications of numerical simulation are the accelerations measured in the axle boxes of a Minden-Deutz type bogie during circulation at a constant velocity on a track section in alignment and crosslevel.

The analysis of the STD accelerations values and of the maximum accelerations calculated against the two axles highlighted: the influence of the amplitude variability of the vertical track irregularities alongside upon the scatter of the values in the bogie accelerations; asymmetry of the behaviour of vibrations in the bogie frame against the two wheelsets; within the frequency range of 0 – 10 Hz, the accelerations in the bogie frame are higher than in the wheelsets, due to the resonance in the bounce and pitch vibrations.

Future directions of this research will be directed on the effect of the vertical track irregularities on the carbody vibration and the ride comfort.

Acknowledgments

The activity of Dorina Fologea, PhD student, in this work has been funded by the European Social Fund from the Sectorial Operational Programme Human Capital 2014-2020, through the Financial Agreement with the title "Scholarships for entrepreneurial education among doctoral students and postdoctoral researchers (Be Entrepreneur!)", Contract no. 51680/09.07.2019 - SMIS code: 124539.

References

[1] Xu L, Zhai W, Gao J, Meacci M, 2017 On effects of track random irregularities on random vibrations of vehicle-track interactions, Probabilistic Engineering Mechanics **50**, pp 25-35.
[2] Pombo J, Ambrósio J 2012 An alternative method to include track irregularities in railway vehicle dynamic analyses, Nonlinear Dynamics **68**, 1-2, pp 161-176.

[3] Cheli F, Corradi R 2011 On rail vehicle vibrations induced by track unevenness: Analysis of the excitation mechanism, Journal of Sound and Vibration, **330**, pp 3744–3765

[4] Lei Y, Tian X, Qi F, Chen D, Tian T 2016 Vertical track irregularity influence on the wheel high-frequency vibration in wheel-rail system, Mathematical Problems in Engineering, Article ID 5082319

[5] Song Y, Liang L, Yanliang Du, Sun B 2020 Railway polygonized wheel detection based on numerical time-frequency analysis of axle-box acceleration, Applied Sciences **10**, 1613

[6] Shan W, Wu P, Wu X, Zhang F, Shi H 2019 Effect of wheel polygonization on the axle box vibrating and bolt self-loosening of high-speed trains, IOP Conf. Series: Journal of Physics: Conf. Series **1213**, Article number 052044

[7] Wang Z, Mei G, Zhang W, Cheng Y, Zou H, Huang G, Li F 2018 Effects of polygonal wear of wheels on the dynamic performance of the gearbox housing of a high-speed train, Proceedings of the Institution of Mechanical Engineers, Part F: Journal of Rail and Rapid Transit 232, 6, pp 1852-1863.

[8] Mazilu T 2016 Geometric model of a railway wheel with irregular contour, Advances in Intelligent Systems and Computing, **356**, pp 155-166

[9] Mazilu T, Dumitriu M, Tudorache C, Sebesan M 2011 Wheel/rail interaction due to the polygonal wheel, UPB Scientific Bulletin, Series D, Mechanical Engineering, 3, pp. 95-108

[10] Mazilu T 2007 A dynamic model for the impact between the wheel flat and rail, UPB Scientific Bulletin Series D: Mechanical Engineering **69**, 2, pp 45-58

[11] Ágh C 2019 Comparative analysis of axlebox accelerations in correlation with track geometry irregularities, Acta Technica Jaurinensis **12**, 2, pp 161-177

[12] Dumitriu M, Sebeşan I 2015 On the wheelset vibration due to the stochastic track vertical irregularities, Applied Mechanics and Materials, **809-810**, pp 1037-1042

[13] Dumitriu M, Cruceanu I C 2019 Effect of vertical track irregularities on the vibration of railway bogie, UPB Scientific Bulletin, Series D: Mechanical Engineering **81**, 1, pp 67-78

[14] Dumitriu M, Fologea D 2019 Numerical analysis on the correlation between bogie dynamic response and vertical track irregularities, UPB Scientific Bulletin, Series D: Mechanical Engineering **81**, 4, pp 99-110

[15] Karis T, Berg M, Stichel S 2020 Analysing the correlation between vehicle responses and track irregularities using dynamic simulations and measurements, Proceedings of the Institution of Mechanical Engineers, Part F: Journal of Rail Rapid Transit **234**, 2, pp 170-182

[16] Sadeghi J, Khajehdezfuly A, Heydari H, Askarinejad H 2020 Incorporating track geometry and rolling stock conditions, Journal Transportation Engineering Part A: Systems **146**, 3, Article number 04020006

[17] Karis T, Berg M, Stichel S, Li M, Thomas D, Dirks B 2018 Correlation of track irregularities and vehicle responses based on measured data, Vehicle System Dynamics **56**, 6, pp. 967-981

[18] Dumitriu M 2015 Analysis of the dynamic response in the railway vehicles to the track vertical irregularities. Part II: The numerical analysis, Journal of Engineering Science and Technology Review, **8**, 4, pp 32 - 39

[19] Sadeghi J, Rabiee S, Khajehdezfuly A 2019, Effect of rail irregularities on ride comfort of train moving over ballast-less tracks, International Journal of Structural Stability and Dynamics **19**, Article number 1950060

[20] Evans J, Berg M 2009 Challenges in simulation of rail vehicle dynamics, Vehicle System Dynamics, **47**, pp. 1023–1048

[21] Dumitriu M 2013 Modeling of railway vehicles for virtual homologation from dynamic behavior perspective, Applied Mechanics and Materials, **371**, pp 647-651.

[22] Dumitriu M 2016 Numerical synthesis of the track alignment and applications. Part I: The synthesis method, Transport Problems **11**, 1, pp. 19 – 28

Design and closed-loop control of a piezoelectric actuator

C Rusu, S Besoiu and M O Tatar

Department of Mechatronics and Machine Dynamics, Technical University of Cluj-Napoca, 28, Memorandumului Ave, 400114, Cluj-Napoca, Romania.

e-mail: sorin.besoiu@mdm.utcluj.ro

Abstract. Piezoelectric elements can produce a high displacement resolution with high force outputs but exhibits large hysteresis nonlinearities. An efficient way to compensate for nonlinearities is using a closed loop control technique. Another limitation for many applications is their relatively short displacement ranges. One method to avoid the above-mentioned shortcoming is to integrate a piezoelectric element with a mechanical displacement amplifier based on a compliant mechanism. This paper presents the design and closed-loop control of such a piezoelectric actuator. Using the integrated strain gauge sensors, a PI controller was designed and tested.

1. Introduction

Piezoelectric elements are almost an ideal displacement actuator and are widely used in many precision systems and devices. They have many advantages including low volume, large blocking force, high resolution, high bandwidth, fast and virtually noise-free response. However, piezoelectric elements can provide only a limited, often insufficient, displacement. To overcome the mentioned disadvantage, it's possible to combine a piezoelectric actuator with a mechanical displacement amplifier usually based on compliant mechanisms with flexure hinges and/or elements [1], [2]. Theoretical models of the displacement amplification ratio for flexure based compliant mechanisms are proposed in [3]. Compliant mechanisms actuated with piezoelectric elements can be found in various mechanical microsystems, such as those presented in [4], [5].

A major drawback of piezoelectric actuators is their hysteresis nonlinearity. The maximum error due to hysteresis can be up to 15% of the actuator stroke and affect their accuracy in many applications. This nonlinearity of piezoelectric actuation can be compensated through mathematical modelling and control approaches [6], [7]. In order to develop a closed-loop control, it is necessary to precisely determine the deformation of piezoelectric element. This can be achieved with different types of sensors, of which the most used are laser-based sensors and strain gauge bridges [8].

This paper presents the design of a piezoelectric actuator with a displacement amplification mechanism. In order to compensate for the hysteresis nonlinearity, we use the feedback signal from a built-in strain gauge sensor.

2. Actuator design

2.1. *The amplification mechanism*

Piezoelectric elements have small displacement values and as mentioned before where more displacement is required an amplifying mechanism must been taken into consideration. One of the most used amplification mechanisms is the bridge-type mechanism, presented in Figure1 in two different topologies.

Figure 1(a) shows a bridge-type mechanism with flexure arms. The vertical output displacement δy is mainly attributed to the bending deformations of the flexure arms, under the action of the force $\vec{F}$ developed by a piezoelectric element. The actuating force is along horizontal direction and forms an angle with arm AB. Figure 1(b) presents the bridge-type compliant mechanism, where the output displacement is due the deformation of flexure hinges and the actuating force is parallel with flexure hinges axis.

Figure 1. Topology of bridge-type amplification mechanisms

Each of the two variants has its own advantages and disadvantages presented in [2]. In this paper we use a bridge-type mechanism with flexure arms.

2.2. Geometrical analysis

Since the amplification mechanism is symmetrical in two coordinates, it's sufficient if we analyze only a quarter of the entire structure. The simplified model is presented in Figure 2. The horizontal displacement δx, due to the force generated by a piezoelectric element, produces a vertical output displacement δy, which modifies the inclination of the element AB.

From the projection equations of the element AB on the x and y axes, it follows:

$$\begin{cases} l\cos\varphi + \delta x = l\cos\alpha \\ l\sin\varphi - \delta y = l\sin\alpha \end{cases} \tag{1}$$

where: l represents the length of AB, φ is the initial value of inclination angle and α denotes its final value.

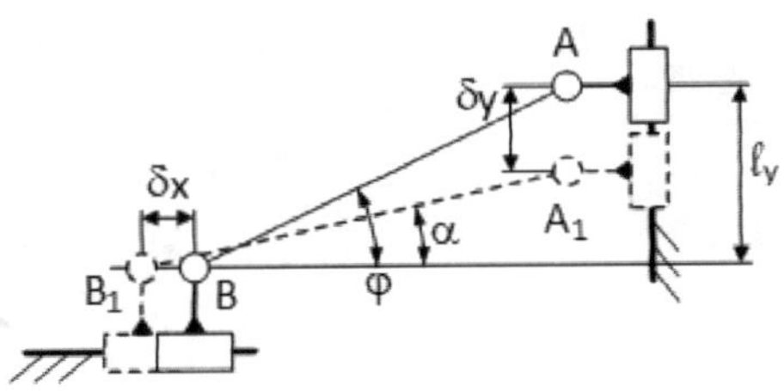

Figure 2. One quarter of mechanism's simplified model

By adding the squares of equations (1) we can eliminate the α variable. The result is a 2nd degree equation with the unknown δy whose solution is:

$$\delta y = l\sin\varphi \pm \sqrt{l^2\sin^2\varphi - \delta x^2 - 2l\delta x\cos\varphi} \tag{2}$$

The ratio between the output displacement and the stroke of the piezoelectric element represents the amplification ratio. The displacement amplification ratio of a flexure type mechanism can be derived analytically by considering the elastic deformation of flexure elements and using the pseudo-rigid body concept [3]. Geometrically it can be written as:

$$A = \delta y/\delta x = l\sin\varphi - \sqrt{l^2\sin^2\varphi - \delta x^2 - 2l\delta x\cos\varphi}/\delta x \tag{3}$$

Theoretically, the maximum output displacement is obtained if $\alpha = 0$. From equations (1) we can write:

$$\delta y_{max} = l \sin\varphi = l_y \quad (4)$$

If the length l and angle φ are known, the amplification ratio A can be seen as a function that depends only of the input displacement δx [3].

2.3. Finite Elements Analysis of the amplification mechanism

In order to estimate the maximum output displacement of the flexure mechanism due to the displacement of piezoelectric element and to prove that an elastic deformation occurs, the finite element analysis method is used. For the purpose of geometric modelling and finite element analysis (FEA) we use CATIA®.

The overall dimensions of the designed amplification mechanism are presented in Figure 3. The flexure elements form an angle $\varphi = 6^o$ with the horizontal direction. The whole mechanism was designed in order to be actuated by a PZS001 piezoelectric stack from ThorLabs. The PZS001 is a co-fired piezoelectric actuator with a 17.4 μm maximum displacement. The active piezo cross section of the actuator is 5 x 5[mm] and the overall dimensions are 6 x 7 x 20[mm].

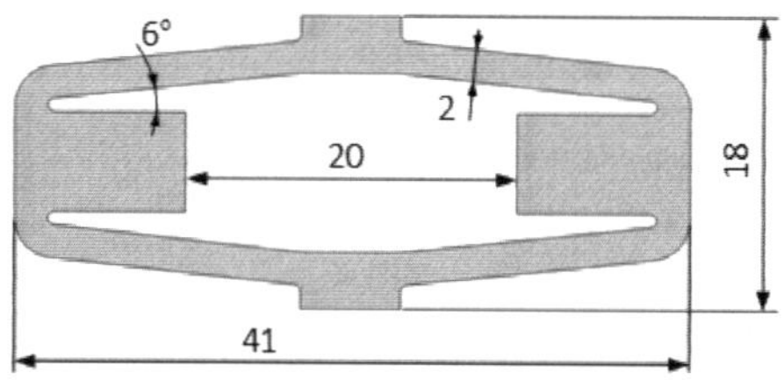

Figure 3. Overall dimensions of the amplification mechanism

We consider that the amplification mechanism is made of aluminum alloy with the modulus of elasticity $E = 69 \cdot 10^9 N/m^2$, yield strength $\tau = 9.5 \cdot 10^7 N/m^2$ and Poisson ratio $\mu = 0.346$. The maximum displacement of the piezoelectric actuator was limited at 16 μm. The result of FEA in terms of translational displacement is presented in Figure 4 (a). The error percent of the FEA is about 0.926%.

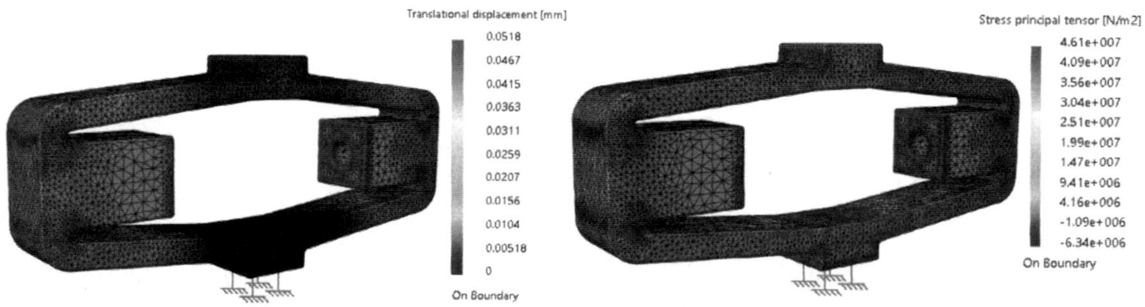

(a) (b)

Figure 4. Results of finite elements analysis

Maximum displacement of the mechanism is 0.0518 mm in -y direction which corresponds to a 0.016 mm displacement of the piezoelectric actuator. Thus, the theoretical amplification ratio can be derived as follows:

$$A_t = \delta y / \delta x = \frac{0.0518}{0.0160} = 3.237 \quad (5)$$

Figure 4 (b) shows the values of resulted stress in the structure of amplification mechanism due to the deformation of the piezo stack. The maximum stress represents 48.53% from the yield strength value, thus an elastic deformation occurs.

IOP Conf. Series: Materials Science and Engineering 1018 (2021) 012002 doi:10.1088/1757-899X/1018/1/012002

3. Experimental results

The bridge-type mechanism was fabricated of a 6 mm thick aluminum plate by laser cutting and has the dimensions presented in Figure 3. The PZS001 piezoelectric stack used to actuate the mechanism, is preloaded with a M2.5 screw in order to eliminate any clearances. The piezoelectric stack has attached a full bridge strain gauge. We intend to use the feedback signal from the strain gauge to provide linear operation of the piezoelectric actuator and determine its displacement without external sensors. Therefore, we configured an experimental testing apparatus, presented in Figure 6.

The displacement of bridge-type mechanism (1) is measured with an external laser displacement sensor (2) from OMRON, with an accuracy of 1μm. The relative position of the sensor to the actuator is maintained using the adjustable device (7). On the controller screen (3) is displayed the measured value of the actuator displacement. The supply voltage for the laser sensor and its controller is provided by the power supply (5). The PZS001 piezoelectric stack is driven by controller (4) from ThorLabs, with a voltage between 0 and 150 V. The DAQ devices (6) are used to measure the feedback signal from the strain gauge and to generate a voltage between 0 and 10V used as an external reference for the piezo controller. These DAQ devices are controlled by a computer running LabVIEW software.

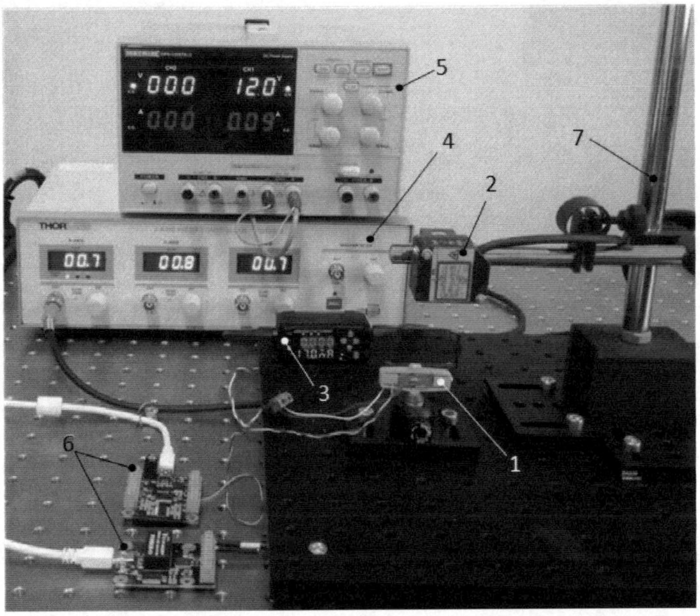

Figure 5. The experimental setup

We performed several series of measurements and based on the average of the measurement values the graphs presented in the Figure 6 were obtained. Thus, in Figure 6 (a) the relationship between the output displacement of the bridge-type mechanism and the input voltage of the piezoelectric stack is presented. Since the amplification ratio is variable in both sections of the hysteresis loop, its exact value is difficult to be determined. We can instead determine the maximum value of the amplification ratio using the measured values. The measured value for the maximum displacement is 0.048 mm and the amplification ratio can be derived as follows:

$$A_{measured} = \delta y / \delta x = \frac{0.0480}{0.0160} = 3 \qquad (6)$$

The difference between the measured and estimated values of the amplification ratio is 7.6 %. This difference is mainly due to the manufacturing errors of the mechanism, the assembly errors of the piezo stack and the fact that the actual material parameters may differ from those considered in the finite element analysis. The functions of the FEA analysis in CATIA® are quite basic, therefore this can be another reason for the obtained difference.

In Figure 6 (b) is presented the relationship between the output signal of the strain gauge bridge and the input voltage of the piezoelectric stack.

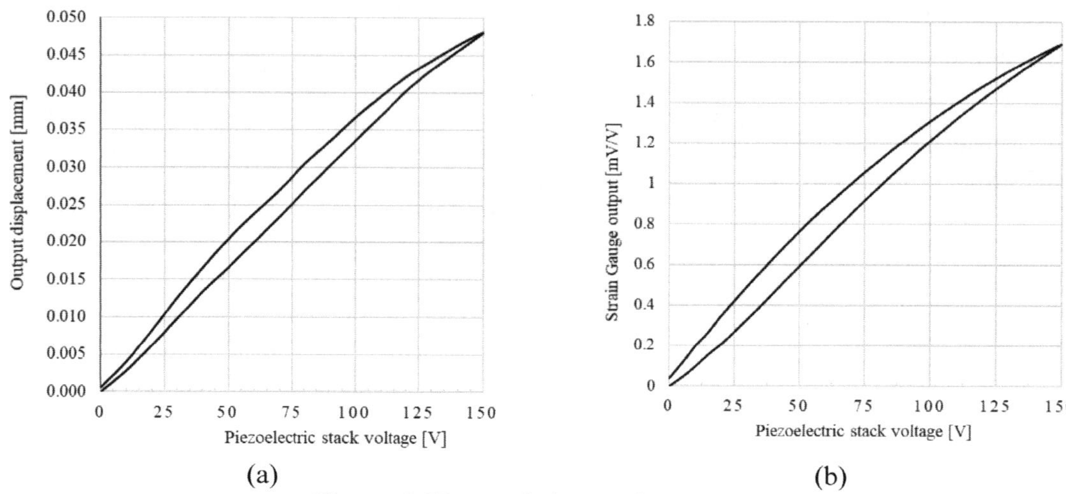

Figure 6. Hysteresis loops of the actuator

In order to implement a closed loop control we need a relation between the output displacement of the mechanism and the strain gauge output signal. This relation was difficult to determine exactly due to the above-mentioned lack of accuracy of the laser sensor. In Figure 7 the relationship between the output displacement of the bridge-type mechanism and the output signal of the strain gauge bridge, is given by a series of points that can be approximated by a line whose equation is $y = 0.0287x - 0.0008$.

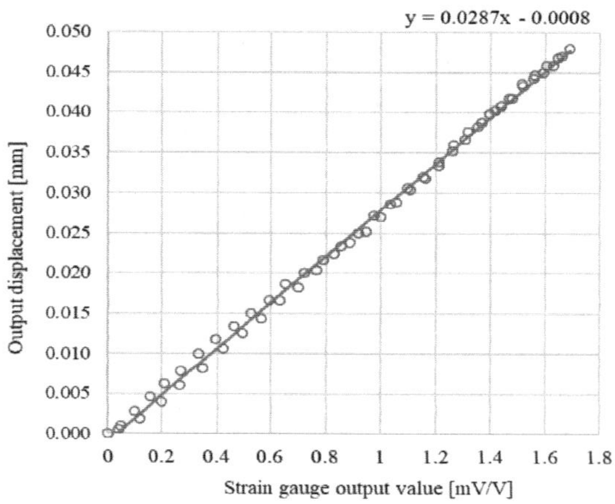

Figure 7. Output displacement vs strain gauge output value

3.1. Closed loop control

The main objectives of the closed loop control are to compensate the hysteresis effect and reach a small steady state error. Although there are many methods to control the piezoelectric actuators [6], we decided to implement a simple Proportional Integrative (PI) controller using LabVIEW software. The controller can be described by the following relation:

$$u = k_P(1 + sT_i^{-1})d \qquad (7)$$

where k_P represents the proportional gain, T_i is the integral time constant, s is the differential operator and d represents the output signal from the strain gauge sensor.

The controller parameters k_P and T_i were determined using the trial and error method. Figure 8 presents the graphical interface of the designed controller that allows us to change the controller parameters and view the system response. In this case, the response of the control system to a sinusoidal input signal with a frequency of 3 Hz and an amplitude of approximately 80% of the total actuator displacement is shown. The tracking error is estimated at 0.378 μm.

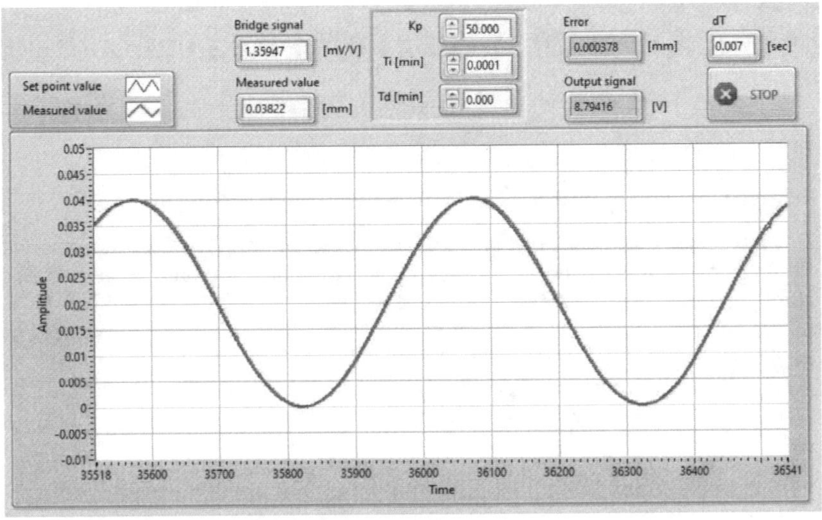

Figure 8. LabVIEW based graphical control interface

4. Conclusions

The paper presents aspects regarding the design and control of a piezoelectric actuator with displacement amplification mechanism. A geometric analysis was performed for the simplified model of the amplification mechanism using the finite element analysis its maximum stroke and amplification ratio was determined. Hysteresis curves were determined experimentally, and the maximum displacement of the actuator was measured using an external laser sensor. The feedback signal from the strain gauge was used to provide linear operation of the piezoelectric actuator and to determine its displacement without external sensors. Using LabVIEW, a Proportional Integral controller was designed and tested.

References

[1] Qian LU 2015 Optimization Design of Flexible Micro-Displacement Amplification Mechanism Based on Parameters *Journal of Applied Science and Engineering* **18** 345-354.

[2] Ling M et al 2016 Enhanced Mathematical Modeling of the Displacement Amplification Ratio for Piezoelectric Compliant Mechanisms *Smart Materials and Structures* **25** 075022.

[3] Xu Q, Li Y 2011 Analytical Modeling, Optimization and Testing of a Compound Bridge-Type Compliant Displacement Amplifier *Mechanism and Machine Theory* **46** 183–200.

[4] Noveanu S, Rusu C, Rancea C, Lateş D 2020 Research Concerning the End-effectors for SiMFlex Microgripper *IOP Conference Series: Materials Science and Engineering* **724** 012055.

[5] Noveanu S, Lates D, Fusariu L, Rusu C 2020 A New Compliant Microgripper and Study for Flexure Hinges Shapes *Procedia Manufacturing* **46** 517-527.

[6] Ziqiang C, Qingsong X 2014 Recent Advances in the Control of Piezoelectric Actuators *International Journal of Advanced Robotic Systems* **11** 182.

[7] Altaher M, Aphale S 2018 High-Precision Control of a Piezo-Driven Nanopositioner Using Fuzzy Logic Controllers *Computers* **7** 10.

[8] Schitter G, Stark RW, Stemmer A 2002 Sensors for Closed-Loop Piezo Control: Strain Gauges Versus Optical Sensors *Measurement Science and Technology* **13** 47-48.

Modeling of nonlinear properties of casing centralizers equipped with axial thrust

I Shatskyi[1], A Velychkovych[2], I Vytvytskyi[3], and M Senyushkovych[3]

[1] Ivano-Frankivsk Branch of Pidstryhach-Institute for Applied Problems in Mechanics and Mathematics, NAS of Ukraine, Department of modelling of damping systems, Mykytynetska 3, Ivano-Frankivsk, Ukraine
[2] Ivano-Frankivsk National Technical University of Oil and Gas, Department of Construction and Civil Engineering, Karpatska 15; Ivano-Frankivsk, Ukraine
[3] Ivano-Frankivsk National Technical University of Oil and Gas, Department of Drilling, Karpatska 15; Ivano-Frankivsk, Ukraine

E-mail: ipshatsky@gmail.com, a_velychkovych@ukr.net

Abstract. The issue of qualitative centering of the casing in wells with complex trajectory is considered. When choosing an effective centralizer, two competing requirements are simultaneously assigned to its mechanical properties. The centralizer must have sufficiently low stiffness in order to provide the descent of the casing into wellbore. At the same time, the centralizer must have high stiffness in order to provide proper clearance gap for high-quality cementing of the annular space. Several new designs of rod centralizers are proposed in the paper. Their main feature is that they are equipped with an additional thrust ring, which allows one to change the mechanical characteristics of the centralizer at high contact loadings. The boundary-value problems for the differential equations of mechanics of the arc rods with boundary conditions in the form of inequalities are formulated and solved. Analytical relations between the clamping force and mutual approach of the casing and the well wall, which characterize the radial stiffness of the centralizer equipped with axial thrust, are determined. It is established that the described designs of the centralizers have essentially nonlinear characteristics that contribute to the efficient centering of the casing. At vertical and slightly deviated from the vertical intervals of the wellbore, the centralizer behaves as soft and does not interfere with the descent of the string; at intervals with large curvature of the wellbore the stiffness of the centralizer increases due to the axial thrusts.

1. Introduction

A careful centering of the casing is the key factor providing a high quality well cementing. Equipping of casing with special centering devices is intended to ensure its equidistance from the walls of the well and thereby improve the quality of cementing of the annular space. The deviation of the casing from the correct trajectory depends on a number of factors: the size of the well, the size of the casing, the type and properties of centralizers, their number and location in the well, etc. One of the actual tasks that petroleum engineers and well completion specialists are required to perform on a regular basis is to choose the optimum type and number of centralizers for each casing. Today, service companies have a wide range of technical solutions to improve the reliability of casing strings [1–4], however the problem of effective casing centering under the complex configuration of the well axis remains relevant. This problem is especially important because of the increase in the volume of directional and horizontal drilling in the world, the development of modern technologies for the re-entry of abandoned and emergency wells, the predrilling of additional lateral wellbores through the casing windows, etc.

Today, elastic rod centralizers [5, 6] are the most used in the practice of well drilling. To evaluate the efficiency of centering devices of this type, to determine their required number and locations, it is recommended to investigate the mechanical properties of the centering devices and their mechanic behavior under contact loading. In particular, the theoretical analysis of the stiffness and strength characteristics will prevent from their excessive consuming or their insufficient use in the centering interval.

The performance criterion used to evaluate the quality of the centralizers is their ability to concentrically place the casing in the well at any position on its axis. There is a problem in choosing an effective centralizer because its two properties are simultaneously competing with each other. The centralizer must have a sufficiently low stiffness not to slow down the descent of casing in the well. At the same time, the centralizer must have a sufficiently high stiffness to provide the necessary clearance for a high-quality cementing of the annular space [6–8]. In previous studies, the authors tried to solve this problem by giving the centralizer an elastic-rigid characteristic by using radial rigid thrusts [6]. In this work, we aim to evaluate the effects of installing axial thrusts in cast and welded rod centralizers. Such additional thrusts give the non-linear characteristics to the centralizers to allow for the descent of the casing and good centering ability of the devices at high clamping forces.

2. Two variants of the centralizer design with axial thrust

To solve the described problem, the authors propose to use the centralizer designs with variable stiffness (Fig. 1). Both centralizers have a rigid connection of the lower ends of the elastic bows *2* with the fixed rings *1*. In the cast centralizer (a) the upper ends of the elastic bows are free, and at some distance from them the centralizer is equipped with a rigid thrust *4*. In the welded centralizer (b) the upper ends of the elastic bows are rigidly connected to the axially movable ring *3*. At a certain distance Δ from the movable ring of the welded centralizer there is a restriction in the form of a rigid thrust *4*.

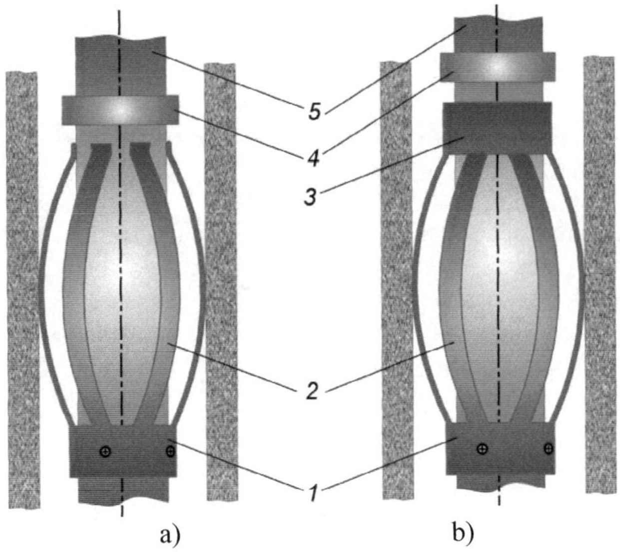

Figure 1. The cast (a) and welded (b) centralizer with an additional axial thrust:
1 – fixed centralizer ring, *2* – elastic bow, *3* – moving centralizer ring, *4* – stiff axial thrust, *5* – casing.

The presence of axial thrusts is intended to constructively ensure the nonlinearity of the elastic characteristics of the centralizers. As long as the rigid thrust is not engaged in the centralizer's operation, the radial stiffness of the device is relatively small. When the distance Δ is out and the thrust comes into play, the stiffness of the centralizer increases.

3. Results of mathematical modeling and analysis

The problems of the interaction of pipe strings with the wall of a well have been the subject of research in [9–15], and more general issues about the interaction of shells with elastic bodies have been considered, in particular, in publications [16–18]. Analytical and numerical approaches to engineering modeling and study of contact phenomena in shell and rod systems using classical theories of rods and cut shells have been proposed in publications [19–24]. In particular, some contact problems for cyclic rod structures were studied in [25–27]. The authors [28] developed a computer model for centering long objects. Experimental and numerical-analytical studies [29, 30] concern the analysis of mechanical properties of centering devices. Analytical models for investigating the contact interaction of elastic linkage of centralizers with well walls are given in [31], where six variants of attaching a centralizer to the casing are considered. In the current paper, we develop a technique for calculating the stiffness of cast and welded centralizers with a thrust, which was proposed earlier in a paper [32] for a hinged centralizer.

For simulation, we used the classical linear theory of hollow arch-shaped rods and consider the deformation of a single link loaded by the contact force (Fig. 2).

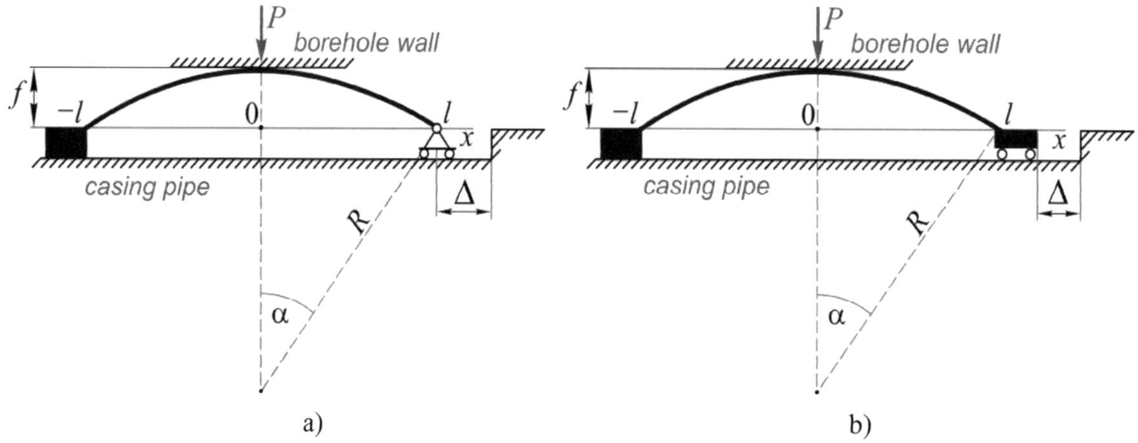

Figure 2. Schemes of contact interaction of the centralizer link with the well wall: (a) – cast centralizer, (b) – welded centralizer.

Let f be the lifting arm of the centralizer's link, $2l$ is the length of the rod projection on the axis of abscissa, 2α is the small slope of the arc and R is the radius of its curvature. The rod is considered flat, it means we assume that the inequilities are realized:

$$(f/l)^2 << 1\,,\ 1/R \approx 2f/l^2\,,\ \alpha \approx 2f/l\,.$$

The equilibrium equations and physical relationships of the model are as follows [33]:

$$\frac{dN}{dx} = 0\,,\ \frac{dQ}{dx} + \frac{N}{R} = -P\delta(x)\,,\ \frac{dM}{dx} - Q = 0\,,\ x \in (-l, l); \tag{1}$$

$$\frac{du}{dx} + \frac{w}{R} = 0\,,\ M = EJ\frac{d\vartheta}{dx} = EJ\frac{d^2w}{dx^2}\,,\ x \in (-l, l). \tag{2}$$

Here N, Q are the axial transverse forces, M is bending moment, P is the contact force, EJ is stiffness of the rod to the bend, u, w are tangential and transverse displacements of the rod, ϑ is the rotation angle, x is the coordinate, and $\delta(x)$ is Dirac function.

The boundary conditions for fastening the rod are as follows.

The anchorage conditions for the left end $(x=-l)$

$$u_x(-l)\equiv u(-l)-w(-l)\alpha=0,\ u_y(-l)\equiv w(-l)+u(-l)\alpha=0,\ \vartheta(-l)=0, \tag{3}$$

and the axial and radial grip conditions of the right end $(x=l)$

$$N_x(l)\equiv N(l)-Q(l)\alpha=0,\ u_x(l)\equiv u(l)+w(l)\alpha\le\Delta \text{ or } u_x(l)=\Delta,\ N_x(l)<0, \tag{4}$$

$$u_y(l)\equiv w(l)-u(l)\alpha=0, \tag{5}$$

are common for both schemes (a) and (b).

At the same time, the constructions under consideration are notable for the freedom of rotation of the cross-section at the right end of the rod, in particular, it is allowed to rotate in case (a)

$$M(l)=0, \tag{6}$$

and prohibbited in case (b)

$$\vartheta(l)=0. \tag{7}$$

Thus, Eqs. (1)–(6) and Eqs. (1)–(5), (7) are the boundary problems for the system of differential equations of the 6th order with boundary conditions in the form of inequalities corresponding to one-sided joints for the construcnions (a) and (b) respectively.

The solution to these problems $\mathbf{z}(x)=(N, Q, M, \vartheta, w, u)^{\mathrm{T}}$ are found in the form:

$$\mathbf{z}(x)=\begin{Bmatrix} P\mathbf{z}^{(1)}(x),\ P<P_*; \\ P_*\mathbf{z}^{(1)}(x)+(P-P_*)\mathbf{z}^{(2)}(x),\ P>P_* \end{Bmatrix}=$$

$$=P\mathbf{z}^{(1)}(x)+(P-P_*)H(P-P_*)\left(\mathbf{z}^{(2)}(x)-\mathbf{z}^{(1)}(x)\right). \tag{8}$$

Here $H(x)$ is the Heveside's function, and P_* is the value of the contact force, for which the axial gap Δ is out.

Moreover, $\mathbf{z}^{(i)}(x)=\left(N^{(i)}, Q^{(i)}, M^{(i)}, \vartheta^{(i)}, w^{(i)}, u^{(i)}\right)^{\mathrm{T}}$ are the solutions of two auxiliary problems: $i=1$ – the problem of the action of a single concentrated force $P=1$ on an arch with a axially-movable fixing (condition in Eq. (4) should be replaced by $N_x(l)=0$); $i=2$ – the problem of the action of a single concentrated force on an arch with a axialy-immovable fixing (it is necessary to take the condition $u_x(l)=0$ in Eq. (4)).

Using the results of [31], we decode the components of vectors $\mathbf{z}^{(1)}(x)$ and $\mathbf{z}^{(2)}(x)$. In particular, the consistent displacements are:

for the scheme (a)

$$w^{(1)}(x)=\frac{1\cdot l^3}{EJ}\left(-\frac{3}{8}\frac{(x+l)^2}{2!l^2}+\frac{11}{16}\frac{(x+l)^3}{3!l^3}-\frac{x^3}{3!l^3}H(x)\right),$$

$$u^{(1)}(x)=1\cdot\frac{2l^2 f}{EJ}\left(\frac{3}{8}\frac{(x+l)^3}{3!l^3}-\frac{11}{16}\frac{(x+l)^4}{4!l^4}+\frac{x^4}{4!l^4}H(x)\right); \tag{9}$$

$$w^{(2)}(x)=\frac{1\cdot l^3}{EJ}\left(\frac{1}{24}\frac{(x+l)^2}{2!l^2}-\frac{17}{48}\frac{(x+l)^3}{3!l^3}+\frac{5}{6}\frac{(x+l)^4}{4!l^4}-\frac{x^3}{3!l^3}H(x)\right),$$

$$u^{(2)}(x)=1\cdot\frac{2l^2 f}{EJ}\left(-\frac{1}{24}\frac{(x+l)^3}{3!l^3}+\frac{17}{48}\frac{(x+l)^4}{4!l^4}+\frac{5}{6}\frac{(x+l)^5}{5!l^5}-\frac{x^4}{4!l^4}H(x)\right); \tag{10}$$

and for the scheme (b)

$$w^{(1)}(x)=\frac{1\cdot l^3}{EJ}\left(-\frac{1}{4}\frac{(x+l)^2}{2!l^2}+\frac{1}{2}\frac{(x+l)^3}{3!l^3}-\frac{x^3}{3!l^3}H(x)\right),$$

$$u^{(1)}(x)=1\cdot\frac{2l^2 f}{EJ}\left(\frac{1}{4}\frac{(x+l)^3}{3!l^3}-\frac{1}{2}\frac{(x+l)^4}{4!l^4}+\frac{x^4}{4!l^4}H(x)\right); \tag{11}$$

$$w^{(2)}(x)=\frac{1\cdot l^3}{EJ}\left(\frac{1}{16}\frac{(x+l)^2}{2!l^2}-\frac{7}{16}\frac{(x+l)^3}{3!l^3}+\frac{15}{16}\frac{(x+l)^4}{4!l^4}-\frac{x^3}{3!l^3}H(x)\right),$$

$$u^{(2)}(x)=1\cdot\frac{2l^2 f}{EJ}\left(-\frac{1}{16}\frac{(x+l)^3}{3!l^3}+\frac{7}{16}\frac{(x+l)^4}{4!l^4}-\frac{15}{16}\frac{(x+l)^5}{5!l^5}+\frac{x^4}{4!l^4}H(x)\right). \tag{12}$$

We evaluate the stiffness of the considering structures by analyzing the axial displacement of the right end of the rod $\Delta_x = u_x(l) \equiv u(l) + w(l)\alpha$ and radial approach of the string with the well wall $\Delta_y = -w(0)$. In accordance with the expressions in Eqs. (8)–(12) we have:

$$\Delta_x = \frac{P}{C_x^{(1)}} + (P-P_*)H(P-P_*)\left(\frac{1}{C_x^{(2)}}-\frac{1}{C_x^{(1)}}\right),$$

$$\Delta_y = \frac{P}{C_y^{(1)}} + (P-P_*)H(P-P_*)\left(\frac{1}{C_y^{(2)}}-\frac{1}{C_y^{(1)}}\right), \tag{13}$$

where $P_* = C_x^{(1)}\Delta$, and stiffness coefficients are

$$C_x^{(1)} = 6\frac{l}{f}\frac{EJ}{l^3},\ C_x^{(2)} = \infty,$$

$$C_y^{(1)} = \frac{96}{7}\frac{EJ}{l^3},\ C_y^{(2)} = 288\frac{EJ}{l^3},$$

as well as

$$C_x^{(1)} = 12\frac{l}{f}\frac{EJ}{l^3},\ C_x^{(2)} = \infty,$$

$$C_y^{(1)} = 24\frac{EJ}{l^3},\ C_y^{(2)} = 384\frac{EJ}{l^3}$$

in the cases (a) and (b) respectively.

The results obtained by Eq. (13) are illustrated in Fig. 3. For calculations we considered: $l / f = 10$ and $C = EJ / l^3$.

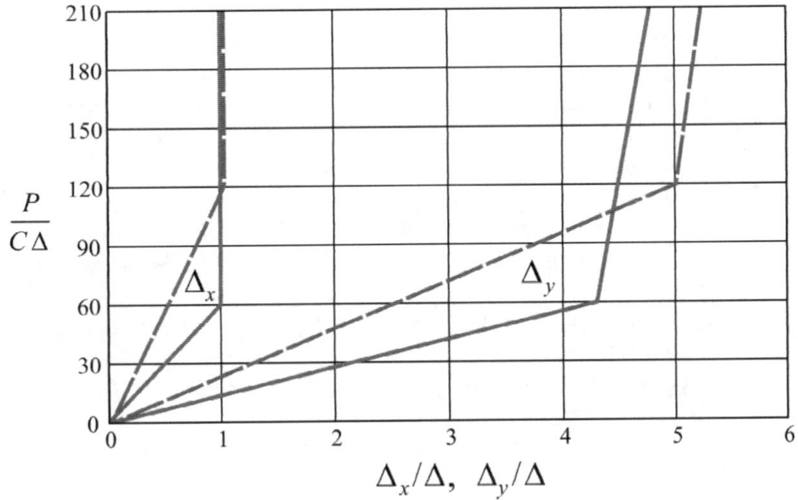

Figure 3. Loading diagrams for centralizers with thrust:
solid lines – cast centralizer (a), dashed lines – welded centralizer (b)

The kinks of the lines in the plots are due to changing boundary conditions in Eq. (4). The ordinates of these kinks are consistent with normalized values P_*.

4. Conclusion

Some new designs of a rod centralizer of a casing column are offered. The key feature of those designs is that, an additional axial thrust ring which allows one to operate the device at large contact loads is presenced.

We are formulated and solved the boundary-value problems for the differential equations of mechanics of the arc rods with boundary conditions in the form of inequalities are formulated and solved. The designed analytical models with a high degree of adequacy describe the autonomous operation of links of cast centralizer (a) for all contact loadings, as well as the operation of links of welded centralizer (b) with limited axial movement. In the case of low load, the calculation according to scheme (b) should be defined more precisely taking into account the combined work of all the rods connected into the cyclic system via the ring.

Both described designs of the centralizers equipped with axial thrust have essentially nonlinear characteristics, which are regulated by the choice of distance Δ. This nonlinearity facilitates the attainment of the aim of effective centering of the casing. At slightly deviated domains of the well with small clamping forces, the centralizer behaves like a soft one and does not impede the descent of the column; in areas with a large curvature the centralizer becomes stiff due to the thrust. In both cases, the radial clearance gap required for high-quality cementing is provided. So the goal of our study has been achieved.

References

[1] Patel D, Thakar V, Pandian S, Shah M and Sircar A 2019 *Petroleum* **5**(1) 1–12
[2] *Casing Accessories: Cement integrity starts here* 2016 (Norway: Weatherford) p 56
[3] Liu G and Weber L D 2012 *Society of Petroleum Engineers* SPE-150345-MS
[4] Kozhevnikov E V, Nikolaev N I, Rozentsvet A V and Lyrchikov A A 2015 *Bull of PNRPU Geology Oil & Gas Eng & Mining* **16** 54–60

[5] Kotskulych Ya S, Seniushkovych M V, Martsynkiv O B and Vytvytskyi I I 2016 *Naukovyi Visnyk Natsionalnoho Hirnychoho Universytetu* **2** 23–30
[6] Vytvytskyi I I, Seniushkovych M V and Shatskyi I P 2017 *Naukovyi Visnyk Natsionalnoho Hirnychoho Universytetu* **5** 29–35
[7] Skadsem H, Saasen A and Havardstein S 2017 *Proceedings of the ASME 2017 36th Intern. Conf.* V008T11A027
[8] Paliichuk I I, Nezamay B S and Malyk H B 2019 *Naukovyi Visnyk Natsionalnoho Hirnychoho Universytetu* **5** 47–54
[9] Gulyaev V I, Gaidaichuk V V, Solov'ev I L and Glovach L V 2007 *Strength of Materials*, **39**(5) 492–501
[10] Andrusenko E N, Gulyaev V I. and Shlyun' N V 2016 *Mechanics of Solids* **51**(2) 234–243
[11] Vlasiy O, Mazurenko V, Ropyak L and Rogal O 2017 *Eastern-European Journal of Enterprise Technologies* **1** 7(85) 25–31
[12] Dutkiewicz M, Gołębiowska I, Shatskyi I, Shopa V and Velychkovych A 2018 *MATEC Web of Conferences* **178** 06010
[13] Andrusyak A, Grydzhuk J, Dzhus A and Steliga I 2017 *Eastern-European Journal of Enterprise Technologies* **1**(7–85) 32–37
[14] Levchuk K G 2018 *Metallofiz. Noveish. Tekhnol.* **40**(1) 45–137
[15] Grydzhuk J, Chudyk I, Velychkovych A and Andrusyak A 2019 *Eastern-European Journal of Enterprise Technologies* **1**(7) 6–14
[16] Mossakowskii V I, Hudramovich V S and Makeev E M 1978 *Contact problems in the theory of shells and rods* (Moscow: Mashinostroienie) p 248
[17] Neto A G, Pimenta P M and Wriggers P 2014 *Int J Numer Methods Eng* **97**(9) 683–706
[18] Jawed M K, Novelia A and O'Reilly O M 2018 *Kirchhoff's theory of an elastic rod, In: A Primer on the Kinematics of Discrete Elastic Rods* (Cham: Springer) p 118
[19] Shatskyi I, Popadyuk I, Velychkovych A 2018 *Springer Proceedings in Mathematics & Statistics* **249** 343–350
[20] Velichkovich A, Dalyak T and Petryk I 2018 *Oil & Gas Science and Technology* **73**(34)
[21] Shats'kyi I P 2005 *Mater Sci* **41**(2) 186–191
[22] Shats'kyi I P and Makoviichuk M V 2005. *Mater Sci* **41**(4) 486–494
[23] Shats'kyi I P and Makoviichuk M V 2009 *Strength Mater* **41** 560–565
[24] Shatskii I P and Makoviichuk N V 2011 *J Appl Mech and Techn Phys* **52**(3) 464–470
[25] Bedzir A A, Shatskii I P and Shopa V M 1995 *Int Appl Mech* **31** (5) 351–354
[26] Bedzir O O and Shopa V M 2010 *J Math Sci* **168**(5) 665–672
[27] Shatskyi I, Ropyak L and Velychkovych A 2020 *Engineering Solid Mechanics* **8**(4) 301–312
[28] Balalayeva E, Kukhar V, Artiukh V, Filatov V and Simonova O 2019 *Advances in Intelligent Systems and Computing* **983** 475–488
[29] Tikhonov V S, Bukashkina O S and Gandikota R 2014 *In: 11th World Congress on Computational Mechanics* 667–675
[30] Money Tuli, Balvinder Singh and Dr Rahul Malhotra 2018 *Int J Acad Res Develop* **13** 1448–53
[31] Shatskyi I, Velychkovych A, Vytvytskyi I and Seniushkovych M 2019 *Engineering Solid Mechanics* **7**(4) 355–366
[32] Shatskyi I, Vytvytskyi I, Seniushkovych M and Velychkovych A 2019 *IOP Conf. Ser.: Mater. Sci. Eng.* **564** 012073
[33] Beer F P, Johnston E R., DeWolf J T and Mazurek D F 2015 *Mechanics of Materials* (New York: McGraw-Hill Education) p 896

IOP Conf. Series: Materials Science and Engineering 1018 (2021) 012004 doi:10.1088/1757-899X/1018/1/012004

Experimental characterisation of a flax fibre - epoxy resin composite

Flavia Artimon[1,2,a)], C Stochioiu[1,b)], H M Gheorghiu[1], Iuliana Sima[3]

[1] University POLITEHNICA of Bucharest, Romania
[2] The National Institute of Research and Development in Mechatronics and Measurement Technique, Bucharest, Romania
[3] Bucharest National University of Arts

a) flavia.artimon@yahoo.ro; b) c.stochioiu@gmail.com

Abstract. In the recent years, the scientific community's attention has gradually started to focus in finding solutions for pollution control and replacement of non-biodegradable products with natural products. In this matter, composite materials with natural fibres are increasingly interesting for different research studies. The present work regards a flax fibre-epoxy resin composite material, which the response is studied under tensile tests until break and repeated loading/unloading cycles. An independent variable under study has been material fibre direction of 0°, 90° and ± 45°, for both types of tests. Samples were extracted from thermo-compression fabricated plates with dimensions following the ASTM D 3039 standard. The experimental setup included an MTS C45 105 Universal Testing Machine, with variable load cells, an extensometer, as well as digital image correlation equipment. The devices allowed the determination of stress-strain curves as well as material constants and Poison ratio. Results show, for the 90° and ± 45° typical responses for composites with these fibre directions, but a particular evolution, bilinear for the 0°.

1. Introduction

The composite materials have long been employed into various fields and the extent of their use is still growing [1, 2]. Since their first use, the goal has been the same: to improve quality and strength. Given their advantages, these materials were implemented in the construction and manufacture of household items (wood and cob) [2], and in the manufacture of battle tools (Mongolian bows, damask swords, or Japanese swords) [3].

Flax fibres use as a material is considerably antique; more than 34.000 years according to a group of archaeologists and paleontologists at Harvard University [4]. Regardless, studies and researches on flax fibres do not completely cover its capabilities.

Flax fibre is distinguished from other composite reinforcement materials, thanks to its high tensile strength and low density [5]. Furthermore, it is a thermal and acoustic insulator and, therefore, it is considered a good replacement for glass fiber [6]. In addition, the plant it is extracted from does not require special conditions for cultivation [7].

Its low density combined with high mechanical properties make it a perfect choice for applications where weight is a critical factor, such as transportation [8]. In the aerospace industry, it is be ideal for lightweight vehicles, such as flying or land drones, ekranoplans, sometimes used in multiagent systems, which would lead to an increase of the entire system performance.

IOP Conf. Series: Materials Science and Engineering **1018**(2021) 012004 doi:10.1088/1757-899X/1018/1/012004

Tests in the literature frequently disregard viscous effects of the material. Often, they are performed with the machine's crosshead speed control [9-11], an indirect parameter, as the material response is also influenced by the specimen's length. At the same time, sample slippage in the machine's jaws occurs during loading, which disturbs the material response. Thus, high variations in material response might appear, through the neglect of the viscous component.

The aim of this article is to determine specific mechanical properties of this material and to analyze its behavior under load. In order to reach this goal, several UD flax fibre – epoxy resin composite specimens, with the dimensions indicated by the standard ASTM D 3039 [12] were used.

First, tensile tests were performed. Further on, loading/unloading tests at constant speed and progressive loading were carried out.

2. Specimens fabrication and experimental protocol

Composite plates (Fig.1) were fabricated from resin pre-impregnated unidirectional fibres, commercially known as LINEO FLAXPREG T-UD 110 [13], through a cycle of thermocompression. The curing cycle (Fig. 2) is performed at a temperature of 130°C and a 0,3 MPa pressure for one hour. A post curing cycle was performed at 80°C, for 60 min, in an industrial oven. The material samples were cut out of the resulting plates using a band saw.

Figure 1. Composite plates for the 3 directions of fibre orientation

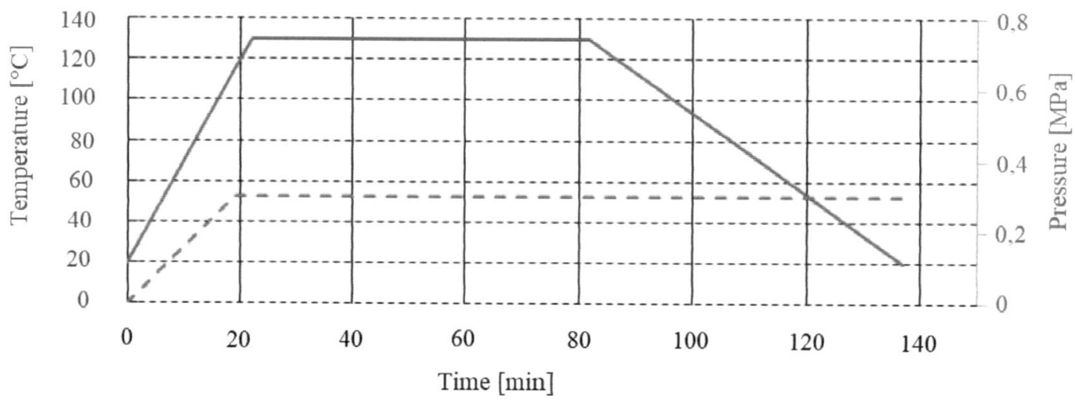

Figure 2. Curing cycle

Several samples were fabricated for different types of fibre orientations in order to determine the mechanical properties of this composite. The dimensions of the samples were chosen according to ASTM D 3039, as presented in Table 1.

Table 1. Specimens dimensions

Fibre orientation with respect to load direction	0°	± 45°	90°
Specimens dimensions [mm]	250 x 25 x 2	250 x 25 x 2,5	175 x 25 x 2

Aluminum tabs were glued at the extremities of each sample, 50 x 2 mm in size, with the purpose of reducing stress concentrators in the grips. The samples were conditioned in a climatic chamber under a relative humidity of 50 % and temperature of 23° C [14]. For humidity control, a saline solution has been prepared by dissolving Potassium Chloride in water and was added in the chamber. The specimens were weighted daily and, while it was found that their weight stabilized in the 3rd day, they were kept for two more days to ensure hygrothermal equilibrium.

The experiments were performed using an MTS C45 105 Universal Testing Machine. It is electromechanically controlled and allows for changeable load cells, a desired feature when characterizing composites with different fibre orientations. Thus, for the directions used, the maximum loads of the cells were: 2.5 kN for 90° samples, 10 kN for ± 45° samples and 100 kN for 0° samples.

The deformation was recorded with a stereo-image correlation camera, for the mechanical characterization and an extensometer, for the cyclic loading tests. For the stereo-image correlation procedure, a surface of each sample was covered with a layer of white paint and black speckles (Fig. 3).

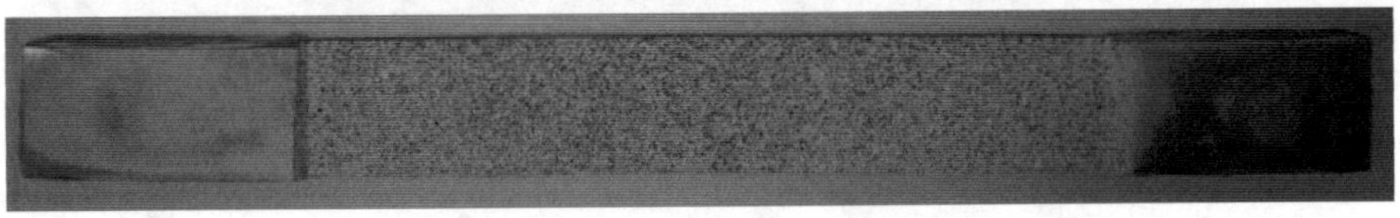

Figure 3. Sample prepared for the stereo-image correlation procedure

3. Results

The tensile tests in the three layouts were carried out with the main objectives of obtaining tensile curves in all directions (Fig. 4), defining the maximum material strength, calculating the modulus of elasticity and identifying their damage evolution with stress.

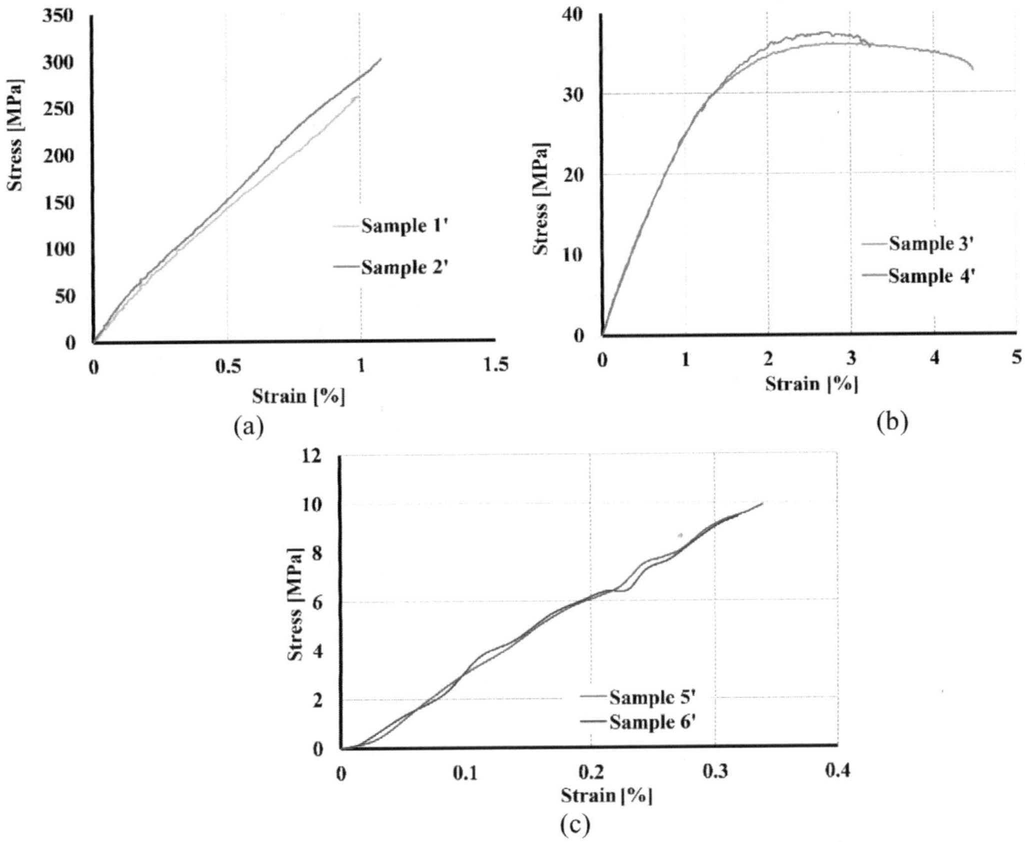

Figure 4. Stress-Strain curves: (a) 0°, (b) ± 45°, (c) 90°

Figure 5. Broken specimens

By analyzing the results, it can be concluded that at ± 45° fibre orientation, the ruptures propagated along the fibre direction; at 0° fibre orientation, the specimen yielded in the grip area, with crack propagations along the length; at 90° fibre orientation, the breaks were sudden, parallel to the transversal section of the sample, with no obvious break region.

The data was analyzed following the norms ASTM D3039, ASTM 3518 [15] for the shear properties and, for the particular bilinear behavior of the composite in the 0° fibre orientation, the method proposed by Shah [16].

The determined tensile stress and elastic moduli are presented in Table 2.

Table 2. Tensile strength and material modulus

Fibre orientation	Tensile strength [MPa]	Elastic modulus [GPa]	
		E1	E2
0°	278,5 ± 29	38,2 ± 2,5	27,7 ± 2,5
± 45°	36,3	1,53	
90°	9,7± 0,4	2,91 ± 0,14	

For the variable loading tests, tensile tests were carried out in the 3 directions under the constant cross speed of 2 mm/min with different loading levels. Although the loading was achieved by controlling the displacement speed, the deformations are measured by the extensometer.

3.1. Specimens at 0°

Various samples were tested under progressive repeated loading, up to 160 MPa and unloading to 0 MPa.

The evolution of the strain determined by the extensometer as a function of time (Fig. 6) reveals that, from 80 MPa onward, residual deformations appear. These deformations can have a viscoelastic and/or plastic nature. This residual strain also appears in the stress-strain curve.

An interesting fact is that, by eliminating the remaining deformations before each cycle, the loading phases of the curves overlap perfectly. The response is nonlinear, approximated by a bilinear evolution (Fig. 7), which allows the identification of two tangent modules of elasticity. The first module $E_1 = 37.14\ GPa$ is calculated up to the stress of 50 MPa and the second module $E_2 = 28.06\ GPa$ is calculated from this stress level onwards.

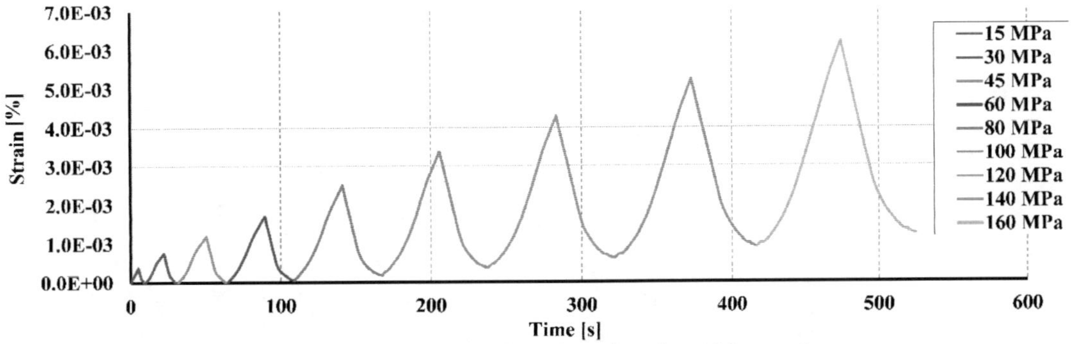

Figure 6. Strain - Time function for a 0° sample

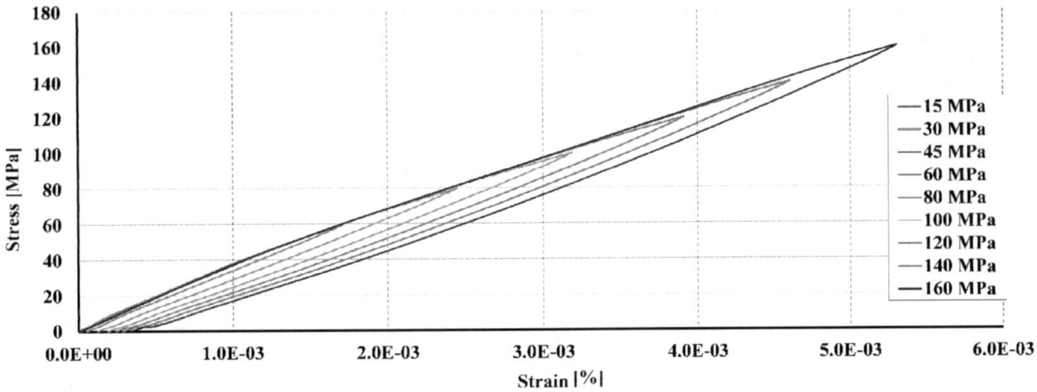

Figure 7. Stress – Strain function depending on the load for the first 0° sample

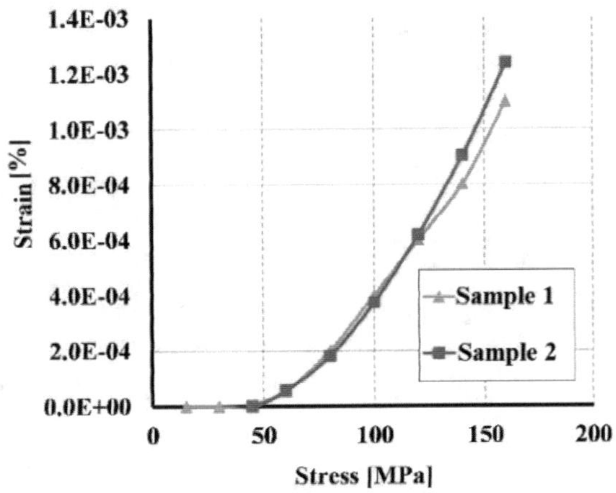

Figure 8. Strain – Stress function for two samples at 0°

Note that the hysteresis loops were almost of the same magnitude for the several tests. This loop depends on the magnitude of the applied stress. Another remark is that all the curves overlap during the loading phase. Residual deformation can be observed starting from 45 MPa (Fig. 8).

3.2. Specimens at 90°

In a similar fashion, several samples with fibre orientation of 90° were subjected to the cyclic loading-unloading. While the one presented in Fig. 9 was tested up to 50% of the failure stress, the loading phases are purely linear, and no residual deformations were recorded at the end of any of the loading cycles.

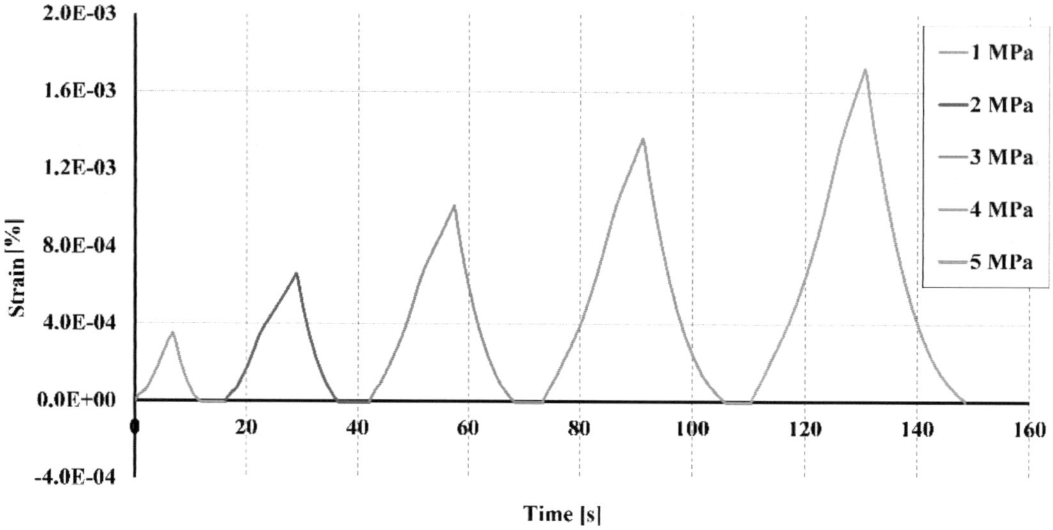

Figure 9. Stress – Time function for a 90° specimen

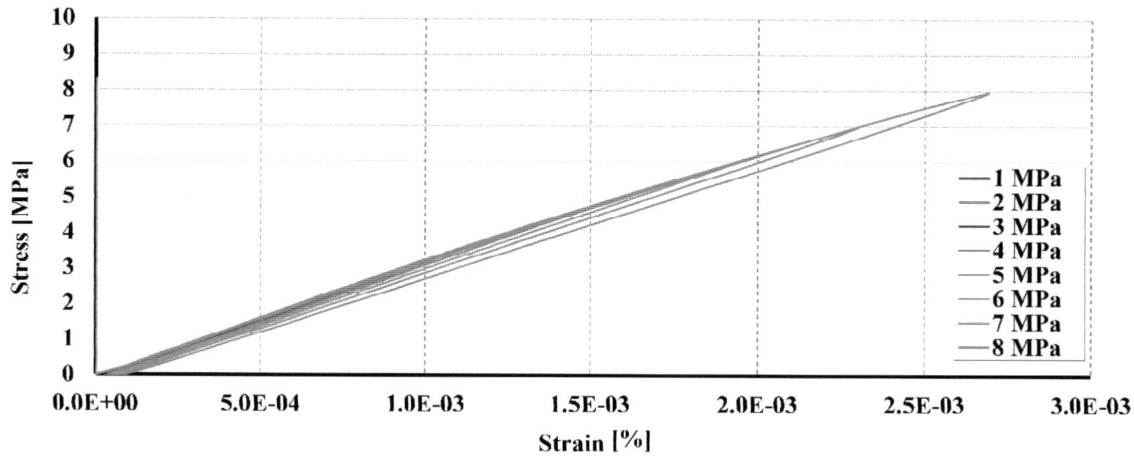

Figure 10. Stress – Strain function depending on the load for a 90° specimen

For the second sample, three more load levels were added, up to 80% of the failure stress. In this case, at the additional loading cycles, a viscous response was indeed recorded, through the appearance of hysteresis loops and residual deformations at the end of the cycles (Fig 10).

3.3. Specimens at 45°

For the ±45° fibre orientation, only one sample was submitted to cyclic loading-unloading, as it was considered enough, by analyzing the previous fibre orientation sample variations. Furthermore, the sample was tested until failure, which occurred at approximately 36 MPa. The sample response in deformation is presented in a time-strain graph in Fig. 11, and in stress-strain coordinates in Fig. 12. A similar response as the sample at the 90° specimens was recorded: an elastic deformation up to 20 MPa (approximately 50% of failure stress) and the appearance of viscoplastic deformations for load levels above this limit.

In the same way as above, a few minutes were waited before the next loading cycle, with the sample mounted in the testing machine. Fig. 11 shows that the residual deformation starts from 20 MPa, which corresponds to approximately 50% of the breaking load of the material.

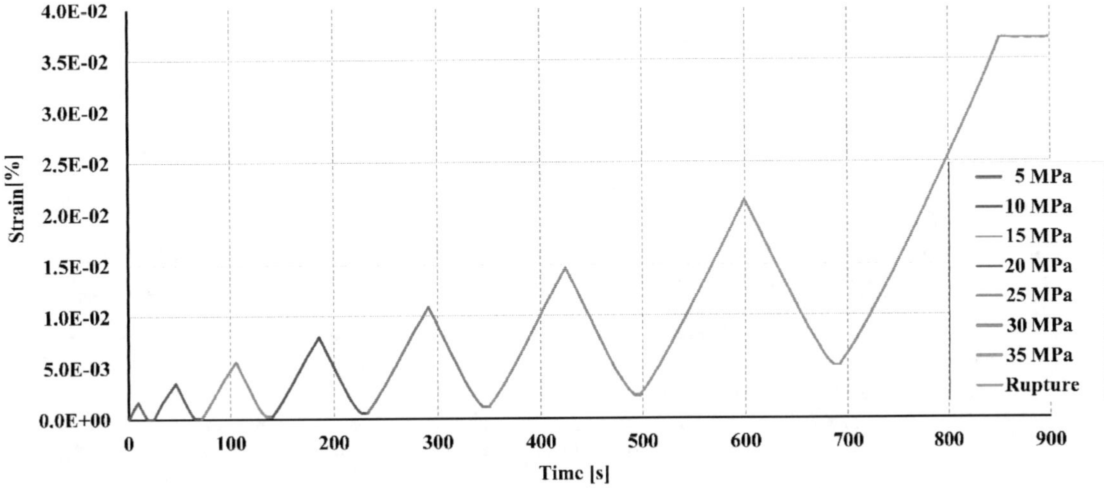

Figure 11. Stress - Time function for a 45° specimen

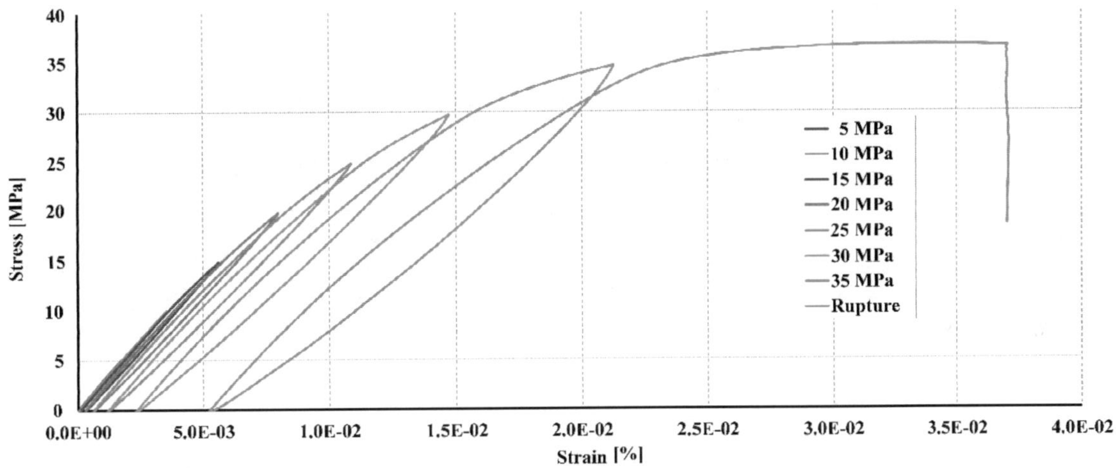

Figure 12. Stress - Strain function depending on the load for a 45° specimen

The loading curves in Fig. 12 have the same evolution, despite the plasticization of the material. There is no loss of modulus of elasticity up to 35 MPa.

4. Conclusions

In this article, two types of tests were carried out, tensile tests to failure and cyclic loading-unloading tests with varying maximum load. The specimens used were fabricated from a resin pre-impregnated unidirectional fibre, with three types of angle orientation, 0°, 90°, ± 45°, and conducted and analyzed according to ASTM D 3039, ASTM 3518 or using the method proposed by Shah [16].

The tensile tests up to failure allowed the determination of the maximum tensile strength in all the three directions as well as the elastic modulus and the linear portion of the characteristic curve. The elastic limit in all directions was determined through the appearance of residual deformations in variable load tests and observed to be 20% for the 0° fibre direction and 50% for the 90° and ± 45° ones.

Results show a complex behavior of the material, in all fibre directions, with a viscous component added to the elastic one. While this behavior has been noticed for other polymer-based composites with off axis fiber directions, it is specific for the 0°, where synthetic fibre composites behave purely linear.

The work was aimed at presenting the flax fiber epoxy resin composites behavior. It opens the way for future research, regarding the quantification of the viscous response in various loading scenarios as well as modeling of the behavior, with the purpose of implementing the material in structural applications.

References

[1] Goutianos S, Peijs T, Nystrom B and Skrifvars M 2006 Development of Flax Fibre based Textile Reinforcements for Composite Applications *Applied Composite Materials* no 13 p 199–215

[2] Nassibeh Hosseini, Samad Javid, Ali Amiri, Chad Ulven, Dean C. Webster and Ghodrat Karami 2015 Micromechanical Viscoelastic Analysis of Flax Fiber Reinforced Bio-Based Polyurethane Composites *Journal of Renewable Materials* no 3 vol 3 p 205-2015

[3] Gay Daniel 2015 *Composite Materials Design and Applications 3rd edition* (Florida, CRC Press) p 598

[4] Berthelot Jean-Marie 2013 *Mécanique des matériaux et structures composites* (Le Mans, Institut Supérieur des Matériaux et Mécaniques Avancés) p 641

[5] Anaële Lefeuvrea, Alain Bourmaudb, Laurent Lebruna, Claudine Morvana and Christophe Baley 2013 A study of the yearly reproducibility of flax fiber tensile properties *Industrial Crops and Products* no 50 p 400– 407

[6] Kestur G Satyanarayana, Gregorio GC Arizaga and Fernando Wypych 2009 Biodegradable Composites Based on Lignocellulosic Fibers-An Overview *Progress in Polymer Science* no 34 p 982-1021

[7] Le Duigou A, Davies P and Baley C 2011 Environmental impact analysis of the pro- duction of flax fibres to be used as composite material reinforcement *Journal of Biobased Materials and Bioenergy* no 5 p 153–165.

[8] Joost R. Duflou, Yelin Deng, Karel Van Acker and Wim Dewulf 2012 Do fiber-reinforced Polymer Composites Provide Enveronmentally Benign Alternatives? A Life-Cycle-Assessment-Based Study *MRS Bulletin* no 37 p 374-382

[9] Gning P B, Liang S, Guillaumat L and Pui W J 2011 Influence of Process and Test Parameters on the Mechanical Properties of Flax/Epoxy Composites Using Response Surace Methodology *Journal of Material Science* no 46 p 6801-6811

[10] Oksman Kristiina 2001 High Quality Flax Fiber Composites Manufactured by the Resin Transfer Moulding Process *Journal of Reinforced Plastics and Composites* no 20 Issue 07 p 621-627

[11] Elias Nassiopoulos and James Njuguna 2014 Thermo-Mechanical Performance of Poly (Lactic Acid)/Flax Fibre-Reinforced Biocomposites *Materials and Design* no 66 Issue 07 p 473-485

[12] D3039/ D3039M 2000 *Standard Test Method for Tensile Properties of Polymer Matrix Composite Materials* ASTM International (US)

[13] Eco-Technilin SAS *Technical Data Sheet - Flaxpreg T-UD* 2019

[14] Davies G C and Bruce D M 1998 Effect of Environmental Relative Humidity and Damag on Tensile Properties of Flax and Nettle Fibres *Textile Research Journal* no 68 Issue 9 p 623–629.

[15] D3518/ D3518M 2001 *Standard Test Method for In-Plane Shear Response of Polymer Matrix Composite Materials by Tensile Test of a ± 45° Laminate* ASTM International (US)

[16] Darshil U Shah 2016 Damage in Biocomposites: Stiffness Evolution of Aligned Plant Fibre Composites During Monotonic and Cyclic Fatigue Loading *Composites Part A: Applied Science and Manufacturing* no 83 p 160-168

Acknowledgments

This work has been funded by the European Social Fund from the Sectoral Operational Programme Human Capital 2014-2020, through the Financial Agreement with the title "Scholarships for entrepreneurial education among doctoral students and postdoctoral researchers (Be Antreprenor!)", Contract no. 51680/09.07.2019 - SMIS code: 124539, and also by a grant of the Romanian Ministry of Education and Research, Research program NUCLEU, contract no. 17N/2019, project number PN 19.24.01.01.

Intelligent control of wheel rail contact noise phenomenon in rail transportation

E Chiba, M Abdelkrim, A Belloufi and I Rezgui

University Kasdi Merbah Ouargla, Faculty of Applied Sciences, Department of Mechanical Engineering, Ouargla 30000, Algeria

chibaelhocine@gmail.com

Abstract. Railway transport systems play a very important role in the future of transport. They offer sustainable solutions that reduce carbon emissions (environmental requirements) and allow more mobility with better energy efficiency, and they generate advantageous economic benefits in the long term. But the interaction between the wheel and the rail is always very difficult to approach. There are a large number of studies in the literature on this problem. However, there is still no comprehensive approach. On this work we propose to study numerical modelling of contact between wheel-rail systems by using fuzzy logic. For this purpose, the experimental collection data reported in literature are used to predict the resonance noise during wheel-rail contact in rail transport. We found a good agreement between the elaborated model and the results of the literature.

1. Introduction

The contact between the Wheel and the rail is important for improving and facilitating the movement of the vehicle especially in bad conditions for better contact without noise, Skating…etc.
Xiaojun. Hu et al. [1] did study experimentally and analyzed the sound field in the cavity of a rotating tire through a custom acoustic pressure sensor attached to the inner surface of the tire, to obtain the distribution and frequency characteristics of the sound field.

Continuous urbanization has resulted in a higher density of population in areas near railway tracks for many years. These factors raise the problem of train noise disruption, which is pushing infrastructure managers to systematically take measures to combat rail noise [2].

In the wheel/rail contact system in Sydney, Zhu. H et al. [3] have done an investigation and control to minimize degradation on mainly squat faults, they discover two types of white layer (Severe Deformation Induced White Etching Layer and Thermally Produced White Etching Layer). Also in Washington; Sound levels were assessed both near the track and at community locations near the subject maintenance yard. Outside the yard, the vibrations were almost undetectable; thus, squeal elimination. After about 3 months of service, however, a reduction of 14 dBA with some squeal was observed; and chronic squeal reappeared after 6 months of service. The rapid contact point wear of the facing treatment was due to this lack of effectiveness [4].

RailCorp. [5] revised the wheel squeal calculation, management and mitigation on the New South Wales Rail Network in 2004. Recent mitigation steps have centered on adding the head of the rail to a friction modifier. On some pages, this has been quite good, but it is now clear that on others it is only partially effective. A rolling touch test rig is used to investigate simple squeal behavior in order to validate wheel squeal prediction models. And, respectively, by experimental effect hammer analysis.

The lateral resonant frequencies and wheel mode shapes are calculated accordingly. Based on this, a dominant mode is defined as the primary peak in the squeal sound spectrum and is used as a measure of squeal incidence and magnitude. The lateral creep curves and wheel vibration amplitudes at different rolling speeds are measured and projected using a strain gauge technique. A simplified model is created involving the interaction of the dominant mode between lateral force and transverse vibration, and the experimental and simulated results show that the wheel's sound pressure level and vibration velocity increase significantly. The wheel squeal sound range has also been tested, they find that the cause of double peaks is due to wheel rotation and, as theoretically expected, the frequency divergence of double peaks increases with rolling [6].

Field measurements of wheel squeal events at a site in Australia in 2013 revealed a growing likelihood of a squeal event occurring as the relative humidity rises. A new method to calculate the lateral and normal force simultaneously on a test rig was created to check these effects, in order to evaluate the curves of friction-creep. The relative humidity within the acoustic enclosure of the test rig was changed under regulated conditions of 50 percent, 70 percent and 90 percent to investigate the influence of relative humidity on squeal and friction creep curves. The test rig results show that with the rise in relative humidity, the lateral adhesion ratio decreases marginally and that squeal is more likely in high relative humidity. The modeling analysis reveals that with the rise in relative humidity, the critical creep decreases, which means that negative damping occurs at a lower angle of attack [7].

Xiaogang L. [8] investigated the effect of various friction modifiers on the incidence of squeal on a rolling touch of the disk test rig under positive friction characteristics. In particular, under different rolling velocities and friction modifiers, friction-creep curves and squeal sound pressure levels were measured. The findings show that friction modifiers can remove or decrease the friction-creep curves' negative slope, but there is still squealing noise. A potential explanation why wheel squeal still persists after the application of friction modifiers is discovered by theoretical modeling of instant creep behaviors. Paul A. Meehan [9] demonstrated that the stiffness mode coupling induces a very high amplitude squeal outside the limits of the simplified creep model under only a limited range of contact angles and closely matched uncoupled modes. Only models of rail mass and damping are shown to reflect stiffness and viscous mode coupling occurrences, respectively, where the optimum amplitudes depend on the complex stiffness and rail damping, respectively. Excessive non-proportional damping and very high squeal amplitudes associated with reverse complete sliding are quantified and shown to be constrained by the reach of the effective analytical model. The analytical model is shown to provide insight into the effects of mode coupling dynamics on the amplitude of wheel squeal noise and can help explain the enigma that squeal tends to occur in the field unreliably [10].

The aim of this work is to develop a precise prediction model of the phenomenon of wheel-rail contact noise in rail transport using the fuzzy logic technique based on the experimental data from literature.

2. Experimental data

The data shown in Table 1 are the results of experimental tests done by Xiaojun Hu [1]. The objective of this study was to research the effects on frequencies of important process parameters (velocity, inflation pressure and the load).

Table 1. Experimental data for the brass electrode [1].

Test	Velocity (km/h)	Pressure (bar)	Load (N)	Frequency (Hz)
1	40	2.5	3000	237
2	40	2.5	3500	237
3	40	2.5	4000	238
4	40	2.5	4500	239
5	40	2.2	3000	238

6	40	2.2	3500	238
7	40	2.2	4000	237
8	40	2.2	4500	239
9	40	1.9	3000	237
10	40	1.9	3500	238
11	40	1.9	4000	239
12	40	1.9	4500	239
13	50	2.5	3000	235
14	50	2.5	3500	235
15	50	2.5	4000	235
16	50	2.5	4500	237
17	50	2.2	3000	236
18	50	2.2	3500	236
19	50	2.2	4000	236
20	50	2.2	4500	236
21	50	1.9	3000	236
22	50	1.9	3500	237
23	50	1.9	4000	237
24	50	1.9	4500	237
25	60	2.5	3000	237
26	60	2.5	3500	237
27	60	2.5	4000	238
28	60	2.5	4500	237
29	60	2.2	3000	237
30	60	2.2	3500	237
31	60	2.2	4000	238
32	60	2.2	4500	238
33	60	1.9	3000	237
34	60	1.9	3500	238
35	60	1.9	4000	238
36	60	1.9	4500	238

2.1. Fuzzy logic

Fuzzy logic is an extension of the Boolean logic that Lotfi Zadeh. [11] developed in 1965based on the mathematical theory of fuzzy sets, a generalization of classical set theory. By adding the idea of a degree of condition verification, thus making a condition of being in a state other than true or false, fuzzy logic offers a very significant versatility in the argument for use, which makes it possible to take into account inaccuracies and uncertainties. As a consequence, the principle of fuzzy logic applied to various manufacturing processes in which tests and expert expertise play an important role [11-12].

2.2. Fuzzy system

In this fuzzy system inference, we chose:

- three variables the input (Velocity, Inflation Pressure and the Load).
- one variable output (we have the Frequency).

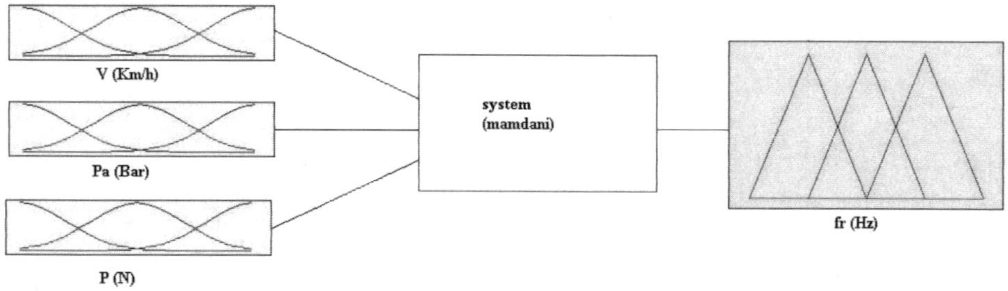

Figure 1. Inputs and outputs of the fuzzy system.

Table 2. Input and output parameters for membership function.

Factor	Symbol	Unit	Levels				
			Very Low (VL)	Low (L)	Medium (M)	High (H)	Very High (VH)
Velocity	V	Km/h	-	40	50	60	-
Pressure	Pa	Bar	-	1.9	2.2	2.5	-
Load	P	N	-	3000	3500	4000	4500
Frequency	fr	Hz	235	236	237	238	239

[40 40 50]	L
[40 50 60]	M
[50 60 60]	H

Figure 2. Linguistic variables for Velocity (V).

[1.9 1.9 2.2]	L
[1.9 2.2 2.5]	M
[2.2 2.5 2.5]	H

Figure 3. Linguistic variables for Pressure (Pa).

[3000 3000 4000]	L
[3000 3500 4000]	M
[3500 4000 4500]	H
[4000 4500 4500]	VH

Figure 4. Linguistic variables for the Load (P).

[235 235 236]	VL
[235 236 237]	L
[236 237 238]	M
[237 238 239]	H
[238 239 239]	VH

Figure 5. Linguistic variables for Frequency (fr).

Membership functions come in many forms figure 2, figure 3, figure 4, figure 5. In this study, triangular membership functions for the input variables in figure 6, figure 7, figure 8and figure 10, explained in Table 2 and the figure 9 for output variables.

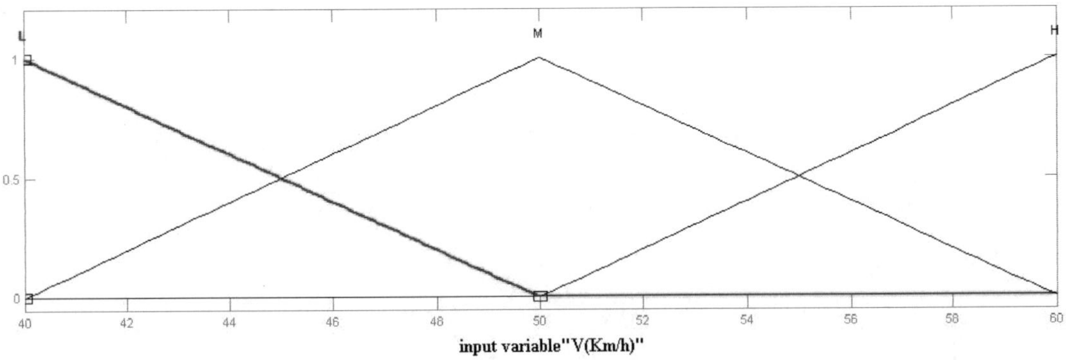

Figure 6. Linguistic variables for Velocity (V).

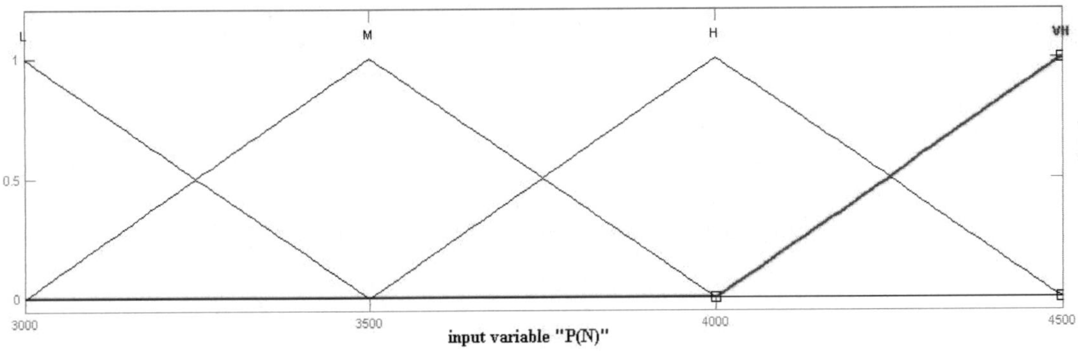

Figure 7. Linguistic variables for the Load (P).

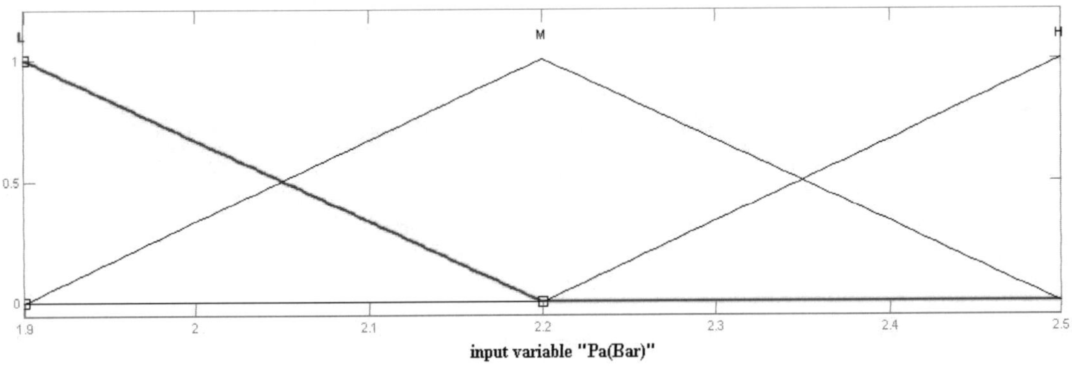

Figure 8. Linguistic variables for Pressure (Pa).

In order to properly define the linguistic variables for the outputs we have used the scatter plot in figure 9, which presents the distribution of the values of the frequency.

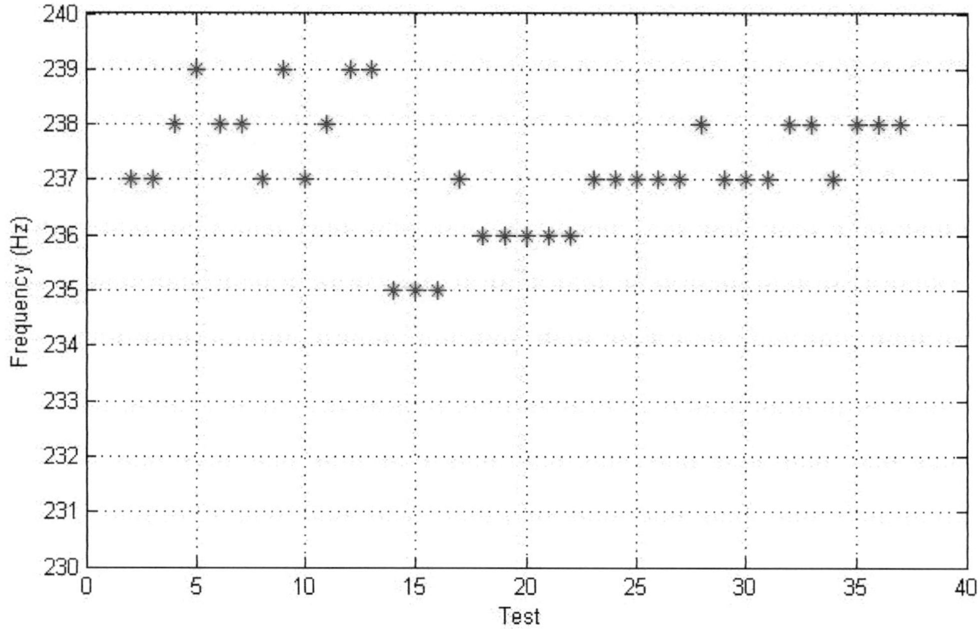

Figure 9. Linguistic variables for Frequency (fr).

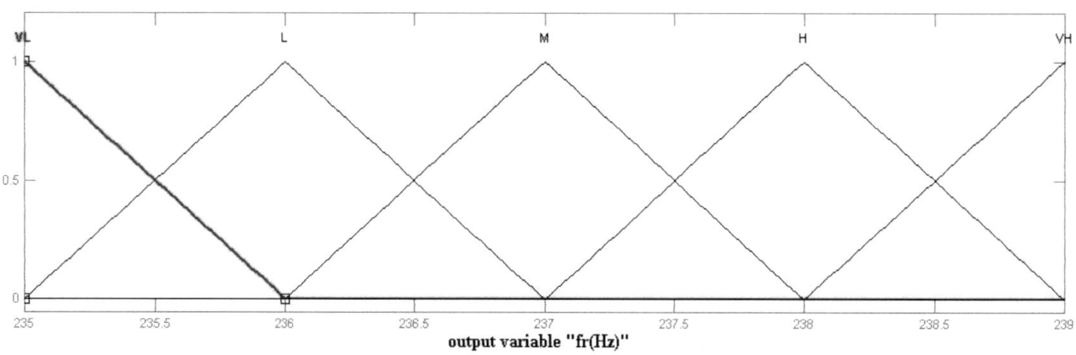

Figure 10. Linguistic variables for Frequency (fr).

2.3. Fuzzy rules

The rule to calculate it's come and represent Table 3:

If a velocity is… and Pressure is… and load is… so Frequency is…

Table 3. Fuzzy rules.

Test	Velocity (km/h)	Pressure (bar)	Load (N)	Frequency (Hz)
1	L	H	L	M
2	L	H	M	M
3	L	H	H	H
4	L	H	VH	VH

5	L	M	L	H
6	L	M	M	H
7	L	M	H	M
8	L	M	VH	VH
9	L	L	L	M
10	L	L	M	H
11	L	L	H	VH
12	L	L	VH	VH
13	M	H	L	VL
14	M	H	M	VL
15	M	H	H	VL
16	M	H	VH	M
17	M	M	L	L
18	M	M	M	L
19	M	M	H	L
20	M	M	VH	L
21	M	L	L	L
22	M	L	M	M
23	M	L	H	M
24	M	L	VH	M
25	H	H	L	M
26	H	H	M	M
27	H	H	H	H
28	H	H	VH	M
29	H	M	L	M
30	H	M	M	M
31	H	M	H	H
32	H	M	VH	H
33	H	L	L	M
34	H	L	M	H
35	H	L	H	H
36	H	L	VH	H

2.4. Defuzzification

The difference between the observed and the forecasted value was determined by calculating the error. The (1) equation can be used to measure errors. A division of the absolute difference in the measurement value was accomplished by dividing the percentage of individual errors.

$$e_i = \left(1 - \frac{|fr_{ex} - fr_p|}{fr_{ex}}\right) * 100\% \qquad (1)$$

Precision is determined using the expected value method to the observed value. The consistency of the model is A in equation (2), and N is the cumulative number of evaluated data sets. The model's precision is average individual precision.

$$A = \frac{1}{N}\sum\left(1 - \frac{|fr_{ex} - fr_p|}{fr_{ex}}\right) * 100\% \qquad (2)$$

3. Results

After using the fuzzy system and the rules of table 3 by equation 1, 2, we obtained the results presented in Table 4.

Table 4. fuzzy results for calculate Ei% and A%.

Test	Frequency (Hz)	Frequency (Hz) SUMI	E_i %	A%
1	237	237	0	100
2	237	237	0	100
3	238	238	0	100
4	239	239	0	100
5	238	238	0	100
6	238	238	0	100
7	237	237	0	100
8	239	239	0	100
9	237	237	0	100
10	238	238	0	100
11	239	239	0	100
12	239	239	0	100
13	235	235	0	100
14	235	235	0	100
15	235	235	0	100
16	237	237	0	100
17	236	236	0	100
18	236	236	0	100
19	236	236	0	100
20	236	236	0	100
21	236	236	0	100
22	237	237	0	100
23	237	237	0	100
24	237	237	0	100
25	237	237	0	100
26	237	237	0	100
27	238	238	0	100
28	237	237	0	100
29	237	237	0	100
30	237	237	0	100
31	238	238	0	100
32	238	238	0	100
33	237	237	0	100
34	238	238	0	100
35	238	238	0	100
36	238	238	0	100
			A= 100%	

3.1 *Schematic representation of the results*

The figures below show the variation of the frequency as a function of the speed and of the load applied during the contact between wheel-rail.

Figure 11 shows the variation of the frequency depending on the speed and pressure.

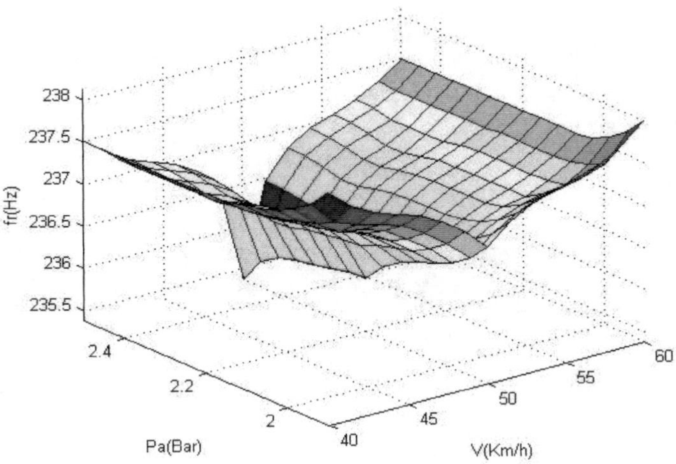

Figure 11. Variation of frequency with speed and pressure.

Figure 12 shows the variation of the frequency depending on the speed and the load.

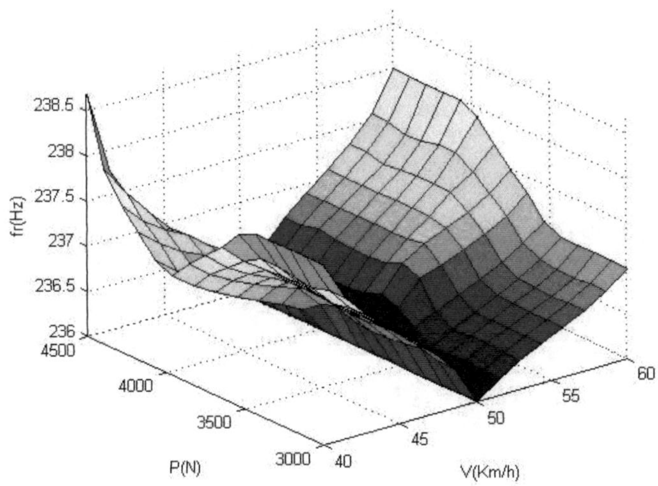

Figure 12. Variation of frequency with speed and pressure.

Both figures show that the maximum frequency Value is obtained for the maximum values of speed and load or pressure.

These figures allowed us to detect the noise zone as a function of the speed and pressure load applied during the contact between the wheel and the rail.

The fuzzy logic method helps us to optimize the input parameters in order to find good solutions for such a phenomenon.

3.2 Validation of results

The following figure shows the comparison between the values predicted by our model proposed by fuzzy logic and the values obtained from the literature.

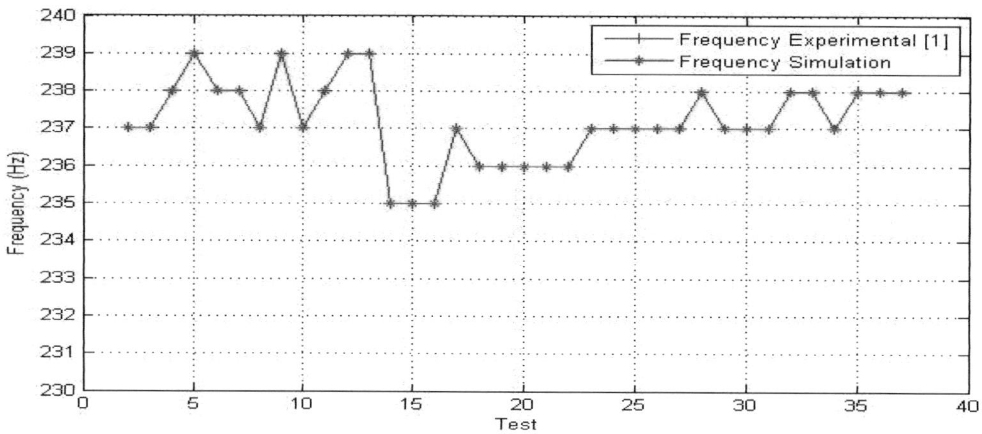

Figure 13. Comparison between the experimental result [1] And fuzzy system result.

From the figure 13, the curves are similar so our fuzzy logic based prediction model works well and with high accuracy and can be used as a solution to predict the wheel rail contact noise phenomenon in rail transportation.

4. Conclusion

The noise resulting from contact between wheel and rail to rail transport has a major effect on the health train drivers and people. The work in this modest research, are part of the objective of the study and development of optimization techniques.

To validate our model developed by the fuzzy logic method proposed to study the wheel rail contact noise phenomenon in rail transportation, a comparison with the experimental study carried out by Xiaojun Hu study [1] was made

The study we presented allowed us to conclude that

- The comparison and validation of the predicted results with the results of the experimental tests confirmed the accuracy of the models developed.
- The maximum values of the frequency are obtained for maximum values of speed and load or pressure.
- The fuzzy logic method is more reliable with an average error of 0 % and the precision of the developed model is 100%.
- The fuzzy modeling technique could be an economical and efficient method to predict the wheel rail contact noise phenomenon in rail transportation.

References

[1] Xiaojun H, Xiandong L, Xiaofei W, Yingchun S and Jiajing Y, Experimental analysis of sound field in the tire cavity arising from the acoustic cavity resonance, *Applied Acoustics*, 2020. **161** p 107172

[2] Matthias A, Matti R, Roger J, Martin H, Combating curve squeal noise, Conference world congress railway research 2016. p. 1-6

[3] Zhu H, Li H, Al-Juboori A,Wexler D,Lu C, Mccusker A, Mcleod J, Pannila S, Barnes J, Understanding and treatment of squat defects in a railway network, *J Wear*. 2020. 442 443 p 203139

[4] Staiano M, Sastry G, Sastry, Control of Wheel Squeal Noise in Rail Transit Cars, *J Transportation Research Record.* 1990(1255)

[5] Dave A, Wheel squeal measurement, management and mitigation on the New South Wales rail network, *Proceedings of Acoustics, Gold Coast, Australia*. 2004 p 233-8

[6] Liu X, Meehan P A, Investigation of the effect of lateral adhesion and rolling speed on wheel squeal noise, *J of Rail and Rapid Transit*. 227(5) 469-480

[7] Liu X, Meehan P A, Investigation of the effect of relative humidity on lateral force in rolling contact and curve squeal, *J Wear*. 2014. 310(1-2) p 12-19

[8] Liu X, Meehan P A, Investigation of squeal noise under positive friction characteristics condition provided by friction modifiers, Journal of Sound and Vibration. 2016. 371 p 393-405

[9] Meehan P A, Prediction of wheel squeal noise under mode coupling, *Journal of Sound and Vibration.* 2020 465 p. 115025

[10] Meehan P A, Liu X, Modelling and mitigation of wheel squeal noise amplitude, *Journal of Sound and Vibration.* 2018 413 p 144-158

[11] Belloufi A, Mezoudj M, Abdelkrim M, Rezgui I, Chiba E, Experimental and predictive study by multi-output fuzzy model of electrical discharge machining performances, The International Journal of Advanced Manufacturing Technology. 2020 109(7) p 2065-2093

[12] Belloufi A, Rezgui I, Mihi A, Experimental investigation of the corrosion resistance of Ni-Al2o3 composite coatings obtained by electrodeposition, 2020

IOP Conf. Series: Materials Science and Engineering **1018**(2021) 012006 doi:10.1088/1757-899X/1018/1/012006

Aspects of the wheel-switch contact studied with the CONTACT software

M C Tudorache[1], V M Dinu[2], R A Oprea[1]

[1] Department of Railway Vehicles, University Politehnica of Bucharest, 313 Splaiul Independenței, Bucharest, Romania
[2] Doctoral School of Transports, University Politehnica of Bucharest, 313 Splaiul Independenței, Bucharest, Romania

E-mail: cristi_tudorache2003@yahoo.com

Abstract. The rapid variation of the wheel and rail profile influences, on one hand, the grip - pseudo-slip - wear characteristics, and on the other hand, the dynamic behavior of the vehicle, because the pseudo-slip forces are significantly influenced by the contact area and pressures from within this area. It must be borne in mind that, in most cases, the rail-road has deviations from the geometric dimensions, which must not affect traffic safety and the demands of the running gear and the track must be within the permissible limits. Railway switches ensure discontinuities on the railway causing high dynamic loads. In operation, the rail-way devices must ensure three basic requirements: safety against derailment, stability while on the move and passenger comfort. In this context the paper presents aspects of the multipoint wheel - rail-way switch contact using the Contact software. Considering the influence of sliding speeds on the contact phenomena at the intersection of the wheel with the running path, the paper presents an analysis of the kinematic behavior of the axle while passing over a point rail.

1. Introduction

The shape and dimensions of the contact area determine the reliability of the wheels and rails, the safety of the steering and the traction and braking characteristics of the vehicle.

It must be borne in mind that, in most cases, the rail-road has deviations from the theoretical geometric dimensions, which must not affect traffic safety and the demands of the running gear and the track must be within the permissible limits.

With what has been mentioned above only the wheel-rail contact was emphasized, but just as important for the safety of the guide is the wheel's contact with the point rail. Thus, the track switches, in this case the point rail, ensure discontinuities on the railway causing high dynamic loads. In operation, the track devices must ensure three basic requirements: safety against derailment, stability while on the move and passenger comfort.

Up until recently the problem of the influence of wheel-rail contact phenomena on steering safety has been done using simplifying hypotheses and approximate methods that have led to illustrative results, on one hand; on the other hand, the analysis of contact phenomena produced over point rail crosses is a field of expertise almost untouched in our country, being limited only to a visual inspection through the use of template checks.

Analysing the specialty literature we can see a focus on the interaction of the vehicle with the tip of the crossing heart, thus imposing the need for a three-dimensional analysis of the wheel's contact with the point rail in terms of traffic safety, [1-12]. This analysis is based on the applicability of the CONTACT software, [13], this being considered an advanced simulation program applied to contact between the wheel and the rail, where in each element of the contact ellipse the law of dry friction of Coulomb is applied. The originality of the work consists in the application of the software in the case of the bi-contact made by the wheel when passing over the track changer, in this case the point rail.

2. The linear kinematics of wheel - railway contact

In France, Levi, a French railway engineer, publishes two papers on the rolling contact, based on experimental investigations. His purely empirical representation can be found in Rocard's monograph on "Stability of locomotive running". The relation established by Levi between the coefficient of friction τ and pseudo-slips is of parabolic form:

$$\frac{1}{\tau}=\frac{1}{\mu}+\frac{1}{\chi\cdot\nu} \tag{1}$$

where $\mu = F_{max} / N$ is the maximum coefficient of friction, and $x = tg\ \varphi = (\tau / \nu)$, $\nu \to 0$ is called the pseudo-slip coefficient.

A laborious treatment of the problem of contact is made by J. Kalker in his PhD, "On the rolling contact of two elastic bodies in the presence of dry friction", held in 1967 in Delft. It tries to establish the delimitation of the adhesion / sliding zones and the unitary efforts within them according to the tangential forces. For this, the longitudinal, transverse and spin pseudo-slips and the local deformations of the two bodies are calculated at each point in the contact area and the normal force resulting from the pressure distribution ellipsoid is considered. It is considered that in the sliding zone the tangential forces in the two directions satisfy Coulomb's law, and in the adhesion zone they are smaller. The size of the areas results from local stresses and deformations, [14].

The first stage in determining the sliding speeds is the study of the contact's geometry of the two surfaces, and based on that the contact points are determined according to the position of the axle on the path.

Assuming that the contact surface, S, is divided into a S_{stick} adhesion zone and a slip zone, S_{slip}, for which the conditions must be met:

$$\begin{aligned}(x,y)\in S_{stick} &\Rightarrow \begin{cases} s(x,y,t)=0 \\ |q(x,y,t)|<\mu\cdot p(x,y,t)\end{cases}\\ (x,y)\in S_{slip} &\Rightarrow \begin{cases} s(x,y,t)\neq 0 \\ |q(x,y,t)|=\mu\cdot p(x,y,t)\end{cases}\end{aligned} \tag{2}$$

where: $q\ (x, y, t)$ - tangential stress in the contact area, μ - Coulomb's coefficient of friction, $p\ (x, y, t)$ - normal pressure in the contact area, $s\ (x, y, t)$ - sliding near the point of contact.

The tangential force at the point of contact has two components, one longitudinal, F_x, and one transverse, F_y, which results from the integration of the tangential force:

$$\begin{aligned}F(t)&=\iint_S q(x,y,t)\\ F(t)&=\{F_x,F_y\}\end{aligned} \tag{3}$$

In kinematic regime the slip is defined by the following relations:

$$s_x(x,y,t) = \nu_x(t) - \phi(t)y + \frac{\partial}{\partial x}u_x(x,y,t) - \frac{1}{V_m}\frac{\partial}{\partial x}u_x(x,y,t)$$
$$s_y(x,y,t) = \nu_y(t) - \phi(t)x + \frac{\partial}{\partial x}u_y(x,y,t) - \frac{1}{V_m}\frac{\partial}{\partial x}u_y(x,y,t) \quad (4)$$

For the stationary regime the term $\partial/\partial t\ u(x,y,t)$ becomes null situation in which the relation (4) becomes:

$$s_x(x,y,t) = \nu_x - \phi \cdot y + \frac{\partial}{\partial x}u_x(x,y)$$
$$s_y(x,y,t) = \nu_y - \phi \cdot x + \frac{\partial}{\partial x}u_y(x,y) \quad (5)$$

Pseudo-slips that occur in relation 1 and which together with the coefficient of friction and the normal force are one of the essential parameters for determining the wheel-railway contact forces.

The longitudinal, transverse and spin pseudo slip are established based on the relationships, [15]:

$$\nu_x = \frac{s_x}{V}, \nu_y = \frac{s_y}{V}, \phi = \frac{r \cdot \omega_s}{V}$$

ω_s – - angular velocity of spin.

Pure rolling in [15] is defined as a rolling without longitudinal and lateral slips and is also called conical rolling. With the mention that the longitudinal or transverse pseudo-slip is defined as the ratio between the relative speed of the wheel in the longitudinal or lateral direction and its forward speed. This rolling can be performed only rarely and for short periods of time; all cases of abnormal rolling of the vehicle on the rails, such as acceleration or braking, lead to the occurrence of pseudo-slips at the wheel-railway interaction. In the case of pure slip, the tangential force between the two bodies reaches the friction limit. There is no slip below this limit. However, specialized studies have found an intermediate state in which the elasticity of the two bodies in contact allows the division of the contact area into an adhesion zone and a sliding zone. Thus, below the friction limit value, there is a transition from the predominance of the adhesion zone (1) to the sliding zone (4), according to fig. 1.

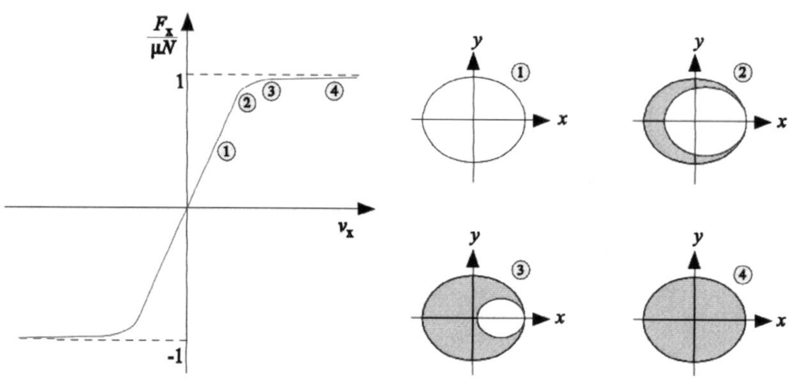

Figure 1. Elliptical contact adhesion area [16]

3. Numerical simulations

The analysis of the kinematic regime of the axle is performed considering a symmetrically loaded axle. For the case presented, the axle is equipped with S78 profile wheels and passes over a point rail with at a speed of 160 km / h in direct and 30 km / h in deviation. The load on the wheel is considered 20kN and 40kN respectively. The attacking wheel comes into contact with the point rail on the lip by making bi-contact between the point rail-railway-wheel, so that later, as the wheels advance, the contact moves to the outer side of the wheel's rolling surface. The coordinates of the two contact points are established by the method of minimum elevation distances. At the same time, the distribution of normal loads on the two contact areas is made based on the fact that the rigidities are equal and the deformation is proportional to the ratio 3/2.

Figure 2 shows the distribution of the relative pseudo-slip velocity in the longitudinal direction considering the zero spin coefficient and the wheel load 40 kN (a,b) - 20 kN (c,d). An intensification of the field of longitudinal pseudo-sliding speeds is observed at the point of contact on the wheel's lip (Fig. 2 a, c), where the sliding area of the contact spot is present. These are reduced as the load on the wheel decreases. At the point of contact on the running surface, point of support, where the sliding area predominates, the sliding speeds show very small intensifications towards the inside of the running surface, fact justified by the takeover of the guide by the lip, see fig. 2 b, d.

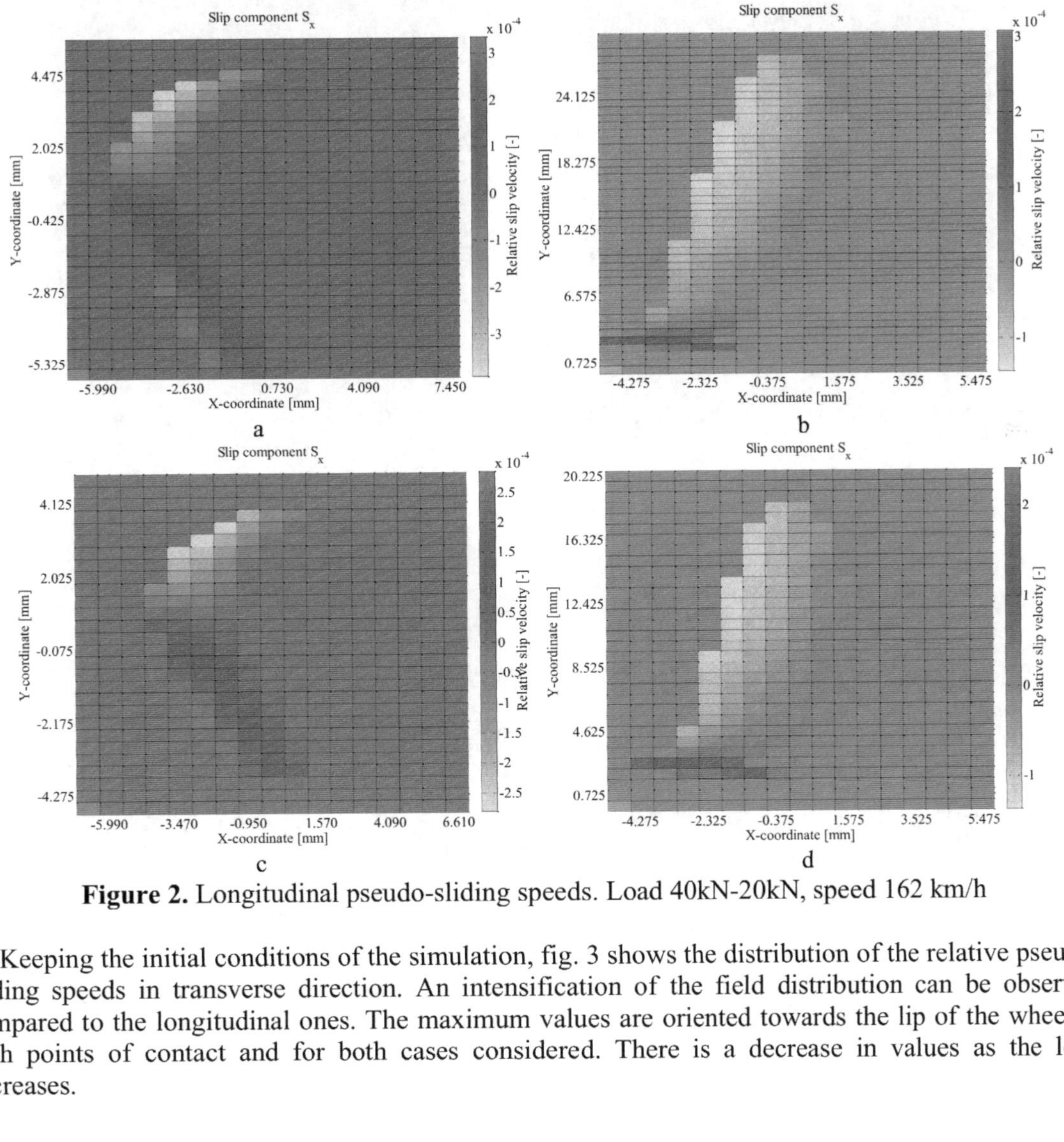

Figure 2. Longitudinal pseudo-sliding speeds. Load 40kN-20kN, speed 162 km/h

Keeping the initial conditions of the simulation, fig. 3 shows the distribution of the relative pseudo-sliding speeds in transverse direction. An intensification of the field distribution can be observed compared to the longitudinal ones. The maximum values are oriented towards the lip of the wheel at both points of contact and for both cases considered. There is a decrease in values as the load decreases.

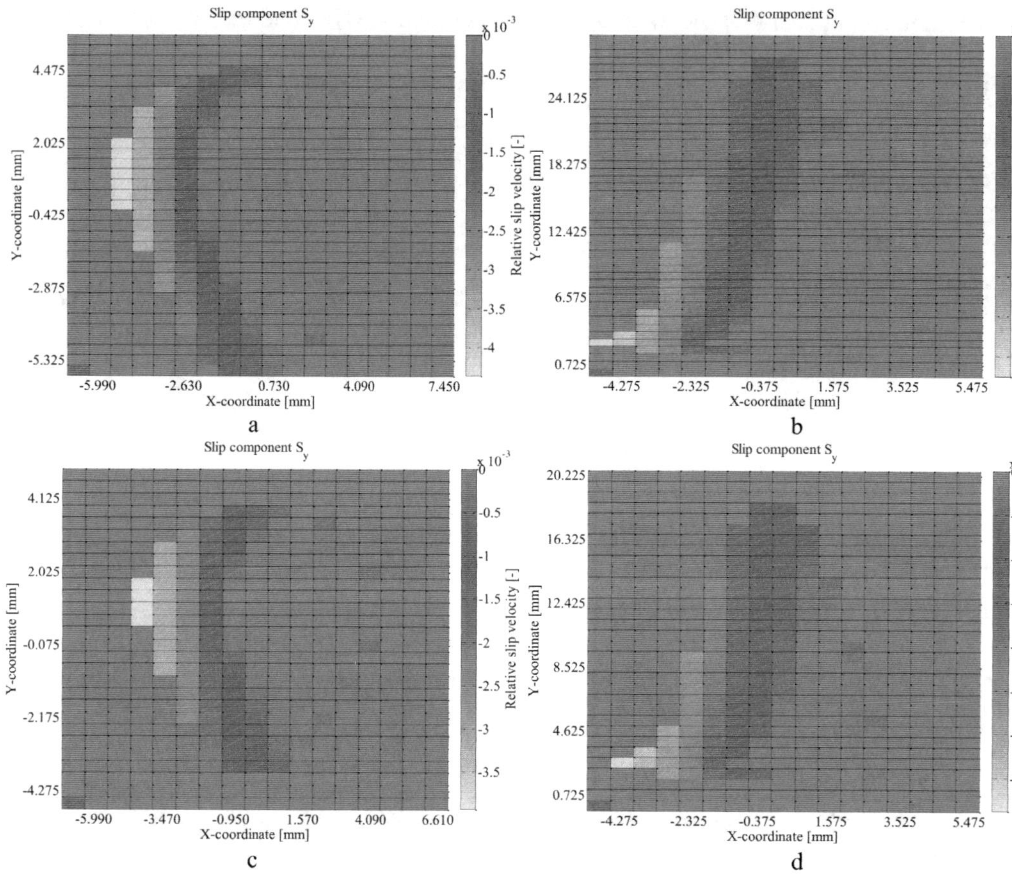

Figure 3. Transversal pseudo-sliding speeds. Load 40kN-20kN, speed 162 km/h

If the traffic speed is reduced to 30 km/h, traffic on deviation, there is no significant change in the distribution of the pseudo-slip speed field.

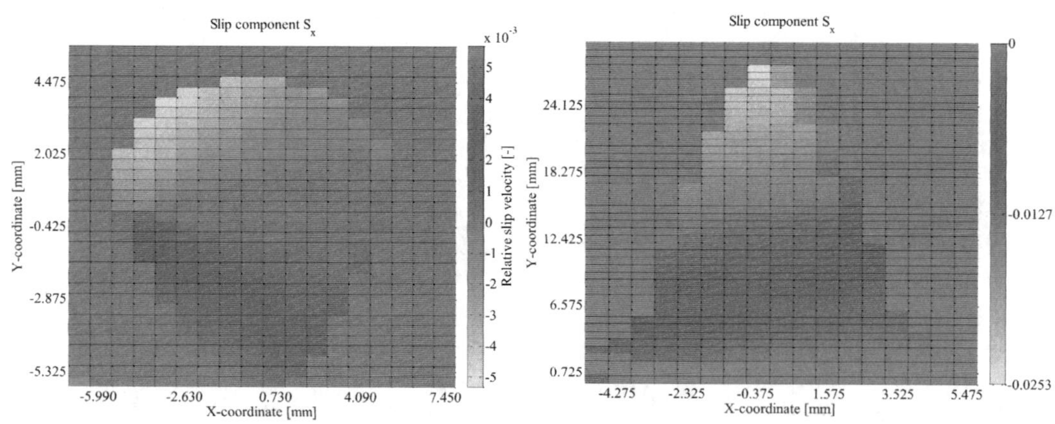

Figure 4. Longitudinal pseudo-sliding speeds, load 40kN, speed 30 km/h, spin 0.000923rad/mm

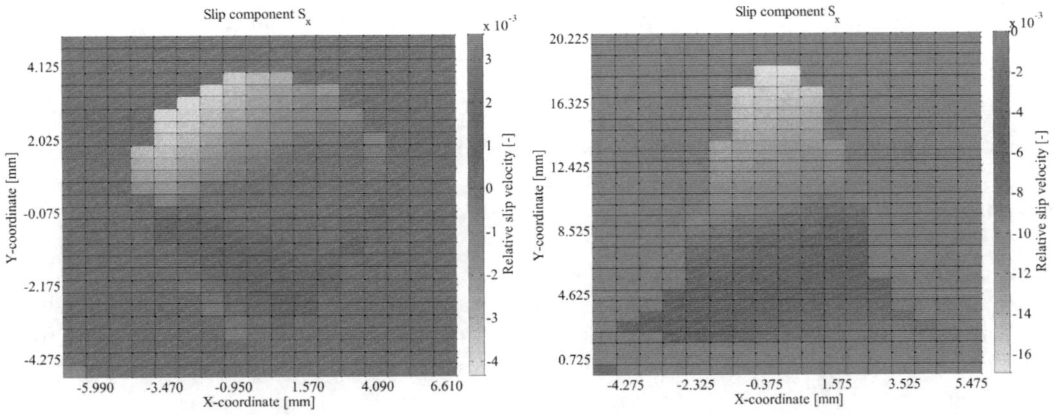

Figure 5. Longitudinal pseudo-sliding speeds, load 20kN, speed 30 km/h, spin 0.000923rad/mm

At the deviation traffic, with a speed of 30 km/h, also taking into account a non-zero value of the spin coefficient, 0.000923 rad / mm, corresponding to the situation in which the axle runs at the adhesion limit, it can be seen in fig. 4 to 6 the distribution of pseudo-sliding speeds in longitudinal and transverse direction.

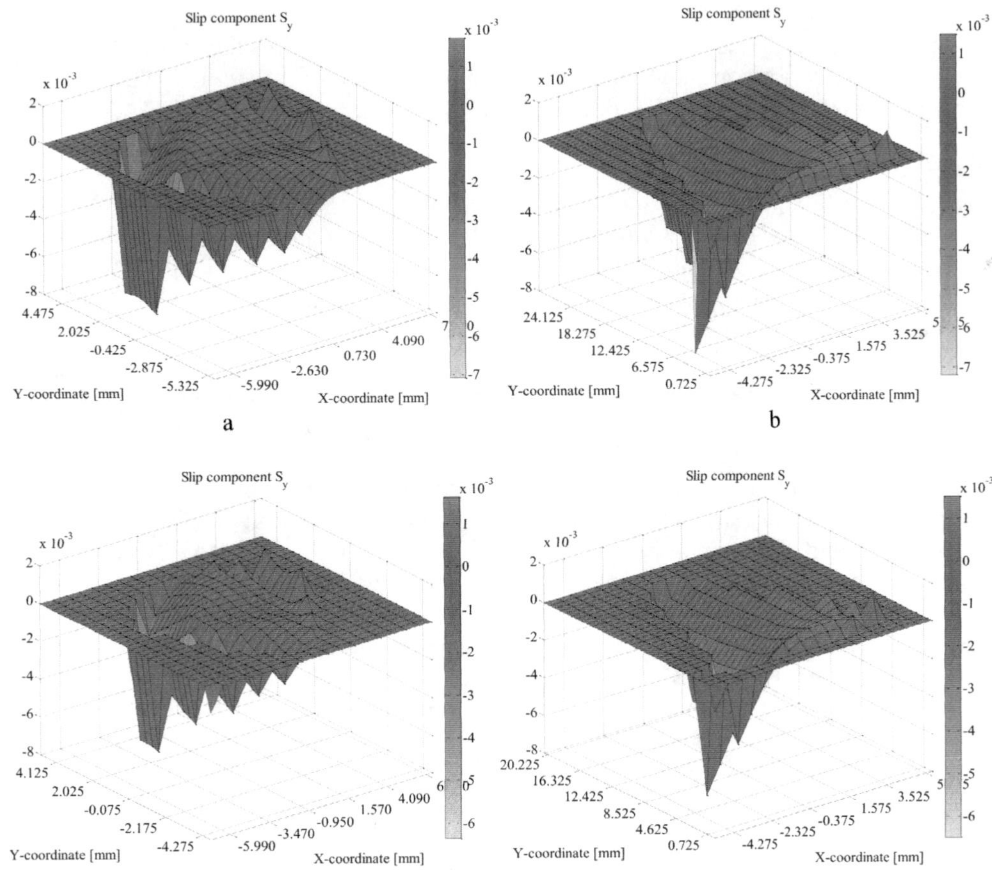

Figure 6. Transversal pseudo-sliding speeds.

There is a significant increase in pseudo-sliding speeds in the longitudinal direction for the situation where the load on the wheel is 40 kN and at the point of contact on the lip, v fig, 4a – 5a. This is justified by the guiding role of the wheel's lip. In the transverse direction, see fig. 6, compared to the situation in which the spin coefficient was considered zero, see fig. 3, there is a slight increase in values for the two cases considered, which is explained by the pivoting effect.

4. Conclusions

In this paper we approached a less studied side from an analytical point of view in our country, namely the analysis of the contact between wheel and point rail and the influence of it on guidance. Also, a three-dimensional approach to the kinematic behaviour of an axle equipped with S78 profile wheels, by adapting the iterative Contact software to bi-contact. The originality of the study consists in the application of the Contact software in case of wheel's interaction when crossing the track changer, in this case the point rail.

The perspective of pseudo-sliding speeds can determine the friction forces and the distribution of stresses in the contact area. At the same time, the problem of tangential contact can be solved and especially as it is observed that of the secant contact. The values of the multipoint sliding speeds are of major importance when the second point of contact is on the lip of the wheel, which is crucial for safety against derailment.

Acknowledgement

This work has been funded by the European Social Fund from the Sectoral Operational Programme Human Capital 2014-2020, through the Financial Agreement with the title "Scholarships for entrepreneurial education among doctoral students and postdoctoral researchers (Be Antreprenor!)", Contract no. 51680/09.07.2019 - SMIS code: 124539.

References

[1] Sugiyama H, Sekiguchi T, 2012 Wheel/rail contact dynamics in turnout negotiations with combined nodal and non-conformal contact approach, Multibody System Dynamics 27:55–**74**, DOI 10.1007/s11044-011-9252-0.

[2] Kassa E, Nielsen J C O, 2008 Dynamic interaction between train and railway turnout: full-scale field test and validation of simulation models, Vehicle System Dynamics **, 46(5)**, pp. 521–534.

[3] Kassa E, Anderson C, Nielsen J, 2006 Simulation of dynamic interaction between train and railway turnout, Vehicle System Dynamics **44**, pp 247–258.

[4] Kono H, Suda Y, Yamaguchi M, Yamashita H, Yanobu Y, Tsuda K, 2005 Dynamic analysis of the vehicle running on turnout at high speed considering longitudinal variation of rail profiles, Proceedings of ASME International Design Engineering Technical Conferences and Computer and Information in Engineering Conference, USA, pp. 2149-2154.

[5] Alfi S, Bruni S, 2009 Mathematical modeling of train-turnout interaction, Vehicle System Dynamics **47**, pp. 551–574.

[6] Ren Z, Sun S, Zhai W, 2005 Study on lateral dynamic characteristics of vehicle/turnout system, Vehicle System Dynamics, **43(4)**, pp. 285–303.

[7] Kassa E, Johansson G , Simulation of train–turnout interaction and plastic deformation of rail profiles, Vehicule System Dynamics, Int. Journal of Veh. Mechanics and Mobility, **44**:sup1, pp. 349-359, http://dx.doi.org/10.1080/00423110600871566.

[8] Shu X, Wilson N, Sasaoka, C, Elkins J, 2006 Development of a real-time wheel/rail contact model in NUCARS and application to diamond crossing and turnout design simulations. Vehicle System Dynamics, **44**(Suppl), pp. 251–260.

[9] Sun Y Q, Cole C, Boyd P, 2011 A numerical method using VAMPIRE modelling for prediction of turnout curve wheel–rail wear, Wear **271**, pp. 482–491, doi:10.1016/j.wear.2010.10.010

[10] Ren Z, Sun W, Zhai S, 2005 Study on lateral dynamic characteristics of vehicle/turnout system, Vehicle System Dynamics: International Journal of Vehicle Mechanics and Mobility, **43:4**, pp. 285-303, DOI: 10.1080/00423110500083262
[11] Wang P, 1997 *Research on wheel/rail system dynamics on tournout*, Phd thesis, Southwest Jiaontong University of China.
[12] Droździel J, Sowiński B, Groll W 1999 The Effect of Railway Vehicle-Track System Geometric Deviations on its Dynamics in the Turnout Zone, Vehicle System Dynamics, 33:sup1, pp. 641-652, DOI: 10.1080/00423114.1999.12063118.
[13] https://www.cmcc.nl/User guide for CONTACT, Rolling and sliding contact with friction.
[14] Kalker J J, 1967 *On the rolling contact of two elastic bodies in the presence of dry friction*, Faculty Electrical Engineering, Mathematics and Computer Science, doctoral thesis.
[15] Sebeşan I,1995 *Dinamica vehiculelor de cale ferată*, Editura Tehnică, pp. 87-85.
[16] Zougar A, 2014 *Estudio del comportamiento vibratorio de vias ferroviarias mediante simulation numerica*, tesis doctorale, Universitat Politecnica de Catalunyia, Barcelona.

Mathematical, detailed and parametric modelling for smart spinal orthoses

R Murzac[1], C V Doicin[1], M E Ulmeanu[1,2] and T Țurcanu[1]

[1] University POLITEHNICA of Bucharest, Splaiul Independentei, No. 313, District 6, Bucharest, Romania

E-mail: mihaela.ulmeanu@upb.ro

Abstract. The paper presents a mathematical model and the design details to manufacture a prototype of a smart orthosis for posture correction while sitting on a chair at your office or at home. The global geometry for developing the mathematical model was obtained by direct measuring a human 3D model, developed by scanning one of the authors' body and 3D printing it. Based on the mathematical model, the detailed design for the entire product was performed in accordance with the product architecture. The stages of detailed design for the critical systems within the final assembly are also presented in the paper. The posture correction orthosis must be customized for each user, thus, once the design for the critical components has been validated, a parametric design has been generated to facilitate the manufacturing process. The technical solutions are generated and analysed for a vertebral element of the orthosis, representing the critical component of the product. The vertebral element is subject to mechanical stresses, so that a finite element analysis was performed. The geometry of the component and its material were validated by using FEA on the CAD model of the vertebral element. The further work will include research regarding the changes needed for transforming the orthosis into a medical device for spinal column illnesses.

1. Introduction

The mathematical modelling and mathematical simulations play an important role in biological sciences, and they are essential for a better understanding of the functions of human body [1].

Physical correction of human posture must use a device able to follow the anatomical shape of the spinal column. Studies show that improper body posture during sitting has detrimental effect to the human spinal column region over an extended period [2]. The paper proposes a mathematical model to represent the spinal column of human body during sitting tasks, based on a two-link kinematic open chain in two dimensional spaces. Thus, all movement joints and their effect to every part of vertebral bodies of human body can be thoroughly analysed.

The global geometry for developing the mathematical model was obtained by direct measuring a human 3D model, developed by scanning one of the authors' body and 3D printing it. A global Cartesian coordinate system was established wherein the X-Y plane is mid-sagittal, the X-Z plane is transverse and tangent to the inferior surface of the ischium, and Y-Z is a frontal plane [3].

The increasing demands have motivated progress in the analysis and simulation of complex mechanical systems. The development of mathematical models of mechanical systems, however, requires both knowledge of the best methods of mechanics as well as practical experience. For

[2] mihaela.ulmeanu@upb.ro

dynamic systems, such as vehicles, mechanisms, machinery, and robots, discrete models have long proved themselves useful. Such multibody models are called mechanical systems even if they also contain devices which are electric, magnetic, hydraulic, or pneumatic in nature [4].

On the other side, due to the advantages of low cost and small size, wearable sensors have been drawing attention. They can meet most needs of human motion monitoring and have been adopted in many applications, such as interactive gaming and learning, animation, health care, personal navigation, and security monitoring [5].

Following all these tendencies, the paper presents the elements that were the basis for the development of a mathematical model for the manufacture a prototype of orthosis for posture correction while sitting on a chair at your office or at home. The detailed design process of the solution that allows the materialization of the mathematical model into a mechatronic product is also described, together with the process of optimizing the design of critical components.

2. Mathematical modelling

To develop a mathematical model for detecting and correcting the user's posture, the anatomy of the spine, its curves and joints were studied. The spine acts as a pillar that supports the human body and protects the spinal cord. Viewed from the side (Figure 1), it normally has the shape of letter `S` due to three natural curves, which is why the spine can be divided into three regions: cervical, thoracic, and lumbar. The spine consists of 24 vertebral bones and two bones in the axial section of the pelvic girdle, sacrum, and coccyx. There are 7 cervical vertebrae, 12 thoracic vertebrae and 5 vertebrae [6].

The mathematical model describes the shape of the column based on the angles read by the sensors on the spinal orthoses prototype (Figure 2).

The input data of the calculation algorithm are:

- coordinates of the initial positions of the sensors: S_1 (x_1, y_1), S_2 (x_2, y_2), S_3 (x_3, y_3) and S_i (x_i, y_i);
- coordinates of the final positions of the sensors: S'_1 (x'_1, y'_1), S'_2 (x'_2, y'_2), S'_3 (x'_3, y'_3) and S'_i (x'_i, y'_i);
- the values of the angles between the rigid segments: α_1, α_2, α_3 and α_i;
- length of rigid segments: l_1, l_2, l_3 and l_i.

The output data of the calculation algorithm are:

- the angular values of rotation for each of the rigid segments: u_1, u_2, u_3 and u_i;

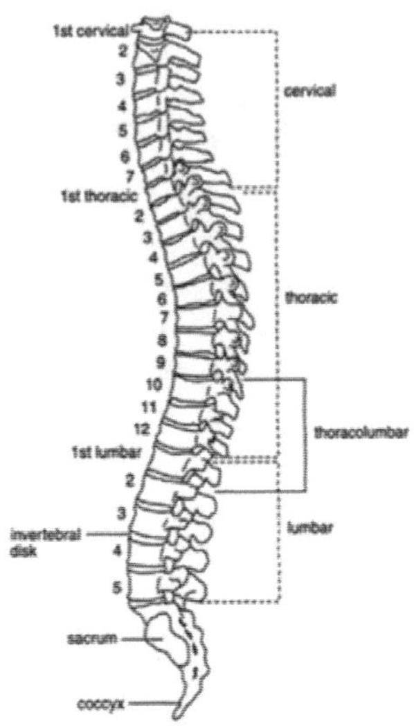

Figure 1. Side view of anatomical Spinal Column [7]

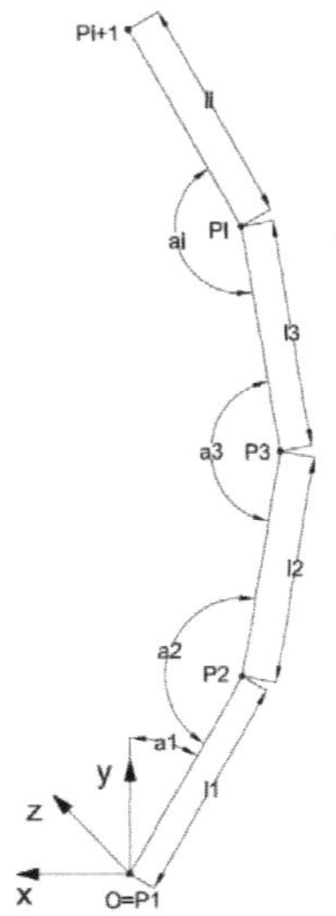

Figure 2. Mathematical model of Spinal Column

IOP Conf. Series: Materials Science and Engineering **1018**(2021) 012007 doi:10.1088/1757-899X/1018/1/012007

2.1. Initial assumptions

Known data for the mathematical model are orientation angles and distances. The chosen approach refers to a mathematical model based on straight segments: the equations needed to calculate the coordinates of the target points are only first-degree relations. Thus, a linear system of equations is reached that is easier and faster to solve.

The following hypotheses are considered:

- The carrier monitoring system will offer the possibility of orientation on the directions of the 3 axes, but only one axis will be studied;
- The proposed mathematical model is based on the approximation of real curves with line segments and geometric functions;
- The model inputs will be the measured distances between the sensors on the flexible frame. Orientation angles will be provided by inertial sensors, as follows:
- The segments formed between two sensors are approximated with segments of length l_1, l_2, l_3 and l_i and the angle between two consecutive segments α_1, α_2, α_3 and α_i;
- The right segments will coincide with each other, and a smooth curve will be obtained.

The model can precisely calculate the coordinates (x, y) of the position of the sensors and can reconstruct the natural curves of the spine.

2.2. Methodology stages

The mathematical model development has the following stages:

- Initialize the coordinates (x_i, y_i);
- Define the centre rotation S_i;
- Calculate the angles of rotation u_i;
- Reset the coordinates (x_i, y_i);
- Repeat the last two steps for the required number of iterations.

2.3. Coordinates initialization

The coordinates for the established points on the spine are initialized. The initialization of the coordinates is done iteratively for each joint as follows:

S_1: (x_1, y_1) - remains fixed.

S_2: A centre rotation S_1 is defined:

$$\begin{pmatrix} x_2 \\ y_2 \end{pmatrix} = l_1 \begin{pmatrix} \cos\alpha_1 & -\sin\alpha_1 \\ \sin\alpha_1 & \cos\alpha_1 \end{pmatrix} \begin{pmatrix} 0 \\ 1 \end{pmatrix} + \begin{pmatrix} x_1 \\ y_1 \end{pmatrix} \tag{1}$$

S_3: A centre rotation S_2 is defined:

$$\begin{pmatrix} x_3 \\ y_3 \end{pmatrix} = \frac{l_1}{l_2} \begin{pmatrix} \cos\alpha_2 & -\sin\alpha_2 \\ \sin\alpha_2 & \cos\alpha_2 \end{pmatrix} \begin{pmatrix} x_1 & -x_2 \\ y_1 & -y_2 \end{pmatrix} + \begin{pmatrix} x_2 \\ y_2 \end{pmatrix} \tag{2}$$

S_{i+1}: A centre rotation S_i is defined:

$$\begin{pmatrix} x_{i+1} \\ y_{i+1} \end{pmatrix} = \frac{l_i}{l_{i-1}} \begin{pmatrix} \cos\alpha_i & -\sin\alpha_i \\ \sin\alpha_i & \cos\alpha_i \end{pmatrix} \begin{pmatrix} x_{i-1} & -x_i \\ y_{i-1} & -y_i \end{pmatrix} + \begin{pmatrix} x_i \\ y_i \end{pmatrix} \tag{3}$$

2.4. Determination of rotation angle u_1

The values of the rotation angles u_1 are determined. With the help of the obtained values the new coordinates are calculated (x'_2, y'_2).

Centre rotation

To determine u_1, a centre rotation $S_1 < u_1$ is defined:

$$\begin{pmatrix} x'_2 \\ y'_2 \end{pmatrix} = \begin{pmatrix} \cos u_1 & -\sin u_1 \\ \sin u_1 & \cos u_1 \end{pmatrix} \begin{pmatrix} x_2 - x_1 \\ y_2 - y_1 \end{pmatrix} + \begin{pmatrix} x_1 \\ y_1 \end{pmatrix} \Rightarrow \begin{pmatrix} x'_2 - x_1 \\ y'_2 - y_1 \end{pmatrix} = \begin{pmatrix} \cos u_1 & -\sin u_1 \\ \sin u_1 & \cos u_1 \end{pmatrix} \begin{pmatrix} x_2 - x_1 \\ y_2 - y_1 \end{pmatrix} \tag{4}$$

$$\Rightarrow \cos u_1 = \left| \frac{\begin{matrix} x'_2 - x_1 & -(y_2 - y_1) \\ y'_2 - y_1 & x_2 - x_1 \end{matrix}}{\begin{matrix} x_2 - x_1 & -(y_2 - y_1) \\ y_2 - y_1 & x_2 - x_1 \end{matrix}} \right|$$

IOP Conf. Series: Materials Science and Engineering **1018**(2021) 012007 doi:10.1088/1757-899X/1018/1/012007

Reset coordinates

Considering the performed centre rotation, the coordinates are reset, thus determining the new coordinates of the points of interest are as follows:

$$\begin{pmatrix} x_i'' \\ y_i'' \end{pmatrix} = \begin{pmatrix} \cos u_1 & -\sin u_1 \\ \sin u_1 & \cos u_1 \end{pmatrix} \begin{pmatrix} x_i - x_1 \\ y_i - y_1 \end{pmatrix} + \begin{pmatrix} x_1 \\ y_1 \end{pmatrix} \qquad (5)$$

2.5. Determination of rotation angle u_i

The values of the rotation angles u_i are determined. With the help of the obtained values the new coordinates are calculated (x'_i, y'_i).

Centre rotation

To determine u_i, a centre rotation $S_i < u_i$ is developed:

$$\begin{pmatrix} x'_{i+1} \\ y'_{i+1} \end{pmatrix} = \begin{pmatrix} \cos u_i & -\sin u_i \\ \sin u_i & \cos u_i \end{pmatrix} \begin{pmatrix} x_{i+1} - x_i \\ y_{i+1} - y_i \end{pmatrix} + \begin{pmatrix} x_i \\ y_i \end{pmatrix} \Rightarrow \begin{pmatrix} x'_{i+1} - x_i \\ y'_{i+1} - y_i \end{pmatrix} = \begin{pmatrix} \cos u_i & -\sin u_i \\ \sin u_i & \cos u_i \end{pmatrix} \begin{pmatrix} x_{i+1} - x_i \\ y_{i+1} - y_i \end{pmatrix} \qquad (6)$$

$$\Rightarrow \cos u_i = \frac{\begin{vmatrix} x'_{i+1} - x_i & -(y_{i+1} - y_i) \\ y'_{i+1} - y_i & x_{i+1} - x_i \end{vmatrix}}{\begin{vmatrix} x_{i+1} - x_i & -(y_{i+1} - y_i) \\ y_{i+1} - y_i & x_{i+1} - x_i \end{vmatrix}}$$

Reset coordinates

Considering the new centre of rotation, the coordinates are reset, thus determining the new coordinates of the points of interest are:

$$\begin{pmatrix} x_j'' \\ y_j'' \end{pmatrix} = \begin{pmatrix} \cos u_i & -\sin u_i \\ \sin u_i & \cos u_i \end{pmatrix} \begin{pmatrix} x_j - x_i \\ y_j - y_i \end{pmatrix} + \begin{pmatrix} x_i \\ y_i \end{pmatrix} \qquad (7)$$

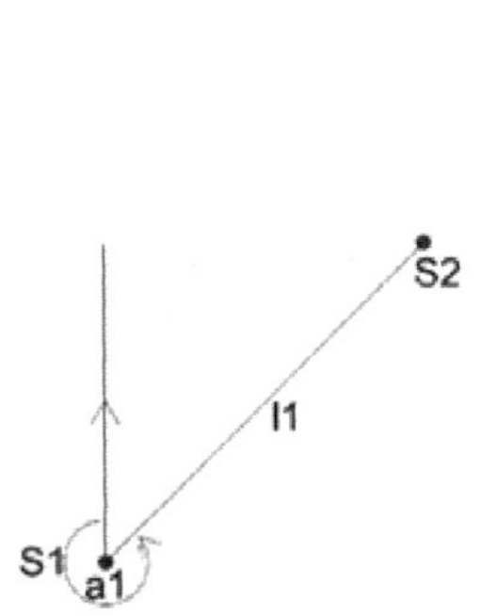

Figure 3.a Geometrical model for S_2 coordinates

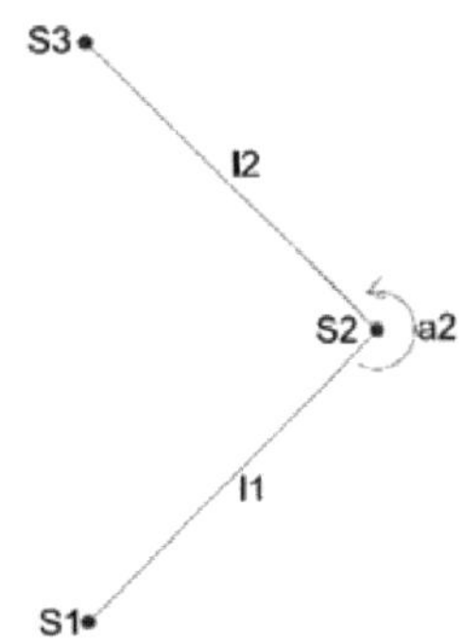

Figure 3.b Geometrical model for S_3 coordinates

3. Detail design

In engineering design, the product architecture is related to the functional elements and physical components of products and is used to define the basic physical building blocks of the product in terms of what they do and what their interfaces are to the rest of the device [8]. One of the most widely accepted definitions of product architecture was given by Ulrich [9]. The spinal orthoses architecture shown in Figure 4 identifies the product's functions, materialised in specific systems and assemblies.

Any engineering design process is based on two factors: (1) the state of technical and technological knowledge and (2) the project theme that includes the initial computer data about the future product [10]. Starting from the established objective specifications and the selected concept, the detailed design has been elaborated (Figure 4). The posture correction orthosis has four main systems, and each system has few assemblies. Further in this paper, the detailed design of the Reading assembly, the most critical component in the final product, is discussed, being marked in green in Figure 4. The other systems are normalized regardless of the user, while the Reading assembly is critical because the user can perform the movements that the system cannot, thus resulting in mechanical stress.

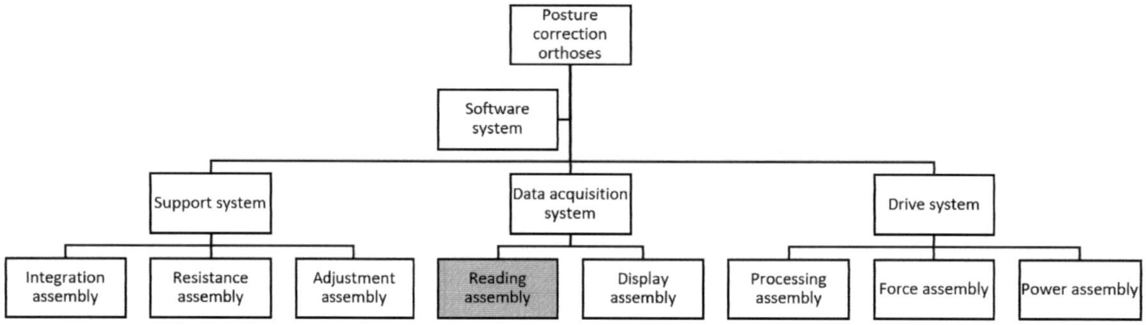

Figure 4. Posture correction orthoses architecture

3.1. Reading assembly detailed design

The reading assembly has five main components, and it represents a mechanical copy of the spine. The Reading system is designed to accurately mimic the movements of the user's spine and read the position in which he is in real time. Thoracic and lumbar joints are represented by a single piece in the final assembly, while the number of the vertebral bodies, sensors and axes are added depending on the desired reading accuracy. The reading accuracy depends on the number of segments added to the system component [11]. Thus, for a 100% accurate reading, 24 segments are needed, one segment for each vertebra. The current design version is constructed with four vertebral bodies (Figure 5.a and 5.b), meaning that the reding accuracy is 1/6 from real user's posture. The final Posture correction orthosis prototype will be manufactured with eight segments (reading accuracy will of 1/3 from real user's posture).

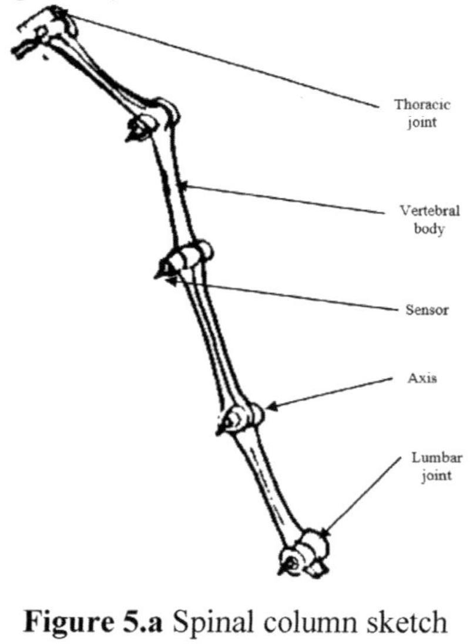

Figure 5.a Spinal column sketch

Figure 5.b Spinal column detailed design

The vertebral element - VE (Figure 6.a) materializes the mathematical model defined in section 2. At each end, the VE has a sensor slot and a joint so that it can be assembled with another VE or with one of the two joints, thoracic or lumbar (Figure 6.b). The body is rigid and has a fixed length, depending on the desired accuracy, the number of segments is kept constant and the length of the VE varies.

During the design process, the assembly tolerance design is an important issue. Assembly tolerances reflect functional requirements of fitting, which can be used to control assembling qualities and production costs. Assembly constraints are essentially constraints between assembly feature surfaces of parts. Assembly tolerances are crucial for many activities in the product's life cycle, which not only affects assembly qualities but also determines manufacturing costs [12].

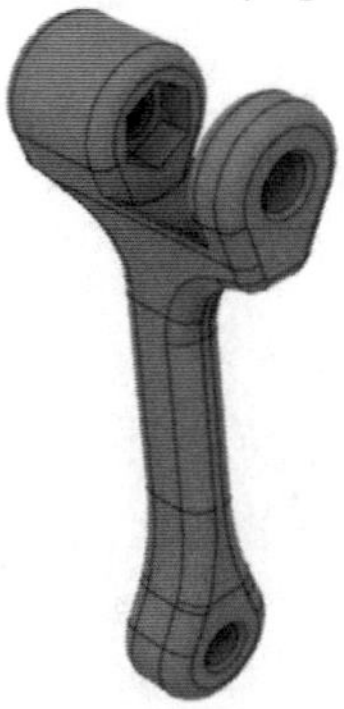

Figure 6.a Vertebral element - VE

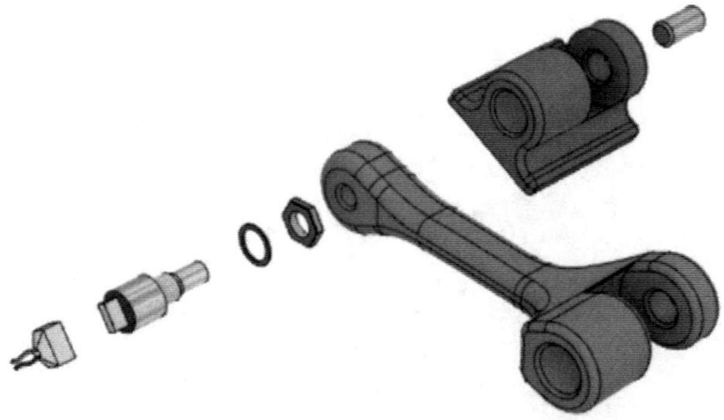

Figure 6.b Components of for one VE

A challenge in the design of the VE was to establish the tolerances of different dimensions, because the VE assembles with most components of the orthosis. Because the VE is the most important component of the posture correction orthosis, it has gone through several design stages. Thus, the final shape of the VE was obtained after several iterations (V1-V5). Some versions of the concept are shown in the figure 7.a-e.

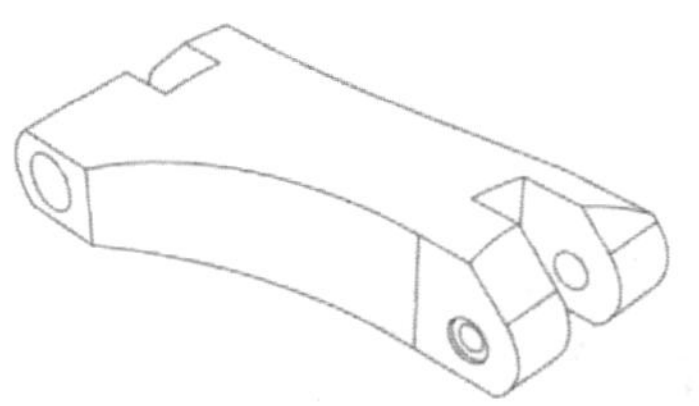

Figure 7.a VE - v1

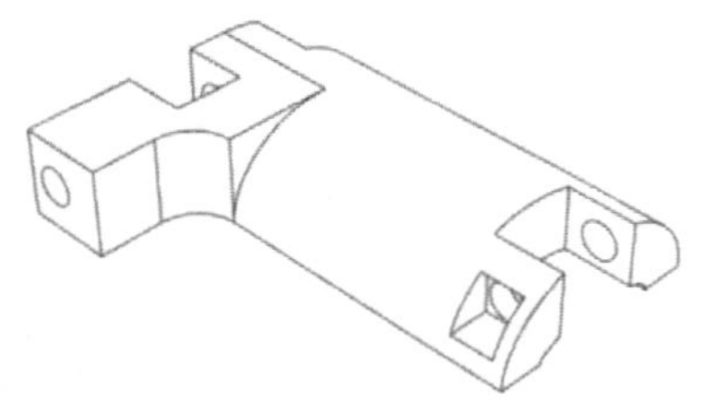

Figure 7.b VE – v2

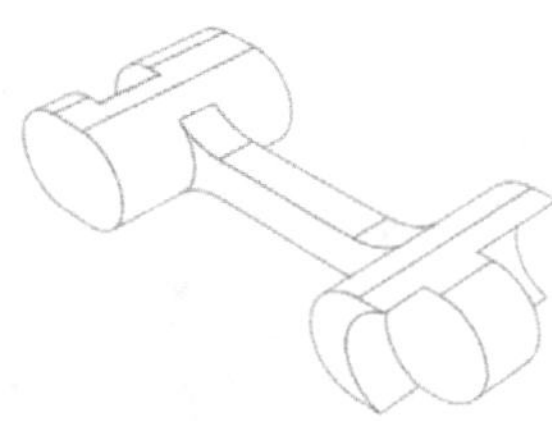

Figure 7.c VE – v3

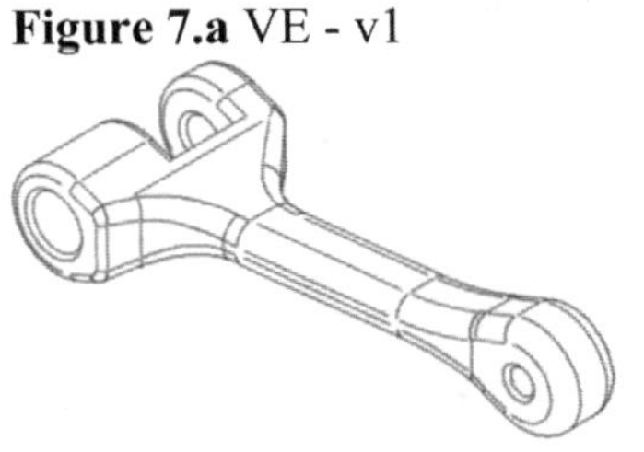

Figure 7.d VE – v4

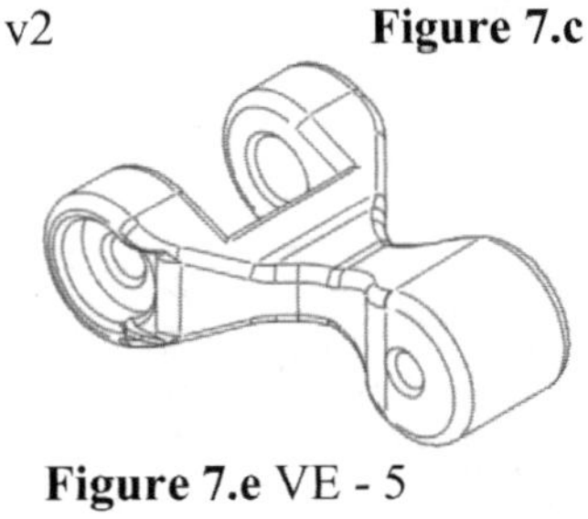

Figure 7.e VE - 5

3.2. Finite Element Analysis

Because the vertebral bodies articulate both with each other and with the thoracic and lumbar joints, they have not been subjected to a finite element analysis on the frontal direction. The working stress scenario was considered when the user tilts to one side. In the lateral plane the components are rigid, thus a finite element analysis (FEA) was performed in this plane. The FEA analysis was performed in same conditions for two plastic materials, namely ABS and PET. The properties of the selected materials are presented in Table 1.

The first simulated material was ABS, and the result of the analysis is presented in Figure 8.a. The safety factor in this analysis is less than 1, ~ 0.41, which indicates that the part will not withstand the action of the applied force F = 100 N.

The second simulated material was PET, and the result of the analysis is presented in Figure 8.b. The safety factor in the case of this analysis is more than 1, ~ 1.2, which indicates that the part will withstand the action of the applied force F = 100 N.

Table 1. Material properties [13]

Properties	ABS	PET
Thermal conductivity [W/(m*K)]	3.000E-01	1.600E-01
Specific heat [J/(g*C)]	2.287	1.500
Thermal expansion [µm/(m*C)]	0.702	87.500
Young's module [GPa]	2.758	2.240
Poisson's ratio	0.420	0.380
Shear modulus [MPa]	1240.000	805.000
Density [g/cm^3]	1.541	1.060
Yield strength [MPa]	54.400	20.000
Tensile Strength [MPa]	55.100	29.600

Because the posture correction orthosis is used by humans, the most important specification is safety during use. For this reason, the safety factor obtained ~ 1.2 is enough for this purpose. In case of emergency, the user will not be blocked by the mechanical spine but will be able to easily damage the component and will be able to move freely. Thus, the safety factor obtained is good both, from a technical point of view and from the safety in use point of view.

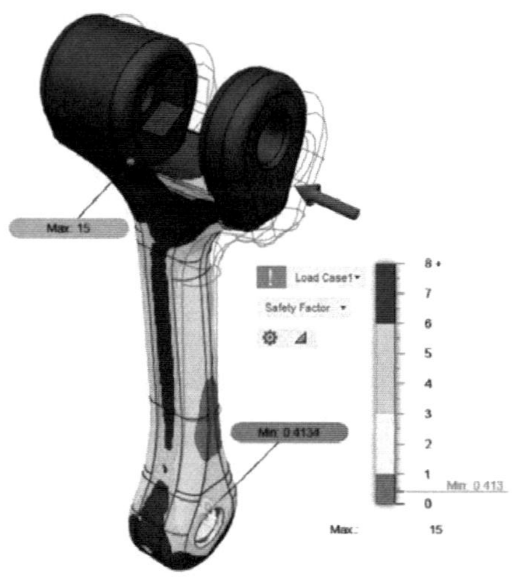

Figure 8.a FEA with ABS

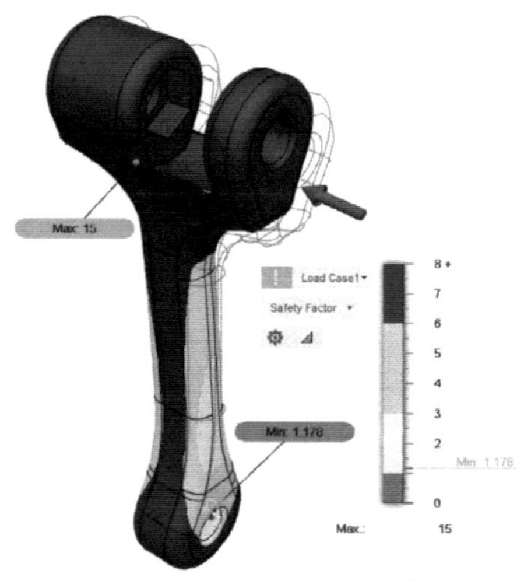

Figure 8.b FEA with PET

4. Parametric design

Design optimization is an engineering design methodology that uses a mathematical formulation of a design problem to support the selection of the optimal design from several generated alternatives. Parametric design is a method in which parameters are used to set the size values of a part or the whole model. In the case of the VE, parametric optimization was performed on the model optimized for 3D printing, as shown in Figure 9. Each element from the initial design has its own parameters and it is important for the components, as they are required to define the final parametric model. Figure 10 shows the elements of the model.

Because the dimensions of the VE bores depend on the dimensions of the components to be assembled, they were left as parameters of the model, thus they were not customized. Firstly, the dimensions that influence the geometry of the vertebral body were parameterized according to the user's height (Figure 11.a,b). Dependencies were created between the dimensions of the model.

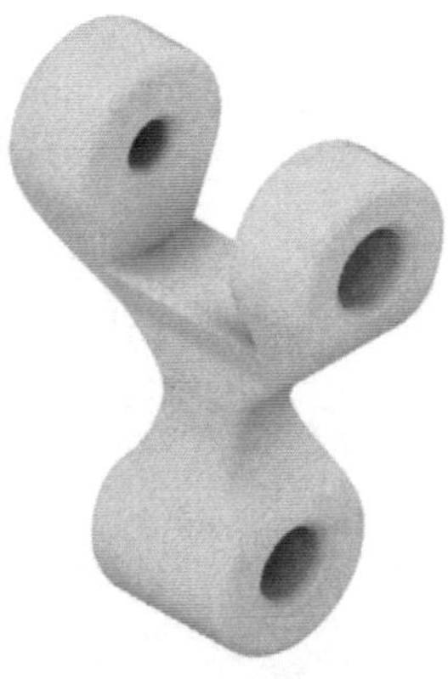

Figure 9 3D model of VE

Parameter	Name	Unit	Expression	Value	Comments
Favorites					
User Parameters +					
˅ Model Parameters					
˅ Corp_Vertebral v1					
› Sketch1					
› Extrude1					
› Sketch2					
› Extrude2					
› Sketch3					
› Extrude3					
› Hole1					
› Hole2					
› Hole3					
› Sketch4					
› Extrude4					
› Sketch5					
› Extrude5					
› Chamfer1					
› Sketch6					
› Extrude6					

Figure 10 Model parameters of VE

Parameter	Name	Unit	Expression	Value	Comments
Favorites					
User Parameters +					
˅ Model Parameters					
˅ Corp_Vertebral_3D_01 v1					
˅ Sketch1					
☆ Diameter Dimension-2	d1	mm	25 mm	25.00	
☆ Diameter Dimension-3	d2	mm	25 mm	25.00	
☆ Linear Dimension-2	d3	mm	57 mm	57.00	
☆ Linear Dimension-3	d4	mm	28.5 mm	28.50	
☆ Linear Dimension-4	d5	mm	5 mm	5.00	
☆ Linear Dimension-5	d6	mm	5 mm	5.00	
☆ Angular Dimension-2	d7	deg	45 deg	45.0	
☆ Angular Dimension-3	d8	deg	45 deg	45.0	
› Extrude1					
˅ Sketch3					
☆ Linear Dimension-2	d20	mm	7.5 mm	7.50	
☆ Linear Dimension-3	d21	mm	11.5 mm	11.50	
☆ Linear Dimension-4	d22	mm	25 mm	25.00	
☆ Linear Dimension-5	d24	mm	16 mm	16.00	
☆ Linear Dimension-6	d25	mm	16 mm	16.00	

Figure 11.a Initial model parameters

Parameter	Name	Unit	Expression	Value	Comments
Favorites					
› User Parameters +					
˅ Model Parameters					
˅ Corp_Vertebral_3D_01 v1					
˅ Sketch1					
☆ Diameter Dimension-2	d1	mm	25 mm	25.00	
☆ Diameter Dimension-3	d2	mm	d1	25.00	
☆ Linear Dimension-2	d3	mm	X * 57 mm	57.00	
☆ Linear Dimension-3	d4	mm	d3 / 2	28.50	
☆ Linear Dimension-4	d5	mm	5 mm	5.00	
☆ Linear Dimension-5	d6	mm	d5	5.00	
☆ Angular Dimension-2	d7	deg	45 deg	45.0	
☆ Angular Dimension-3	d8	deg	d7	45.0	
› Extrude1					
˅ Sketch3					
☆ Linear Dimension-2	d20	mm	7.5 mm	7.50	
☆ Linear Dimension-3	d21	mm	11.5 mm	11.50	
☆ Linear Dimension-4	d22	mm	25 mm	25.00	
☆ Linear Dimension-5	d24	mm	(d3 - 25 mm) / 2	16.00	
☆ Linear Dimension-6	d25	mm	d24	16.00	

Figure 11.b Customized model parameters

The next step was to create parameters that define the dimensions of the VE depending on the height of the user. Two user parameters were defined (Figure 12.a). The first parameter was the user's height in mm, and this is the only parameter that is modified by the product manufacturing team. The second parameter is the height factor that proportionally converts the dimensions of the VE according to the height of the user (Figure 12.b).

Parameter	Name	Unit	Expression	Value	Comments
Favorites					
˅ User Parameters +					
☆ User Parameter	h	mm	170 mm	170.00	Înălțime utilizator
☆ User Parameter	X	mm	h / 170	1.00	Factor de înălțime

Figure 12.a Initial model parameters

˅ Model Parameters					
˅ Corp_Vertebral_3D_01 v1					
˅ Sketch1					
☆ Diameter Dimension-2	d1	mm	25 mm	25.00	
☆ Diameter Dimension-3	d2	mm	d1	25.00	
☆ Linear Dimension-2	d3	mm	X * 57 mm	57.00	
☆ Linear Dimension-3	d4	mm	d3 / 2	28.50	
☆ Linear Dimension-4	d5	mm	5 mm	5.00	
☆ Linear Dimension-5	d6	mm	d5	5.00	
☆ Angular Dimension-2	d7	deg	45 deg	45.0	
☆ Angular Dimension-3	d8	deg	d7	45.0	
› Extrude1					
˅ Sketch3					
☆ Linear Dimension-2	d20	mm	7.5 mm	7.50	
☆ Linear Dimension-3	d21	mm	11.5 mm	11.50	
☆ Linear Dimension-4	d22	mm	25 mm	25.00	
☆ Linear Dimension-5	d24	mm	(d3 - 25 mm) / 2	16.00	
☆ Linear Dimension-6	d25	mm	d24	16.00	

Figure 12.b Customized model parameters

In figures 13.a-c several dimensional variations of the VE, depending on the height of the user, are developed and presented. For a user with a height of 1.5 m, each of the eight vertebral bodies has a length between the centres of the joint points of 50.3 mm (Figure 13.a). For a user with a height of 2.0 m, the distance between the joint points of the model is 67 mm (Figure 13.c).

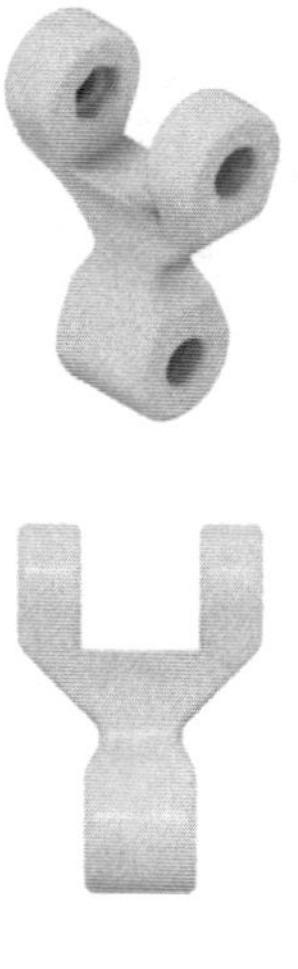

Parameter	Name	Unit	Expression	Value
Favorites				
User Parameters				
User Parameter	h	mm	1500 mm	1500.00
User Parameter	X		h / 1700 mm	0.882353
Model Parameters				
Corp_Vertebral_3D_01 v1				
Sketch1				
Diameter Dimension-2	d1	mm	25 mm	25.00
Diameter Dimension-3	d2	mm	d1	25.00
Linear Dimension-2	d3	mm	X * 57 mm	50.294
Linear Dimension-3	d4	mm	d3 / 2	25.147
Linear Dimension-4	d5	mm	5 mm	5.00
Linear Dimension-5	d6	mm	d5	5.00
Angular Dimension-2	d7	deg	45 deg	45.0
Angular Dimension-3	d8	deg	d7	45.0
Extrude1				
Sketch3				
Linear Dimension-2	d20	mm	7.5 mm	7.50
Linear Dimension-3	d21	mm	11.5 mm	11.50
Linear Dimension-4	d22	mm	25 mm	25.00
Linear Dimension-5	d24	mm	(d3 - 25 mm) / 2	12.647
Linear Dimension-6	d25	mm	d24	12.647

Figure 13.a Parametric model for user's height of 1.5 m

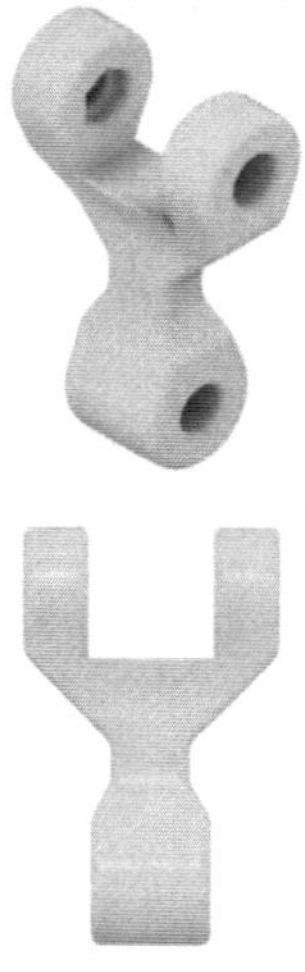

Parameter	Name	Unit	Expression	Value
Favorites				
User Parameters				
User Parameter	h	mm	1700 mm	1700.00
User Parameter	X		h / 1700 mm	1
Model Parameters				
Corp_Vertebral_3D_01 v1				
Sketch1				
Diameter Dimension-2	d1	mm	25 mm	25.00
Diameter Dimension-3	d2	mm	d1	25.00
Linear Dimension-2	d3	mm	X * 57 mm	57.00
Linear Dimension-3	d4	mm	d3 / 2	28.50
Linear Dimension-4	d5	mm	5 mm	5.00
Linear Dimension-5	d6	mm	d5	5.00
Angular Dimension-2	d7	deg	45 deg	45.0
Angular Dimension-3	d8	deg	d7	45.0
Extrude1				
Sketch3				
Linear Dimension-2	d20	mm	7.5 mm	7.50
Linear Dimension-3	d21	mm	11.5 mm	11.50
Linear Dimension-4	d22	mm	25 mm	25.00
Linear Dimension-5	d24	mm	(d3 - 25 mm) / 2	16.00
Linear Dimension-6	d25	mm	d24	16.00

Figure 13.b Parametric model for user's height of 1.7 m

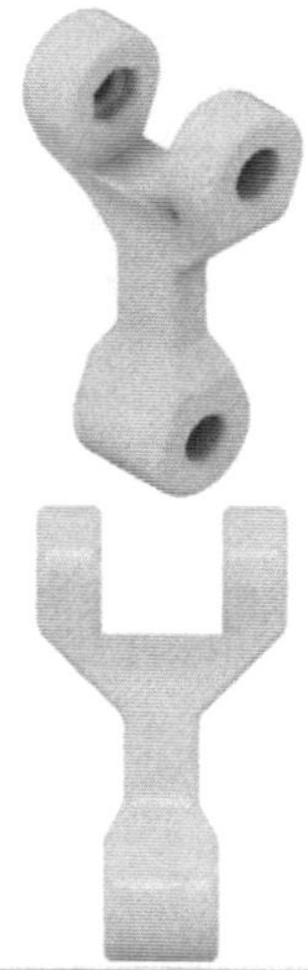

Parameter	Name	Unit	Expression	Value
Favorites				
User Parameters				
User Parameter	h	mm	2000 mm	2000.00
User Parameter	X		h / 1700 mm	1.17647
Model Parameters				
Corp_Vertebral_3D_01 v1				
Sketch1				
Diameter Dimension-2	d1	mm	25 mm	25.00
Diameter Dimension-3	d2	mm	d1	25.00
Linear Dimension-2	d3	mm	X * 57 mm	67.059
Linear Dimension-3	d4	mm	d3 / 2	33.529
Linear Dimension-4	d5	mm	5 mm	5.00
Linear Dimension-5	d6	mm	d5	5.00
Angular Dimension-2	d7	deg	45 deg	45.0
Angular Dimension-3	d8	deg	d7	45.0
Extrude1				
Sketch3				
Linear Dimension-2	d20	mm	7.5 mm	7.50
Linear Dimension-3	d21	mm	11.5 mm	11.50
Linear Dimension-4	d22	mm	25 mm	25.00
Linear Dimension-5	d24	mm	(d3 - 25 mm) / 2	21.029
Linear Dimension-6	d25	mm	d24	21.029

Figure 13.c Parametric model for user's height of 2.0 m

Apart from the commercial components, all components went through several design iterations until the optimal solution was chosen. For all components that are subjected to mechanical stresses during operation, FEA was performed. With the help of this analysis the geometry and the right material for the given application were chosen (Figure 14.a,b).

Figure 14.a Final design front view

Figure 14.b Final design back view

IOP Conf. Series: Materials Science and Engineering **1018**(2021) 012007 doi:10.1088/1757-899X/1018/1/012007

5. Conclusions

For the development of the smart posture correction orthosis prototype a mathematical model was generated considering the principle of anatomical functioning of the human body and especially the particularities of the spine. Regardless of the user, the operating principle remains the same and the technical solution can be adapted for each individual user. Because the technical solution is customized for each user, along with the detailed design, a parametrical design was developed. This allows the use of the same technological production process for each device depending on the needs of the given user.

The current orthosis is designated for posture correction of the healthy subjects. As a future work we consider the process of transforming the product into a medical device for correcting the spinal column illnesses, including the cervical vertebrae, not only lumbar ant thoracic. In this case the subject's axial rotation and frontal movement will also be considered.

References

[1] Nedoma J, Stehlik J, Hlavacek I, Danek J, Dostalova T and Preckova P 2011 Mathematical and computational methods in biomechanics of human skeletal systems *Wiley* **3-5** 565

[2] Ismail K N, Basah S, Omar N, Murugappan M and Yaacob S 2012 Mathematical modelling of human body for lifting task *IEEE International Conference on Control System Computing and Engineering (Penang, Malaysia)* pp. 149-154

[3] Roberts S B and Chen P H 1970 Elastostatic analysis of the human thoracic skeleton *J. Biomech.* **3**(6) 527-545

[4] Kreuzer J E 1987 Mathematical modelling of complex mechanical systems *Math. Model.* **8** 37-42

[5] Xu C, He J, Zhang X, Yao C and Tseng P H 2018 Geometrical kinematic modeling on human motion using method of multi-sensor fusion *Inf. Fusion* **41** 243-254

[6] Whitwam R 2018 The Vertebral Column *A&P Labs* Mississippi University for Women

[7] Watkins J 2009 Structure and function of the musculoskeletal system *Champaign IL Human Kinetics*

[8] Ulrich K T and Eppinger S D 2000 *Product design and development* (New York: McGraw-Hill)

[9] Ulrich K T 1995 The role of product architecture in the manufacturing firm *Res Policy* **24**(3) 419-440

[10] Dolga V 2007 *Design of mechatronic systems* (Timisoara: Ed. Politehnica)

[11] Ćuković S, Devedzic G, Ivanovic L, Zecevic L T and Subburaj K 2010 Development of 3D kinematic model of the spine for idiopathic scoliosis simulation *Comput. Aided Des. Appl.* **7**(1) 153-161

[12] Zhang Y, Li Z, Gao J, Hong J, Villecco F and Li Y 2012 A method for designing assembly tolerance networks of mechanical assemblies *Math. Probl. Eng.* **2012** 513958

[13] Autodesk Fusion 360 On-line Retrieved 2020 https://www.autodesk.com/products/fusion-360/overview

IOP Conf. Series: Materials Science and Engineering 1018(2021) 012008 doi:10.1088/1757-899X/1018/1/012008

Chemical and mechanical tests on the SiO_2-ZnO materials for protective coatings of industrial use

A M Mocioiu[1], D I Băilă[2*] and O C Mocioiu[3*]

[1]National R&D Institute for Non-ferrous and Rare Metals,102 Biruinţei Blvd, Pantelimon, Ilfov 077145, Romania
[2]University Politehnica of Bucharest, Blv. Splaiul Independentei, no. 313, sector 6, cod 060042, Bucharest, Romania
[3]Ilie Murgulescu Institute of Physical Chemistry of Romanian Academy, 202 Splaiul Independenţei, Bucharest, Romania

E-mails: baila_d@yahoo.com, oana.mocioiu@yahoo.com

Abstract. Many industrial processes, such as chemical process and pharmaceuticals, use aggressive acidic and caustic solution which may result in an early failure of the components of plant. The various corrosion processes in harsh environments could be uniform corrosion, localized corrosion, galvanic corrosion, etc. In order to extend the life of components, a proper material selection is necessary. This direction is intended to address the behaviour of materials due to exposure in harsh environments. The main objective is to develop new vitreous coatings for different kind of materials as metals or glasses in order to protect them against acid attack. SiO_2-ZnO materials were prepared by chemical route in order to be used as protective coatings. The SiO_2-ZnO materials were investigated by infrared spectroscopy and differential thermal analysis in order to determine their structure and thermal characteristics. SEM and EDS measurements were performed to investigate their structure and morphology. The chemical resistance tests in acid media (HCl and HF) were followed based on STAS and ISO standards and the extraction solutions were tested by ICP-MS in order to identify elements lost during tests. Mechanical tests were performed.

1. Introduction

Many industrial processes, such as chemical process and pharmaceuticals, use aggressive acidic and caustic solution which may result in an early failure of the components of plant [1,2].

In the industry are used reactors made from various materials as steel, glass, cement, titanium, etc. [3,4,5]. Each of them is used according with reactions that are performed inside.

Glass reactors are transparent equipment that is used in the production of chemicals, polymers, and pharmaceuticals. They comprise strong glass walls that shield the interior contents from any exposure to or contact with the outside world. In doing so, they facilitate the reactions of substances. Glass reactors are mainly used for synthetic reaction, distillation and concentration of different types of materials. In the pharma industry are used glass reactors in order to prevent the metal contamination for high pure drugs and biological materials.

There are few types of glass reactors used in the Pharma Industry as:

- *The batch glass reactor* is mainly used for those small-scale applications. Though simple, this kind of reactor has the ability to accommodate many kinds of gears at a time. Indeed, it will receive all the materials until such a time that the capacity is packed to the brim [3].

- *The continuous stirred-tank reactor model* where one or more liquid reagents are introduced into the tank segment. This mixture is subsequently stirred using an impeller to mix the reagents fully and uniformly. In this way, the reagents produce great outcomes indeed. This kind of reactor is larger than the batch glass reactor. It finds applicability in those areas that demand industrial-size applications; that include the brewing, pulping, and handling of various biological materials [3].

- *The continuous tubular reactor* accepts one or more fluid reagents from a side to side tube or pipe. In this reactor, the chemical reactions will transpire simultaneously with the movements of the reagents from the tubes into the tubular reactor. As these reagents mix, the rates vary significantly in such a way as to generate a gradient that is further characterized by added admiration. This reactor is mainly useful for mass production where many ingredients have to be mixed and handled at a time [3].

- *Catalytic Cracker.* Catalytic cracking is a chemical process that entails the extensive use of carbon ions. The cracker is mainly used to convert the high molecular weights of the hydrocarbons into more valuable small molecular hydrocarbons [3].

The various corrosion processes in acid environments could be uniform corrosion or localized corrosion. In the case of hydrochloric acid corrosion on reactors it is uniform corrosion and it can cause contamination of products. Glass reactors can be used in industries were small quantities of contaminants have influence on the properties of product as pharmaceutics.

In order to extend the life of components of reactors and industrial plant, a proper selection of the material is necessary. Other method used in industry is coating the inside part of the reactor with an anticorrosive material. The chemical stability of the glasses can be improved by depositing protective layers of metal oxides (SnO_2, TiO_2, CuO, ZnO) or films of organic compounds [6-10].

The main objective of this paper is to develop new vitreous coating for glass reactors in order to protect them against acid attack. The SiO_2-ZnO composition was choosing in order to maintain the vitreous character of protective film and also due to well-known antibacterial role of zinc.

The literature data report obtaining of glasses synthesized by classical route of melt-quenching in the SiO_2-ZnO system in compositional domain between 0-35 mol % ZnO [11, 12]. The classical route imply high melting temperature (>1400°C) and the deposition on other glass cannot be performed.

In order to obtain SiO_2-ZnO vitreous films on glass substrate, a precursor solution was prepared by sol-gel and dip-coating method was used for deposition.

2. Experimental parts

Coatings with 7% ZnO and 93% SiO_2 were obtained by deposition of thin films by dip-coating of precursor solutions obtained by sol-gel method. The method to prepare similar precursor solution was presented in a previous paper [12]. Precursor solution with molar ratio: C_2H_5OH: TEOS: H_2O: HCl = 10: 1: 3: 0.03 was prepared. Doping with 7% zinc oxide was achieved by adding zinc acetate dehydrate at precursor solution and mixing for 2 hours. Thin films were obtained by dip-coating of solution on glass substrates and thermally treated at 500°C.

STA/TG measurement on gel was made with a Setsys Setaram equipment with 10 degrees/minute between 20-900°C.

The chemical resistance tests in acid media (HCl 37% and HF 35%) were performed based on STAS 598/3-76; STAS 2413-87 and ISO 1776:1985 [13, 14, 15]. In order to simulate a long period of contact between acid media and reactor wall, the samples were boiled at 100°C in acid time to 2 hours. The extraction solutions were tested by ICP-MS in order to identify elements lost during tests.

Morphology and composition of samples before and after acid attack were studied with a scanning electron microscope (SEM) Quanta 250, produced by FEI Company, incorporated with Energy Dispersive X-Ray Spectrometer, produced by EDAX, consisting of ELEMENT Silicon Drift Detector Fixed, Element EDS Analysis Software Suite.

MHT-10 Microhardness tester, produced by Anton Paar, with Vickers indenter was used for mechanical properties. The tester was bonded with electronic microscope with polarized light Carl Zeiss Microimaging Gmbh, equipped with a digital camera for photo acquisitions and soft AxioVision Release 4.8.1 for image processing.

3. Results

The 93% SiO_2-7% ZnO gel was investigated by STA/TG and FT-IR in order to determine their thermal behaviour and structure. Gel obtained after drying of precursor solution at room temperature was studied. Figure 1 present the FTIR spectrum of 93% SiO_2-7% ZnO gel. Figure 2 show the STA/TG curves of 93% SiO_2-7% ZnO gel.

In order to record FTIR spectrum, the sol with composition 93% SiO_2-7% ZnO was dried at room temperature and 1 mg powder was mixed in a pellet with 200 mg KBr. The recorded spectrum shown the bands characteristic to Si-O vibrations similar with those reported in literature [16-21]. Main bands at 1223 cm^{-1}, 1088 cm^{-1}, 795 cm^{-1} and 463 cm^{-1}, are assigned to Si-O-Si bonds in network, chains and band at 950 cm^{-1} is characteristic to isolated [SiO_4] tetrahedra; while the small band at 555 cm^{-1} can be assigned to vibration of the Si-O-Zn bonds. The FTIR spectrum show shape of vitreous material due to wide bands. The band position of vibration of the bonds between Si-O and Zn-O appear in the spectrum but they cannot be assigned to a crystalline form. So, the film on the glass substrate has also vitreous structure. The bands at 3478 cm^{-1} and 1643 cm^{-1} are assigned to vibration of the O-H in the absorbed water.

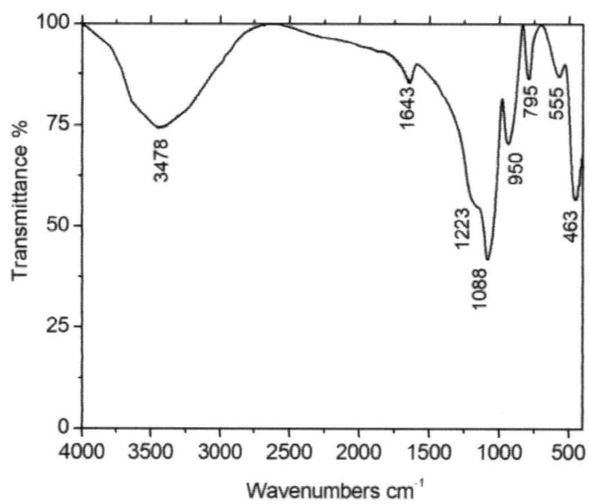

Figure 1. FTIR spectrum of 93% SiO_2-7% ZnO gel

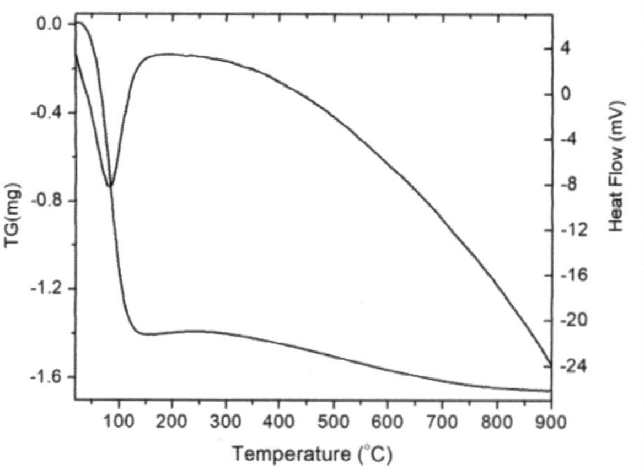

Figure 2. STA/ TG curves of 93% SiO_2-7% ZnO gel

STA curve shown main endothermal effect between 32°C and 126°C corresponding to elimination of water and alcohol from obtained sol. Mass loss was of 12 %. As can be seen in the figure after 500° C the loss mass is zero. Thermal treatment at this temperature was established because the elimination of residual solvents was complete and glass substrate is not modified by treatment temperature.

The film deposited on glass substrate from precursor solution and thermally treated at 500°C was investigated from structural and morphological point of view using SEM / EDAX methods.

Figure 3 show SEM image of the film with morphology characteristic for glasses and EDAX of 93% SiO_2-7% ZnO film. The material have a uniform and homogeneous morphology. Identified position of Si and Zn in EDAX of the film are similar with literature data [8, 14].

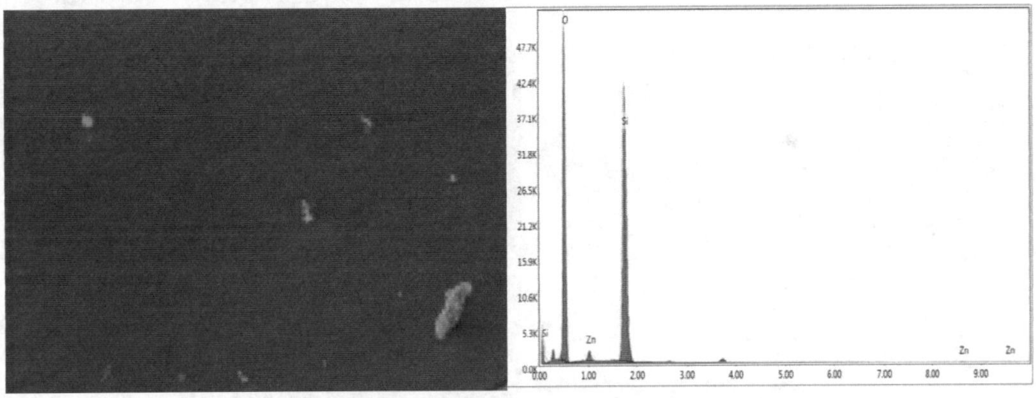

Figure 3. SEM/EDAX of SiO_2-ZnO film

The chemical resistance tests in acid media (HCl and HF) were followed based on STAS 598/3-76; STAS 2413-87 and ISO 1776:1985 [9, 10, 11] and the extraction solutions were tested by ICP-MS in order to identify elements lost during tests.

Figure 4 show SEM images of SiO_2-ZnO film after acid attack with HCl and HF. The morphology of the films surfaces after corrosion tests was studied by scanning electron microscopy. Figure 4 displays both morphologies. In case of HCl attack a network of mottled material, as usually happens in hydrated silica layers was evidenced while in case of HF attack can be seen a change of the morphology. In case of glasses are well known 5 corrosion mechanism. In case of acid attack of the film two mechanism can be observed. In case of HCl attack the formation of a silica-rich film without destruction of the silicate network due to the strength of the acid is observed. In case of HF attack the concentration of ions in solution is high enough and the loss of silica atoms from glass network cannot be prevent until the equilibrium of reaction.

By exposing hot glass surfaces to the action of volatile compounds with flour, an increase in chemical resistance occurs without forming visible reaction products on the surface thus treated.

Chemical resistance is increased due to the replacement of O^{2-} and OH^- ions on the surface glass, by F^- ions. Fluor ions penetrate deep into the silico-calco-sodium glasses and not are abundant on the surface of the glass [6]. Improving chemical and mechanical strength can be achieved by treatment with fluorine-based compounds [7].

Although random wet etching with a hydrogen-fluoride (HF)-chemical solution is a simple and inexpensive method, a non-uniformly-etched surface is produced due to the remains of the insoluble product present in the chemical solution [22,23]. The HF-solution-based etching also leads to an increased haze value, which has an adverse effect on the improvement of the transmittance. Park [22] reported a geometrical hemisphere-array structure developed with a textured-glass substrate using the HF-H_2SO_4 wet-etching method. The height profiles glasses corroded with solutions of 10%, 25% and 35% concentration correspond to texture etched sizes. The morphology was modified in order to increase optical properties and chemical resistance of the glasses.

ICP-MS measurements after corrosion tests shows extraction of 20 ppb Zn from film after HCl attack and 40 ppm zinc and 20 ppm Si from film after HF attack. In the case of HCl attack occur the extraction of zinc ions at level of ppb and silica extraction is not observed. This result is in agreement with previous observation in SEM image and conform formation of silica-rich film without destruction of the silicate network. In case of attack of HF in the solution are extracted higher quantities of both ions Zn and Si and the result confirm the glass network destruction observed in SEM image. The attack of HF was much higher than attack of HCl. In case of HCl the equilibrium is obtained and the silica-rich film act as a barrier and corrosion is stopped. In case of HF the destruction of silica network will continue and new atomic layers will be dissolute in solution, but if the time set until the equilibrium of the corrosion reaction is reached, then the changed morphology will probably have a higher chemical resistance.

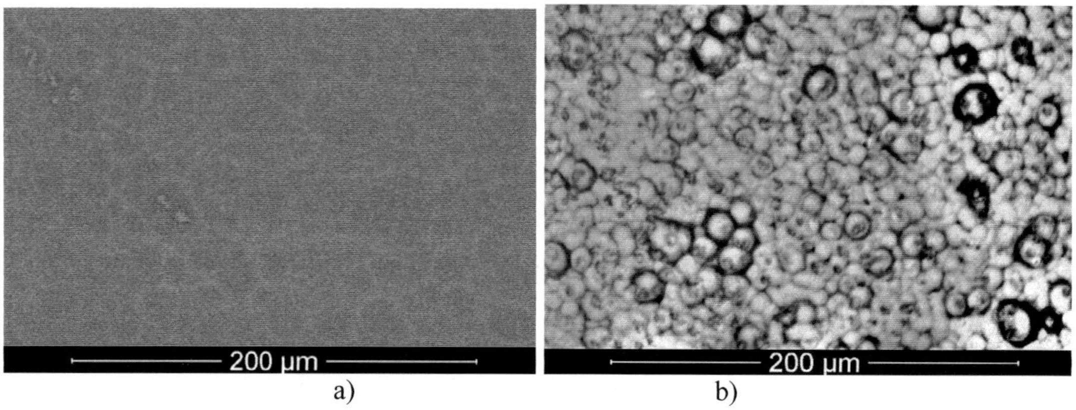

Figure 4. SEM images of SiO_2-ZnO film after acid attack: a) HCl and b) HF

Microhardness tests were performed on the studied film deposited by dip-coating on glass. Figure 5 show the indentation prints were small. When the crack appeared in the corner of the indentation prints they propagated in the film and not to interface. The force used in determination was 2N, time to 2s, slope 0,6N/s.

The measurements were repeated three times and the results were between 765 and 780.

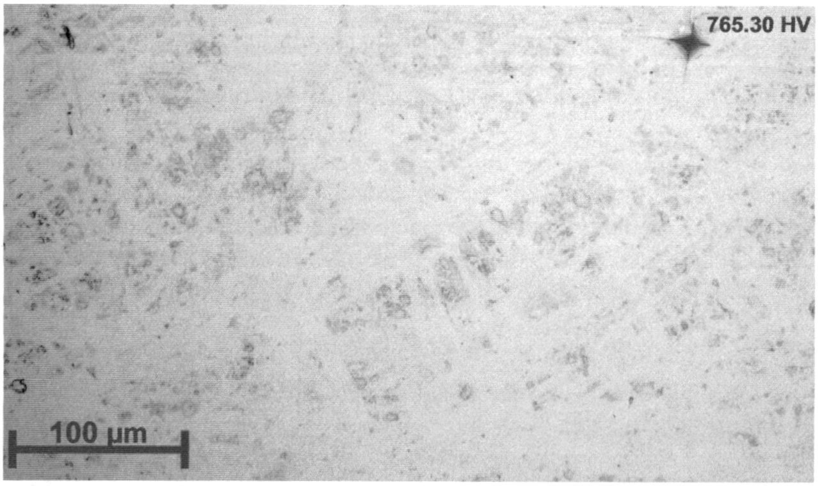

Figure 5. Micro hardness prints on SiO_2-ZnO film deposited on glass

4. Conclusions

The chemical resistance of 93% SiO_2-7% ZnO film deposited on glass in acid media (HCl 37% and HF 35%) was different.

The way in which the surface of the film deposited on the glass reacts to the attack of acids is greatly influenced by the strength of the acid used.

The mechanical resistance HV was between 765 and 780 and it is a good resistance for glasses.

The studied film can be used as protective film for glass reactors that work in acid media as some polymer synthesis.

References

[1] Solé A, Barreneche C, Martorell I and Cabeza L F 2016 *Appl. Therm. Eng.* **94,** 355-363.
[2] Suri S, PHARMACEUTICAL ENGINEERING, Reactors and Fundamentals of Reactors Design for Chemical Reaction.
[3] https://wkielab.com/glass-reactors-in-pharma-industry
[4] Espino-Estévez R, Fernández-Rodríguez C, González-Díaz O M, Navío J A, Fernández-Hevia D, and Doña-Rodríguez J M 2015 *Chem. Eng. J.* **279**, 488-497.
[5] Lin C H, Lee J W, Chang C Y, Chang Y J, Lee Y C, and Hwa M Y 2010 *Surf. Coat.Technol.* **205**, S341-S344.
[6] Solacolu S, 1968 *Chimia Fizica a Silicatilor*, Ed. Tehnica, Bucuresti.
[7] Baltă P, Baltă E, 1971 *Introducere în chimia fizică a stării vitroase*, Ed. Academiei, Bucureşti.
[8] ISO 12543-4:1998 Glass in building - Laminated glass and laminated safety glass - Part 4:Test methods for durability
[9] Holland L, 1966 *The properties of Glass Surfaces*, ed. Chapman & Hall, London.
[10] Bunting E. N., J. 1930 *Am. Ceram. Soc.* **13**, 5-10.
[11] Xia L, Liu Z, and Taskinen P A 2015 *J. Eur. Ceram. Soc.* **35**, 4005-4010.
[12] Mocioiu O C, Atkinson I, Cusu-Pandele J, Bratan V, Petrescu S, Baila D I, and Mocioiu A M 2018 *Rev. Roum.Chim.* **63**, 419-424.
[13] STAS 598/3-76 Sticlă. Determinarea stabilităţii faţă de acizi.
[14] ISO 1776:1985 Glass — Resistance to attack by hydrochloric acid at 100 degrees C. Flame emission or flame atomic absorption spectrometric method.
[15] STAS 2413-87 Ambalaje de sticlă. Butelii pentru produse chimice şi farmaceutice.
[16] Mocioiu O C, Mocioiu A M, Marin A, and Zaharescu M 2017 *Ceram. Int.* **43**, 77–83.
[17] Zaharescu M, Mocioiu O C, and Andronescu C 2008 *J. Opt. Adv. Mater.* **10,** 1315-1319.
[18] Mocioiu O C, Atkinson I, Pandele J, Boroica L, Sava B, Mocioiu A M,and Zaharescu M 2014 *Rev. Roum. Chim.* **59,** 267-272.
[19] Zaharescu M and Mocioiu O C 2013 Infrared spectroscopy, *Chemical Solution Deposition of Functional Oxide Thin Films*, Ed T Schneller, R Waser, M Kosec, D Payne (Wien: Springer-Verlag) chapter 9, pp. 213-230.
[20] Mocioiu O C, Zaharescu M, Atkinson I, Mocioiu AM, Budrugeac P 2014 *J. Therm. Anal. Calorim.* **117**, 131-139.
[21] Pruna A, Cembrero J, Pullini D, Mocioiu A M and Busquets-Mataix D 2017 *Appl. Phys. A* **123**, 792-797.
[22] Park H, Kim D, Jung J, Pham DP, Le A H T, Cho J, Hussain S Q, Yi J 2018 *Heliyon* **4,** e00835.
[23] Iliescu C, Jing J, Tay F E H, Miao J, Sun T 2005 *Surf. Coat. Technol.* **198**, 314.

IOP Conf. Series: Materials Science and Engineering **1018** (2021) 012009 doi:10.1088/1757-899X/1018/1/012009

Manufacturing of SiO_2-ZnO materials with optical properties

O C Mocioiu[1], D I Băilă[2]*, I Atkinson[1], V Brătan[1] and A M Mocioiu[3]*

[1] Ilie Murgulescu Institute of Physical Chemistry of Romanian Academy, 202 Splaiul Independenţei, Bucharest 060021, Romania
[2] Politehnica University of Bucharest, 303 Splaiul Independentei, Bucharest, 060042, Romania
[3] National R&D Institute for Non-ferrous and Rare Metals,102 Biruinţei Blvd, Pantelimon, Ilfov 077145, Romania

Emails: baila_d@yahoo.com, ammocioiu@imnr.ro

Abstract. Antireflection coatings have been widely used to increase the absorption of solar collectors and to reduce surface reflection. Many techniques have been explored to prepare thin coatings, such as sputtering, chemical etching, chemical vapour deposition, and sol-gel method. SiO_2-ZnO coatings were prepared on the surface of glass substrates via the sol-gel dip-coating process. The structural, morphological and optical properties of the coatings were characterized. The vitreous structure was identified by infrared spectroscopy (FTIR) and X-ray diffraction (XRD). Morphology of materials was observed by scanning electron microscopy (SEM). The reflectance spectra were investigated and the optical performance of the structure was determined.

1. Introduction

The wavelength of the light from sun that reaches the Earth's surface conventionally ranges from 300 to 2500 nm. The human eye is sensitive to only a part of the electromagnetic spectrum. Antireflective coatings of glasses are used in many domains as buildings, solar cells, spectacles, etc. The purpose of reflective coatings is to keep objects cooler than they would be normally. The decrease of reflectance will be directly related to the increase of transmittance and the enhancement of conversion efficiency for solar cells [1-3]. Heat is a direct consequence of visible or infrared radiation incident on an object. The heat-producing region of the infrared radiation ranges from 700 to 1100 nm [1-3]. In order to prepare thin coatings, new techniques such as sputtering, chemical etching, chemical vapor deposition, and sol–gel method were used. Surface anti-reflection treatment can be categorized to 4 different techniques including surface texturing, single-layer interference coating, multi-layer coating and moth-eye structure forming [1,2]. Li [1] reported obtaining of bilayer porous coatings with refractive index of SiO_2 layer in the range from 1.10 to 1.45. The nanoporous ZnO/SiO_2 coatings preparation was made in two steps:1) deposition of ZnO bottom layer and 2) deposition of SiO_2 top layer. Due to SiO_2 top layer, the coatings have nanoporous structure. Han [2] obtained the reflection lower to around 5.3% at the wavelength of 400 nm for moth-eye pattern.

In this paper, ZnO-SiO_2 coatings with antireflective properties were prepared on the surface of glass substrates using sol–gel dip-coating process. The structural, morphological and optical properties of single layer coatings were investigated. The influence of different molar ratio, pH, and thermal treatment on the optical properties was studied.

2. Experimental parts

The sol was synthesized using tetraethyl orthosilicate [$Si(OC_2H_5)_4$] (TEOS) and zinc acetate dehydrate [$Zn(CH_3COO)_2\ H_2O$] as sol-gel precursors, HCl as catalyst and ethanol (C_2H_5OH) as solvent and distillated water for hydrolysis. All chemicals were reagent grade. Coatings with 10% ZnO and 90% SiO_2 were obtained by deposition of thin films on glass substrates by dip-coating at a controlled speed of 60 mm/min. Two precursor solutions with different molar ratio of were prepared as can be seen in table 1. The influence of water quantity on the properties of coatings was followed. The obtained coatings were noted S10ZA3 and S10ZA6. The coatings were consolidated at 200°C and then they were thermally treated at 500°C.

Table 1. Precursor solution preparation.

	Molar ratio TEOS:C_2H_5OH:H_2O	pH
S10ZA3	1: 4: 3	1
S10ZA6	1: 4 : 6	1.5

Obtained coatings were characterized structural, morphological and optical. X-ray diffraction experiments were performed with Rigaku Ultima IV diffractometer in parallel beam geometry equipped with CuKα radiation (wavelength 1.5406 Å). The XRD patterns were collected in 2Θ range between 10 to 70 with a speed of 5°/min and a step size of 0.02°. Structures of films were investigated by FT-IR Spectroscopy with a 6700 Nicolet FTIR Spectrometer in 400-1600 cm^{-1} domain; with sensibility of 4 cm^{-1}. Morphology of coatings was observed with a scanning electron microscope (SEM) Quanta 250, produced by FEI Company. Samples preparation was minimal and consisted in immobilizing on a double-sided carbon tape, without coating. Diffuse reflectance UV-VIS spectra were recorded using a spectrophotometer Perkin Elmer Lambda 35. The certified reflectance standard has been used and the measurements were carried out in the range 900-200 nm.

3. Results and discussions

Figure 1 presents XRD patterns of the coatings after thermal treatment at 500°C. Both coatings have vitreous state, without crystallites. In order to see if vitreous state was a response of substrate - glass, or a response of coatings, we made X-ray diffraction measurements on coatings on glass substrate and two powders resulted from gels dried and thermally treated at 500°C. In all cases the materials have vitreous state.

Figure 2 presents the FTIR spectra of SiO_2-ZnO coatings after treatment at 200°C and Figure 3 presents FTIR spectra after thermal treatment at 500°C.

The wide bands in 850–1300 cm^{-1} range, in 700–840 cm^{-1} range, and below 600 cm^{-1} were observed. The band with maximum at 1088 cm^{-1} is attributed to ring or double chains of tetrahedra in silicates. In the spectra of S10ZA3-200°C and S10ZA6-200°C coatings two bands at 950 $cm^{-1}$and 573 cm^{-1} are presented. Both bands evolve as shoulders in S10ZA3-500°C and S10ZA6-500°C coatings. The band at 950 cm^{-1} is assigned to Si-O vibration in chains of structural units. The bands at 573 cm^{-1} and 788 cm^{-1} can be assigned to Si-O-Zn or O-Zn-O bonds, in agreement to literature data [4].

The strong band at 458 cm^{-1} is also characteristic of Si-O-Si bonds in agreement with literature data [4-7]. The shoulder at 448 cm^{-1} present in spectrum of S10ZA6-200°C is characteristic to Si-O-Zn vibration. Fausta et all [8] reported vibration of Zn-O at 427 cm^{-1} in $60TeO_2$- (30-x) ZnO- $5Bi_2O_3$- $5TiO_2$- xB_2O_3 (where x = 0, 2.5, and 5 mol%). The bands at 482 cm^{-1} assigned to the stretching vibrations of ZnO and at 735 cm^{-1} due to the Zn–O bonding were observed by Slimani et al in ZnO films deposited on FTO glass substrate [9].

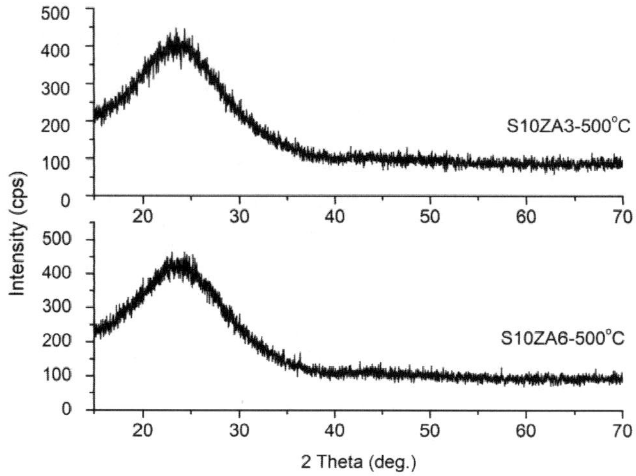

Figure 1. XRD patterns of SiO_2-ZnO coatings thermally treated at 500°C

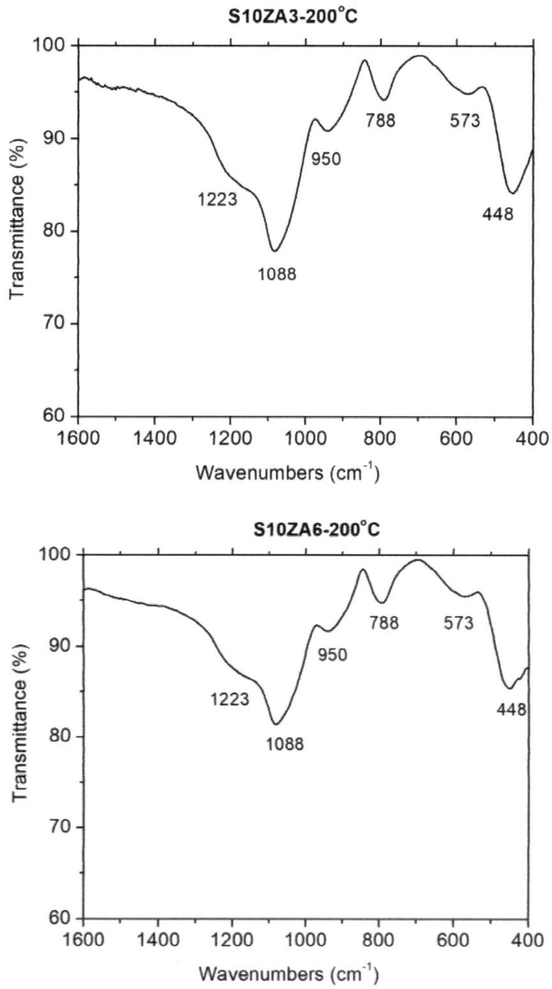

a)

b)

Figure 2. Infrared spectra of SiO_2-ZnO coatings thermally treated at 200°C

The thermal treatment at 500°C leads to an ordering of the silica network as in silicates. From FTIR spectra can be concluded that the higher transmittance of coatings treated at 200°C is correlated with depolymerized silica network. In the coatings treated at 500°C the silica network is structured and structural units are more arranged. The bands below 600 cm^{-1} characteristic Zn-O vibrations are smaller due to strong bond energy of silica network.

Coatings treated at 500°C show high transmittance in infrared so they have good optical properties.

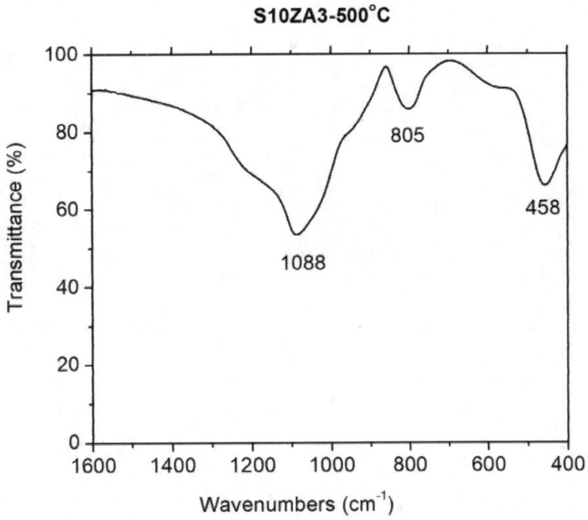

a)

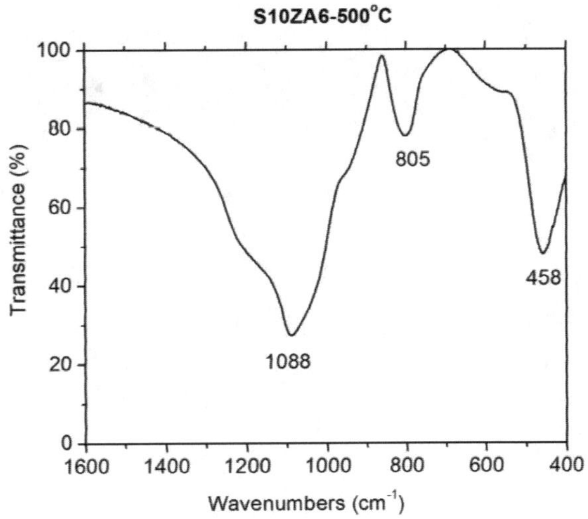

b)

Figure 3. Infrared spectra of SiO_2-ZnO coatings thermally treated at 500°C

Figure 4 show morphology examined by SEM of SiO_2-ZnO coatings after treatment at 500°C. The S10ZA3-500°C is uniform and homogeneous specific of vitreous coatings. The SEM image of the S10ZA6-500°C coating show small white separations. The observation is in agreement with FTIR results that show a network arrangement into coating.

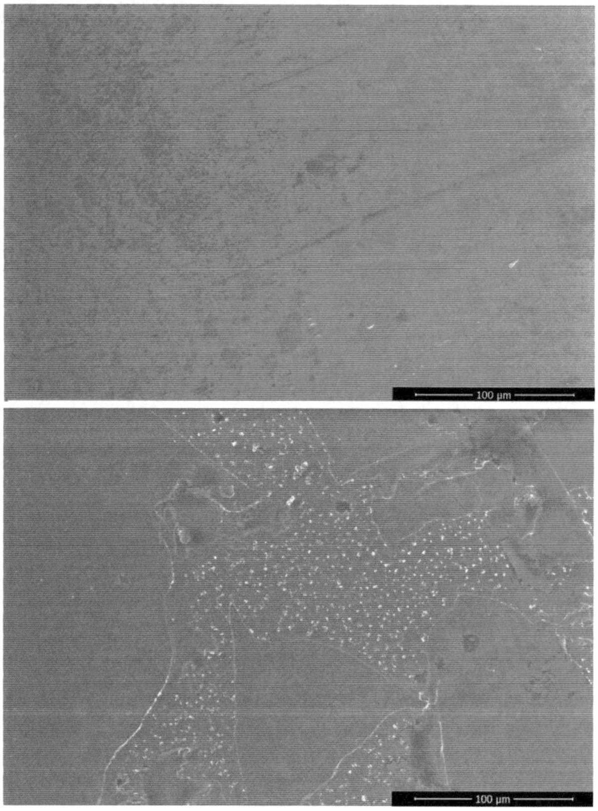

a)
b)

Figure 4. SEM images of SiO_2-ZnO coatings: a) S10ZA3-500°C b)S10ZA6-500°C

The reflectance spectra of the coatings were recorded and compared with the reflectance spectrum of the spectralon etalon. The results of the SiO_2-ZnO coatings are shown in Figure 5.

In case of coatings consolidated at 200°C appear supplementary bands in UV zone at 272 nm and in Visible zone at 343 nm. ZnO thin film synthesized by sol-gel spin coating method shown a sharp decrease in UV-VIS absorption in the wavelength range of about 310 – 430 nm [10]. Motlan et all reported this absorption also in the visible light range and therefore suitable for optoelectronic application [10]. In the visible zone of the spectra, the baselines of diffuse reflectance curves exhibit an almost horizontal area below 10%.

The reflectance of the coatings after thermal treatment at 500°C increased in visible domain to 60% in case of S10ZA6-500°C and to 90% in case of S10ZA3-500°C.

In vitreous coatings the refraction index can be calculated with formula $n=v_0/v$ where v_0 in air and v is wavelength in vitreous material. The refraction index in studied coatings is situated in blue zone and have value of n=1.15.

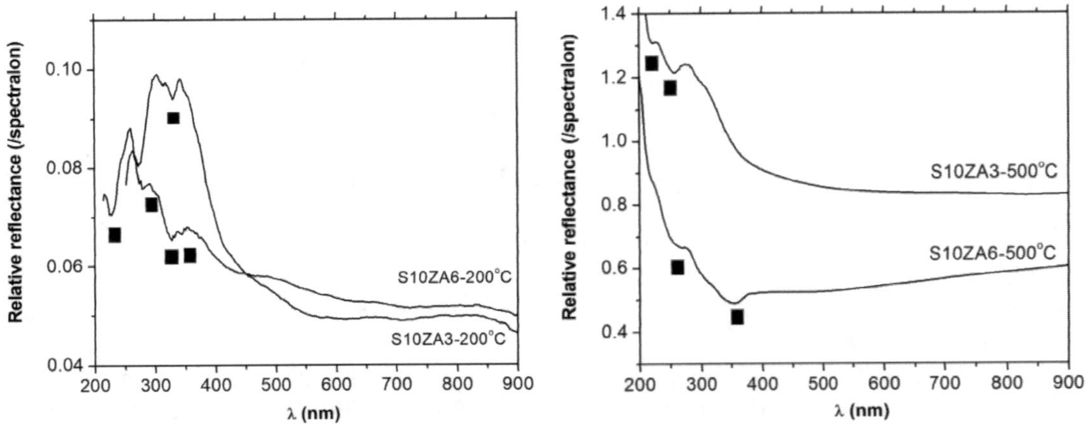

Figure 5. UV-VIS spectra of SiO_2-ZnO coatings

4. Conclusions

SiO_2-ZnO coatings present refractive properties in UV-Vis and infrared domains. Water content used in synthesis has influence on structure and morphology of the coating. Thermal treatment at 500°C leads to improvement of optical properties. The refraction index in studied coatings is situated in blue zone and have value of n=1.15. The studied materials can be used as protective and refractive coatings for glasses in applications as solar cell and in the optoelectronic devices.

References

[1] Li D, Huang F, Ding S 2011 *Appl Surface Sci* **257,** 9752– 9756
[2] Han K-S, Shin J-H, Yoon W-Y, Lee H 2011 *Solar Energy Mater & Solar Cells* **95,** 288 291
[3] Coser E, Moritz V F, Krenzinger A, & Ferreira C A 2015 *Polímeros* **25**, 305-310,
[4] Mocioiu O C, Atkinson I, Cusu-Pandele J, Bratan V, Petrescu S, Baila D I, Mocioiu A M 2018 *Rev Roum Chim* **63**, 419-424.
[5] Mocioiu O C, Mocioiu A M, Marin A, Zaharescu M 2017 *Ceram. Int.* **43**, 77–83.
[6] Zaharescu M, Mocioiu O C 2013 Infrared spectroscopy, Chemical Solution Deposition of Functional Oxide Thin Films, Ed T Schneller, R Waser, M Kosec, D Payne(Wien: Springer-Verlag) chapter 9, pp. 213-230.
[7] Mocioiu OC, Zaharescu M, Atkinson I, Mocioiu AM, Budrugeac P 2014 *J. Therm. Anal. Calorim.***117**, 131-139.
[8] Fausta D E, Marzuki A, Cari 2020 *J. Phys.: Conf. Ser.* **1511**, 012076.
[9] Slimani H, Bessous N, Dagher S, Hilal-Alnaqbi A, El Gamal M, Akhozheya B, Mohammed M 2020 *Mater. Res. Express* **7** 025026.
[10] Motlan M, Siregar N, Panggabean J 2020 *J. Phys.: Conf. Ser.* **1428** 012065.

IOP Conf. Series: Materials Science and Engineering 1018(2021) 012010 doi:10.1088/1757-899X/1018/1/012010

The role of the friction process in abrasive grain micro cutting technology

S Mazuru and S Scaticailov

Department of Machines Construction Technology, Faculty of Mechanical Engineering and Transport, Technical University of Moldova, 9/8, Studentilor str., block of study nr. 6, Chisinau, Moldova

sergiu.mazuru@tcm.utm.md

Abstract. The process of micro-cutting with abrasive grain - complicated tribological phenomenon. The removal of the chip-shaped material from the surface of the semi-finished product is accompanied by elastic and plastic deformations of the surface, with heat releases and phase changes. One of the features of the grinding process and other types of abrasive processing, is that the removal of the chip especially takes place with negative front angles. According to [1] the average value of the front cutting angle values is within the limits - (46.6… 56.9). To the geometric parameters of the distant layer that characterizes the cutting process can be attributed: the thickness and width of the cut and the shrinkage of the chip. A series of scientific and manual works on cutting materials can be objected to the fact that in the process of deformation at cutting which is characterized by the coefficient of shrinkage of the chip a physical-mechanical influence has the characteristic of the processed material, the value of advance, depth of cutting and other parameters.

1. Introduction

Even in the middle of the last century, the authors [1, 2] remarked that when analyzing the mechanics of cutting materials, there is a wrong physical explanation of the phenomena and an incorrect determination of the main influencing factors. The authors assumed that the main role in the value of plastic deformation belongs to the coefficient of friction. The influence of cutting regimes and physical-mechanical properties of materials on the value of plastic deformation of the removed layer, determined by the shrinkage of the chip, can be expressed only by the corresponding influence on changing the coefficient of friction of the chip on the cutting surface of tools.

Experimental research [3, 4] confirms the above. In the development of the above we can add that the local value of the coefficient of friction within the contact area "tool-chip" is not constant. Depending on the purpose of the research, we can examine only the most requested areas of contact of the abrasive grain with the chip. Because, the discussion takes place about the formation of the chip, so the main role is given to the plane of separation of the front and back surface of the granule, ie - the cutting edge of the tool.

The results of experimental research related to the determination of the shear force at grinding are known [5]. Models of cutting forces are known [6, 7] they have an essential value in determining the friction force, and can explain the processes of friction, cutting and the formation of wrinkles on the surface of the part (figure 1) [8]. The increase of the cutting depth is accompanied by the increase of the cutting forces, as a result tempering increases and there are changes in the structure and hardness of the part, attempts have been made to describe mathematically [9] the value of the influence of cutting forces on the process of forming residual stresses. .

Decreasing the cutting forces and the cutting depth leads to an increase in the influence of friction [10]. Consequently, the volume of heat in the cutting area increases, which leads to the appearance of thermal stresses, which can be determined by the expression [10, 11]:

$$\sigma = \frac{E}{1-\mu}\lambda(t_{\phi} - t) \quad (1)$$

where: E - Yung module; μ - Poisson's ratio; λ - thermal coefficient; t_{φ} - average temperature; t - current surface temperature.

It is known that along the cutting line the normal stresses reach the maximum values and the tangential ones are equal to zero, ie the friction forces are missing.

After the transformation the expression (2) will have the form:

$$\frac{\mu_1+\mu_2}{1-\mu_1\mu_2} < \frac{\eta cos\gamma}{1-\eta cos\gamma}. \quad (2)$$

2. **Mathematical model**

In the paper [3] it was demonstrated that the rotating hyperboloid is a figure that better fits the shapes of the abrasive grain and can be used to model geometry. This is also mentioned in the authors' work [3]. Then the value of the front angle on the surface of the abrasive granules and in the plane of the main sections at the level of the semi-finished section can be determined from the following expression [4]:

$$\frac{1}{\cos\gamma} = \pm\sqrt{\frac{\rho^2}{a^2(2\rho + a_z \mathrm{tg}^2\gamma_3)} + \mathrm{tg}^2\gamma_3 + 1}, \quad (3)$$

where: ρ - radius of rounding of the cutting edge (radius of the tip of the hyperboloid); a_z- depth of purchase with a single grain; γ_3 - the front angle of the abrasive granule, determined by the position of the asymptotes of the hyperboloid - the model of the granule section ($\gamma_3 = \varepsilon / 2$, ε - the angle between the asymptotes).

By transforming expression (2), it follows that,

$$\pm\sqrt{\mathrm{tg}^2\gamma + 1} = \pm\sqrt{\frac{\rho^2}{a^2(2\rho + a_z \mathrm{tg}^2\gamma_3)} + \mathrm{tg}^2\gamma_3 + 1},$$

$$\frac{1}{\cos\gamma} = \pm\sqrt{+\mathrm{tg}^2\gamma + 1}:$$

$$\mathrm{tg}^2\gamma = \frac{\rho^2 + 2\rho a_z \mathrm{tg}^2\gamma_3 + a_z^2 \mathrm{tg}^4\gamma_3}{a_z(2\rho + a_z \mathrm{tg}^2\gamma_3)}, \quad (4)$$

$$\mathrm{tg}\gamma = \frac{\rho + a_z \mathrm{tg}^2\gamma_3}{\sqrt{a_z(2\rho + a_z \mathrm{tg}^2\gamma_3)}}, \mathrm{tg}^2\gamma = \frac{(\rho + a_z \mathrm{tg}^2\gamma_3)^2}{a_z(2\rho + a_z \mathrm{tg}^2\gamma_3)}, \gamma = -\mathrm{arctg}\left(\frac{\rho + a_z \mathrm{tg}^2\gamma_3}{\sqrt{a_z(2\rho + a_z \mathrm{tg}^2\gamma_3)}}\right),$$

$$\text{where } \gamma = \in \left(\frac{\pi}{2}; 0\right) \quad (5)$$

By replacing the expression (5) in the inequality (2) we obtain the possibility, depending on the cutting depth a_2 and the groove geometry, to determine the maximum possible value of the generalized friction coefficient, at which the displacement deformations take place.

As a criterion, which characterizes the existence or lack of chip formation in the process of interaction of abrasive granules and chip processing can serve the ratio a_z/ρ - the parameter of chip formation, the value of which in different works is measured in a wide range - from 0.01 to at 1.0. The formulas for determining this value in the literature sources are missing, and the data presented are as a rule the results of experimental research. We will try to solve this problem.

We return to expression (4) and highlight the parameter of chip formation $x = a_z/\rho$, dividing the denominator and the numerator by ρ^2. Preparation:

$$\mathrm{tg}^2\gamma = \frac{1 + 2x\mathrm{tg}^2\gamma_3 + (x\mathrm{tg}^2\gamma_3)^2}{2x + (x\mathrm{tg}^2\gamma_3)^2},$$

After transformations we obtain:

$$2x\mathrm{tg}^2\gamma + x^2\mathrm{tg}^2\gamma\ \mathrm{tg}^2\gamma_3 = 1 + 2x\mathrm{tg}^2\gamma_3 + x^2\mathrm{tg}^4\gamma_3,$$

$$\mathrm{tg}^2\gamma_3(\mathrm{tg}^2\gamma_3 - \mathrm{tg}^2\gamma)x^2 + 2(\mathrm{tg}^2\gamma_3 - \mathrm{tg}^2\gamma)x + 1 = 0.$$

We will note differently $\mathrm{tg}^2\gamma_3 - \mathrm{tg}^2\gamma = K$: then $K\mathrm{tg}^2\gamma_3 x^2 + 2Kx + 1 = 0$.
We solve this quadratic equality with respect to x.

$$x = -\mathrm{ctg}^2\gamma_3\left(1 - \frac{\mathrm{tg}\gamma}{\sqrt{-K}}\right),$$

The necessary root is found in the expression:
that is, the chip formation parameter can be expressed by the angle ε at the top of the abrasive grains (because $\gamma_3 = \varepsilon/2$) and current front angle γ following relationship:

$$\frac{a^2}{\rho} = -\mathrm{ctg}^2\gamma_3\left(1 - \frac{\mathrm{tg}\gamma}{\sqrt{\mathrm{tg}^2\gamma - \mathrm{tg}^2\gamma_3}}\right) \quad (6)$$

At the denominator on the left side of the expression (6) we have the parameter ρ - the radius of the hyperboloid vertex. But in this case it is more correct to examine not the radius at the top of the hyperboloid, but the radius of curvature r at the point at the level of the semi-finished plane. There is a dependency between them:

$$r = \frac{\rho}{\cos^3\alpha}, \quad (7)$$

where α - the angle between the normal at the exaninate point and the radius of the vector, descended from the "focus" of the hyperboloid to the given point.

Taking into account this fact, expression (6) will have the form:

$$\frac{a^2}{\rho} = -\mathrm{ctg}^2\gamma_3\left(1 - \frac{\mathrm{tg}\gamma}{\sqrt{\mathrm{tg}^2\gamma - \mathrm{tg}^2\gamma_3}}\right) \times \cos^3\alpha. \quad (8)$$

The value of the front angle in relation (8) can be determined from expression (1) according to the formula [5]:

$$\gamma \mathrm{acrsin}\left[\frac{1}{1+\mu^2}\left(\frac{\mu^2}{\eta} - \sqrt{1 + \mu^2 - \left(\frac{\mu}{\eta}\right)^2}\right)\right] \quad (9)$$

Using (9) and the expression (6) or (8) we will have the possibility to determine the chip formation parameter. In fig. 1 shows the influence of the generalized friction coefficient μ and the chip contraction coefficient η at the value a_z/ρ - the results of the calculations according to the relations (9) and (7). The top angle of the granule is equal to the angle between the asymptotes of the hyperboloid and accepted as $1*10^9$.

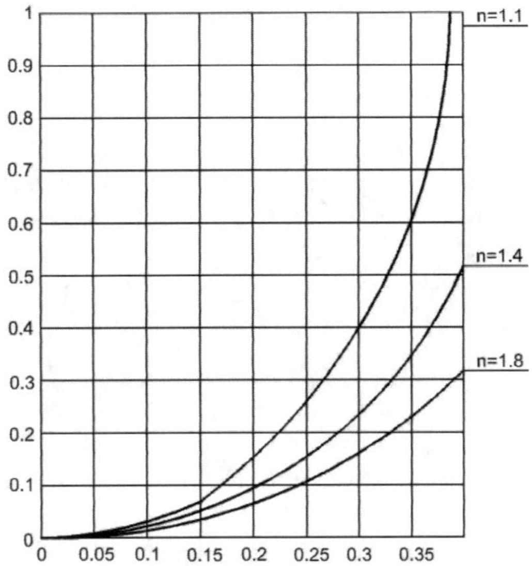

Figure 1. The influence of external and internal friction coefficients and chip compression on the chip formation parameter

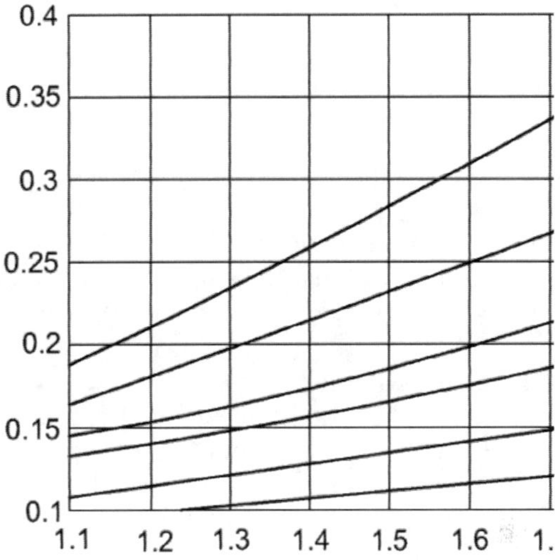

Figure 2. The curves of the equal level of the chip volume dependence V (mm), removed by an abrasive grain, depending on the coefficient of friction μ and the compression coefficient of the chip η.

The presented curves show that the value of the chip formation parameters can vary in a very wide spectrum. Obviously, that in the literature sources the data are obtained for concrete processing conditions and have a connection with the friction coefficients, the cutting speed, the processed material, which in turn determines the compression value of the chip. The influence of external and internal friction coefficients on the chip formation parameter as seen is extremely high within the values of 0.1… 0.4 when grinding [6]. This fact can be clarified by the fact that, for the abrasive granule, the external and internal friction - this is an externally directed load, which determines the possibility of exceeding it depending on the direction of the applied force vector.
Some authors propose that [6] the internal friction angle equals the texture angle, which is logical and correct. Which allows all forms of basic loads on abrasive grains to be presented in angular units. Since the projection of the resultant of the cutting force following the texture plane must be positive [7], then the inequality must be fulfilled:

$$\Sigma = \beta_1 + \xi + \psi - \gamma < \frac{\pi}{2}, \tag{10}$$

where: Σ - load on the abrasive grain in angular units; β_1 - travel angle; ξ - external friction angle; ψ - internal friction angle (texture angle); γ - the front angle next to the cutting edge line in the plane of the main section (figure 2).

The value of the internal friction coefficient in angular units μ_2 = tgψ can be determined from formulas, presented in [7]:

$$\mu_2 = \frac{2}{\Delta + \sqrt{\Delta^2 + 4}}, \tag{11}$$

where: $\Delta = ctg\beta_1 + tg(\beta_1 - \gamma)$ – displacement relative to the transformation of the cut layer into chips.

$$\mathrm{tg}(\beta_1-\gamma)=\frac{\mathrm{tg}\beta_1-\mathrm{tg}\gamma}{1+\mathrm{tg}\beta_1\mathrm{tg}\gamma},\mathrm{tg}\beta_1=\frac{\cos\gamma}{\eta-\sin\gamma},$$

Given the fact that after these transformations we get:

$$\Delta=\frac{\eta^2+1}{\eta\cos\gamma}-2\mathrm{tg}\gamma.$$

From the point of view of the interaction of contact at atom level, the external and the internal friction have a similar nature. The idea of this theory is based on the fact that each discrete elementary act of sliding atomic layers towards each other is accompanied by a sudden (inevitable) rupture of the bonds between the atoms of contacted bodies and elsewhere, as well as the formation of new bonds. “Additive lava”. Internal friction - the dissipation of wave energy elastic at the expense of internal processes in the solid body, has its peculiarities and can be caused by different forms of processes [8].

3. The results obtained

The reduction of internal and external friction is a perspective direction in tribology, especially applied for mechanical processing of abrasive grain materials. It will allow the reduction of the energy consumption quota, consumed for the plastic deformation on the given surface of the abrasive grain and to increase the energy quota for the movement within the front plane. The position of the cutting edge line, of course, has changed, because the micro-cutting process - synergistic process.

The influence of friction in the cutting area on the chip formation process is considerable, especially when a thin layer is removed from the surface of plastics. This refers to different types of abrasive processing, and for grinding in particular. A maximum influence on the chip formation process is exerted by the coefficient of friction on the front plane of the tool near the cutting edge.

The paper [10, 11] presents the mechanism of influence of the coefficient of friction on the position of the cutting-edge line at micro-cutting with unitary abrasive granule of hyperboloid shape. The spatial configuration of the edge determines the real shape and surface of the front abrasive grain, and correspondingly the geometric parameters of the chip, which ultimately affects the volume of the metal removed and the productivity of the process.

The volume of the chip, removed from a single grain, taking into account the plastic deformation process, can be calculated according to the formulas obtained in the works [12, 13]. Based on the regressive analysis, the presented formulas can be substantially simplified, if to examine particular cases, ie to link them to concrete processing conditions, and as variable values to choose the most essential ones. As an example, depending on the influence of the coefficient of friction (μ) and the coefficient of compression (η) on the volume (V) of the chip can be presented as follows:

$$V=0{,}074\frac{0{,}033}{\mu}-\frac{0{,}253}{\eta}. \tag{12}$$

Expression (12) obtained by a nonlinear regression for the initial parameters: diameter of the abrasive disc - 250 mm; accepted cutting depth - 0.01 mm; the apex angle of the abrasive granules - 83… 143°; peak radius - 7 - 50 µm; cutting speed 35.5 m / s, longitudinal feed - 10 m / min.

The curves of the dependence of the volume of chips, removed by a single chip, on the coefficient of friction at different compression coefficients of the chip are shown in figure 2.

The figure shows that if the compression coefficient of the chip has small values, then the coefficient of friction obtains high values, in this case the chip will not be possible to be removed ($V<0$ - the field in which only the plastic deformation accompanied by friction takes place without metal removal).

When grinding in the contact area of the abrasive and semi-finished disc, complicated physical-mechanical and chemical processes take place, among them we can highlight the adhesion, the elastic and the plastic deformation, and the phase transformations. According to the law of conservation of kinetic energy, the energy of the granules of the abrasive disc and the semi-finished product is transmitted to the chips and is transformed into thermal energy, including by elastic deformation and the mechanism of plastic deformation - double sliding and displacement between granules. Some of the energy goes to change the internal state of the semi-finished product, which is manifested by the deformation of the crystal lattice, the accumulation of internal stresses and the structure becomes unstable.

In the conditions when high required contact voltages take place, the thermal energy momentarily causes different forms of phase transformations, even until the metal melts. The momentary high temperature, on the one hand, increases the plasticity of the deformed metal and improves the conditions for chip formation by abrasive grains, on the other hand - causes structural transformations in the surface layer and promotes the appearance of tensile stresses in the given layer [13]. At the same time, the main mechanism of plastic deformation becomes slipping. Simultaneously, the processes of diffusion, polygonization, recrystallization and other similar effects take place, which tend to increase the stability of the structure [14].

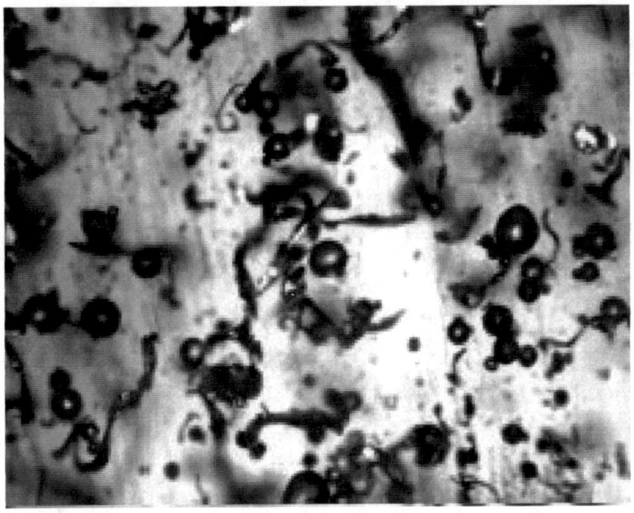

Figure 3. The shape of the sludge under the microscope (you can see the chips and molten metal balls).

The temperature in the cutting area, in the local sounds, can reach 1100… 1500°C, and this is much higher than the temperature of the critical points (as for example for steel 35 - it is 730°C and 810°C). In other areas the temperature can be much lower, which leads to the non-uniformity of the structure of the processed material and the appearance of a number of phases with different physical-mechanical properties. On the outside it is expressed by the appearance of burns, the formation of micro cracks (in case of critical internal stresses), the formation of unstable thermodynamic phases (such as the formation of austenite with large granules), which pass into other phases (eg pearlite) or subsequent destruction. worsen the quality of the outer layer [15-19].

To control the processes, caused by the non-uniformity of the thermal fields, at rectification it is practically impossible, as well as to measure the temperature in the local points in the conditions of the rapid thermodynamic processes. But the measures, which allow to reduce the temperature in the cutting area, in some cases are extremely effective, refer to the rational choice of cutting regimes.

We can talk about the fact that the thermal tension takes place in the cutting area due to the analysis of the chip composition, collected after grinding the semi-finished products, composed of deformed chips, molten metal elements, in the form of balls of various sizes (figure 3), abrasive dust and others dispersion product (ground granule, binder, chipped metal) [20-25].

IOP Conf. Series: Materials Science and Engineering **1018**(2021) 012010 doi:10.1088/1757-899X/1018/1/012010

4.Conclusions

The respective mathematical apparatus and the possibilities of the current calculation technique allow the modelling of different technological processes and the processing of experimental data according to known methods. At the same time, the time for carrying out scientific research is reduced, but there is also the possibility of finding an optimal solution, which is not obvious after carrying out the research.

A considerable influence on the section plane of the sharpening element of the abrasive grain is exerted by its penetration depth, at the same time the influence of its rounding radius and the angle at the top is much smaller but approximately equal to each other; The influence of different factors on the value of the plastic deformation of the removed layer can be expressed by the corresponding influence on the modification of the coefficient of external friction of the chip on the outer surface of the cutting tool and internal friction of the processed material near the cutting edge line; The a_z/ρ chip formation parameter in reality can vary in a very wide spectrum of values, determined, to a large extent, by the value of the external and internal friction coefficient and by the compression of the chip; As a result of decreasing the external friction coefficient increases the number of granules that remove the chip and decreases the number of granules that deform the surface layer, which leads to reduced grain wear, increases tool life, productivity and machining accuracy. Using the results of the sludge analysis, we can determine the thermal stress of the grinding process and optimize the processing regimes (for example, the advance value) so as to lower the temperature in the cutting area, which will have a positive effect on reducing unwanted tensile stresses and increase fatigue resistance of the obtained surface.

References

[1] Klocke F and Eisemblatter G 1997 *CIRP Annals* **46**(2) 519-526

[2] Brinksmeier E, Mutlugunes Y, Aurich JC and Shore P 2010 *CIRP Annals* **59**(2) 652-671

[3] Aurich JC, Brinksmeier E, Blum H and Brecher C 2009 Modelling and simulation of process: machine interaction in grinding *Prod. Eng.* **3**(1) 111-120

[4] Bostan I, Mazuru S and Scaticailov S 2013 Technologies for precessional planetary transmissions too thing generation *TEHNOMUS journal* **20** p 226-233

[5] Botnari V and Mazuru S 2014 *Appl. Mech. Mater.* **657** 147-153

[6] Mazuru S and Casian M 2014 *Appl. Mech. Mater.* **1036** 286-291

[7] Bostan I and Mazuru S 2009 *Influence of the grinding parameters on the characteristics of gears teeth outer layer* Proceedings of the13th Int. Conference on Modern Technologies, Quality and Innovation IASI & Chisinau

[8] Bostan I Dulgheru V Gluşco C and all 2011 *Antologia invenţiilor* Vol 2 *Transmisii planetare precesionale* (Chişinău: Bons Offices)

[9] Mazuru S 2010 *Mechanism of training component kinematics error gears in operation technology hardening chemical – heat* (Buletinul Institutului Politehnic din Iași, Universitatea Tehnică „Gheorghe Asachi" din Iași Tomul LVI (LX) Fasc 2a)

[10] Bostan I, Mazuru S and Botnari V 2011 *Cinetic process of teeth grinding* (The 15 th International Confercence Modern Tehnologies, Quality and Innovation Vadul lui Voda Moldova România

[11] Bostan I, Vaculenco M and Scaticailov S *Issues technology manufacturing precessional gears with nonstandard profile generating* IX international congress "Machines Technologies Materials 2012" Varna Bulgaria Vol I

[12] Casian M and Scaticailov S 2017 *Adv. Mat. Res.* **112** 01026

[13] Mazuru S, Vlase A and Scaticailov S 2014 *Tehnologii de prelucrare pe mașini de danturat* (Chișinău: Tehnica-UTM)

[14] Mazuru S and Scaticailov S 2018 *Tehnologii și procedee de danturare a roților dințate* Univ. Tehn A Moldovei (Chișinău: Tehnica-UTM)

[15] Bostan I 2004 *Aprecierea calităţii organelor de maşini la etapa de pregătire tehnologică a producţiei.* Buletinul Institutului Politehnic Iaşi tomul LIV Fascicula Vc Iaşi 749–752

[16] Scaticailov S 2002 *L'efficatite de la rectification de la force et de la vitesse.* Buletinul Institutului Politehnic Iaşi, tomul XLVIII, Supliment I Iaşi 237 – 240
[17] Laurenţiu S, Coteaţă M, Pop N and Coelho A 2009 *Nonconventional Tech. Review* **1** 96-99
[18] Botnari V 2014 *Appl. Mech. Mater.* **657** 147-153
[19] Kukhar V, Balalayeva E, Prysiazhnyi A, Vasylevskyi O and Marchenko I 2018 *MATEC Web of Conferences* **178** 02003
[20] Anishchenko O, Kukhar V, Grushko A, Vishtak I, Prysiazhnyi A and Balalayeva E Yu 2019 *Mater. Sci. Forum* **945** 531–537
[21] Kukhar V, Grushko A V and Vishtak I V 2018 *Solid State Phenom.* **284** 408–415
[22] Slatineanu L, Dodun O and Coteata M 2008 *Theoretical Model of the Surface Roughness at the End Milling with Circular Tips* Annals of DAAAM for 2008 & Proceedings of the 19th International DAAAM Symposium Editor B. Katalinic Vienna Austria 1273-1274
[23] Bostan I and Botnari V 2011 *Cinetic process of teeth grinding* The 15th International Conference Modern Technologies Quality and Innovation Vadul lui Voda 121-124
[24] Casian M and Mazuru S 2014 *Adv. Mat. Res.* **1036** 286-291
[25] Casian M and Mazuru S 2014 *dv. Mat. Res.* **1036** 292-297

IOP Conf. Series: Materials Science and Engineering 1018(2021) 012011 doi:10.1088/1757-899X/1018/1/012011

Some aspects of the nitriding process of parts in machine construction

S Mazuru, N Trifan and A Mazuru

[1] Department of Machines Construction Technology, Faculty of Mechanical Engineering and Transport, Technical University of Moldova, 9/8, Studentilor str., block of study nr. 6, Chisinau, Moldova

sergiu.mazuru@tcm.utm.md

Abstract. Nitriding is the most common and effective surface hardening method. Such chemical-thermal treatment is capable of increasing surface hardness, contact endurance, wear and seizure resistance, as well as heat resistance and corrosion resistance of a wide range of machine parts. This process of surface hardening has found its application in many branches of modern mechanical engineering. The operational requirements for the parts led to the need to replace high-temperature methods of chemical-thermal treatment (carburizing, high-temperature nitrocarburizing, etc.) for hardening processes at lower temperatures (500-650° C), namely nitriding. This replacement was facilitated by the latest technological developments in the field of various nitriding methods. The scientific developments obtained to date make it possible to gradually eliminate such disadvantages of nitriding as a significant duration of the process, increased fragility of the surface layer, insufficient values of contact endurance, and labor intensity of the process [1, 2].

1. Introduction

Nitriding is the most common and effective method of surface hardening. Such chemical-thermal treatment is capable of increasing surface hardness, contact endurance, wear and seizure resistance, as well as heat resistance and corrosion resistance of a wide range of machine parts. This process of surface hardening has found its application in many branches of modern mechanical engineering. The operational requirements for the parts, came to the conclusion to replace high-temperature methods of chemical-thermal treatment (carburizing, high-temperature nitrocarburizing, etc.) for hardening processes at lower temperatures (500 - 650 ° C), namely nitriding. This replacement was facilitated by the latest technological developments in the field of various nitriding methods. The scientific developments obtained to date make it possible to gradually eliminate such disadvantages of nitriding as a significant duration of the process, increased fragility of the surface layer, insufficient values of contact endurance, labor intensity of the process [1 - 3]. Despite the fairly widespread use of the nitriding process in the practice of various areas of mechanical engineering, there are still a lot of unresolved issues associated mainly with the mechanism of structure formation in the formation of a diffusion layer [2 - 6]. It is the structural features of the nitride layer and the matrix itself that determine the performance of machine parts and, consequently, the choice of steels, pretreatment technologies and the nitriding technology itself. Therefore, today the prediction and modeling of the formation of structure and properties in the diffusion layer during nitriding is a rather urgent problem [7 - 9].

2. Modeling the processes occurring during nitriding

Modeling the processes occurring during nitriding is the most effective method for developing a technological process. The use of computational methods makes it possible to obtain rather promptly information about the main properties of the hardened layer, about the rate of diffusion and structures of forming processes, about the formation of nitride and carbon nitride layers with varying temperature and time regimes. Simulation of various nitriding processes makes it possible for a simple and prompt solution of many technological problems to control and manage saturation modes of diffusion layers, predict the final results and the possibility of developing innovative processing modes. At the same time, taking into account the complexity and high cost of carrying out a large number of experiments makes mathematical modeling of nitriding processes a particularly promising research method. Mathematical models describing the dependences of the depth of the nitride layer and surface hardness on the values of temperature and duration of chemical-thermal treatment of steel during ionic nitriding can be represented as a polynomial of the second degree [1, 3]:

$$y_i = b_0 + a_1x_1 + a_2x_2 + a_3(x_1^2 - \beta) + a_4(x_2^2 - \beta) + a_5x_1x_2 \quad (1)$$

where - a_i are the estimated coefficients, β – is the parameter calculated depending on the number of points of the core of the compositional plan 2n-p, the shoulder of "star" points α and the number of points in the plan by the formula:

$$\beta = \Sigma_{j=1}^{N} \quad (2)$$

The procedure for estimating the coefficients of the model, checking its adequacy and statistical analysis of accuracy are given in [1, 2]. The resulting model of the depth of the hardened layer, depending on the normalized values of temperature and duration of chemical heat treatment, has the form:

$$y = 295{,}55556 + 42{,}5085 \cdot x_1 + 280{,}056 \cdot x_2 - 9{,}211 \cdot x_1^2 + 33{,}289 \cdot x_2^2 + 13{,}75 \cdot x_1 \cdot x_2$$

The procedure for estimating the coefficients of the model, checking its adequacy and statistical analysis of accuracy are given in [1, 2]. The resulting model of the depth of the hardened layer, depending on the normalized values of temperature and duration of chemical heat treatment, has the form:

$$y = 295{,}55556 + 42{,}5085 \cdot x_1 + 280{,}056 \cdot x_2 \quad (3)$$

The mathematical model describing the effect of the temperature and duration of nitriding on the values of the surface hardness of the nitride layer, in general form, is represented as follows

$$y = 9{,}8 - 0{,}8335 \cdot x_1 + 1{,}30026 \cdot x_2 + 0{,}09853 \cdot x_1^2 - 0{,}80147 \cdot x_2^2 - 0{,}75 \cdot x_1 \cdot x_2$$

Checking the significance of the coefficients showed that the coefficient $a3$ can be considered insignificant, therefore, the model is converted to the form:

$$y = 9{,}8 - 0{,}8335 \cdot x_1 + 1{,}30026 \cdot x_2 - 0{,}80147 \cdot x_2^2 - 0{,}75 \cdot x_1 \cdot x_2 \quad (4)$$

Response surfaces in factor space in a given planning area are presented in [2, 3].

3. Wear on the nitride surface

One of the reasons for the failure of many parts is the wear of the working surfaces during operation in friction units. To solve this problem, expensive materials are used, which are necessary for the manufacture of parts, with high hardness, wear resistance, etc. Experimental data on the determination of the hardness and thickness of nitride surface layers of steels of different grades by the nanoindentation method are presented. The dependence of the change in the layer hardness on the current power during nitriding has been determined. For the study were taken samples of steel of different grades, the composition of which is presented in [10, 11]. Determination of the thickness and hardness of the modified layer was carried out using a Nano hardness tester "Nano Hardness Tester" company CSEM by the Vickers method. The Vickers method uses a diamond pyramid with an apex angle between opposite faces equal to 136°. Hardness is defined as the ratio of the load to the surface area of the print or its projection. Since nanoindentation examines the layer in the nanometer range, and the indenter size is 5 μm, the surface roughness, cracks, and inhomogeneity of the chemical composition have a significant effect on the measurement result; therefore, the samples are pre-polished and the area with the least number of defects is selected during indentation. The maximum load on the indenter was 50 mN. An example of indentation is shown in Figure 1.

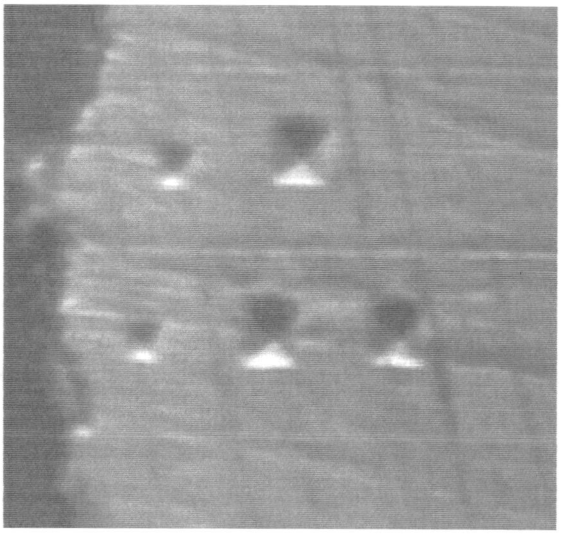

Figure 1. Matrix indentation on the sample nr. 3.

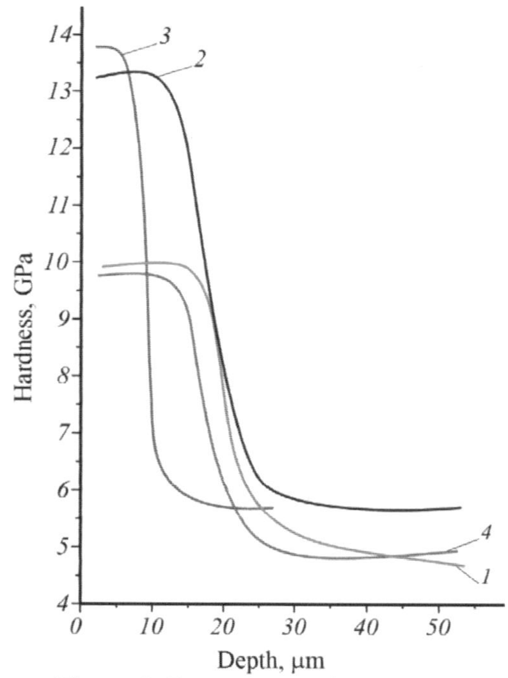

Figure 2. Dependence of the change in hardness on the depth of the nitride layer.

The primary obtained dependence of hardness on penetration depth has peaks and valleys that arise from the errors described above. But the hardness in the nitride layer changes smoothly, therefore, during data processing, an interpolation operation was used in order to reduce the error. Since the samples were polished, the roughness error is small. The instrumental error is determined by the standard. For this, the indenter pierces a quartz sample, then the device independently adjusts this error. The dependences of hardness on depth are shown in Figure 2, and a comparison of the initial and maximum hardness in the table [10, 11]. From the obtained dependences it follows that on the surface of the nitride layer the maximum value of hardness, when moving along the cross section of the sample in the direction from the nitride surface, the value of hardness decreases. At a certain depth of the layer, the hardness does not change, and from this it can be concluded that this value corresponds to the initial hardness of the sample, and this depth is the depth of the nitride layer.

4. Gas-cyclic and thermo-gas-cyclic nitriding

Nitriding was subjected to ARMCO iron, 105MnCrW11 alloy steel, the samples of which were subjected to wear tests. Standard processing for steels of this type - hardening and low tempering, but they can also be hardened by nitriding, because this type of chemical-thermal processing allows to increase the resistance to wear, thermal, corrosion of products [12 - 14]. Gas-cyclic and thermo-gas-cyclic nitriding was performed on an experimental installation, which includes the control and temperature maintenance system in the work space (oven). Gas supply system, control of ammonia consumption and dissociation value, gas cleaning and drying. At the same time, two solenoid valves for gases, controlled and controlled to order, and the possibility in automatic mode to act on the technological parameters of the nitration process.

The factor that has accelerated the enrichment process is the diffusion coefficient, as it is better known is influenced by temperature and concentration gradient. Therefore, in order to speed up the process and the possibility to regulate it, a thermocycle was proposed instead of isothermal treatment at a normal nitriding by gas flow. The temperature range was determined in accordance with the iron-carbon diagram, where there are critical points, which divide the domain into phases by content. The correct choice of the temperature range for the thermocycle process is a condition for researching the influence of cyclic gas and thermo-gas on the nitriding process. Therefore, the temperature range was chosen so that the process when nitrogen diffuses into iron (the first stage of the nitriding cycle) takes place below the temperature of eutectoid transformations according to the "iron-nitrogen" state diagram, iron-nitrogen at a temperature below 591°C. When researching the nitride samples, the temperature was in the limits of 490 - 540°C. Then the dissociation of the nitride layer was performed (the second stage of the cycle), the supply of ammonia is interrupted, the pressure in the furnace for a period of time remaining constant.

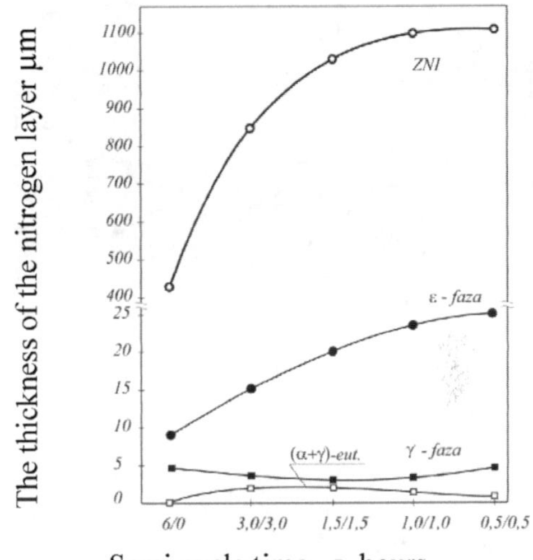

Figure 3. Dependence of the thickness of the nitride layer on the duration of the half-cycles at the nitriding and dissociation stage.

The upper limit of the nitriding cycle was 620°C. The value of ammonia dissociation in each first half-cycle was checked every 15-20 minutes. using the water disociometer. At the enrichment temperature 520°C the dissociation value with ammonia was 25-30%, at temperatures of 620°C - 40-45%. At each second half-cycle, the dissociation value with ammonia increased and constituted 68-70%, at a temperature of 520°C and 94-98%, at a temperature of 620°C.

The microstructure research after nitriding was performed on the "VEGA TS 5130" microscope, and the distribution of nitrogen on the nitriding depth was obtained by scanning with X-ray spectral analyzer. The wear resistance research was performed on the installation, which models the cutting process. The road traveled is equal to 200 m [14, 15]. Thermo-gas-cyclic nitriding technology is based on the periodic alternation of cycles when nitrogen diffuses into the iron during nitriding by gas flow and dissociation of the nitride layer to a maximum possibility of reducing the atmospheric saturation property.

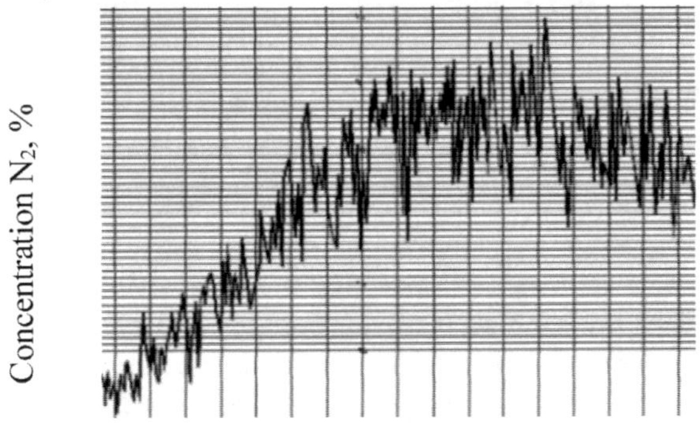

Figure 4. Distribution of nitrogen over the thickness of the nitrogen layer.

At the same time, the additional intensification takes place at the expense of carrying out the process in two phases - at the thermo-gas-cyclic nitriding the nitrogen enrichment process takes place, at the second dissociation which is carried out at different temperatures. This technology ensures an effective regulation of the phase structures of the diffusion layer, with the help of a new technological parameter cycle - and allows reducing gas consumption at the expense of maximum use at each gas consumption volume at the stage when nitrogen diffuses into iron. At the same time, the requirements for ecology are improved, which are advanced compared to the chemical-thermal processing processes, by reducing the harmful gases released into the environment.

Figure 5. Microstructure of diffusion layers.

Thermo-gas-cyclic nitriding technology is based on the periodic alternation of cycles when nitrogen diffuses into the iron during nitriding by gas flow and dissociation of the nitride layer to a maximum possibility of reducing the atmospheric saturation property. At the same time, the additional intensification takes place at the expense of carrying out the process in two phases - at the thermo-gas-cyclic nitriding the nitrogen enrichment process takes place, at the second dissociation which is carried out at different temperatures. This technology ensures an effective regulation of the phase structures of the diffusion layer, with the help of a new technological parameter - the duration of the half-cycle - and allows reducing gas consumption at the expense of maximum use at each gas consumption volume at the stage when nitrogen diffuses into iron. At the same time, the requirements for ecology are improved, which are advanced compared to the chemical-thermal processing processes, by reducing the harmful gases released into the environment.

The increase of the total thickness of the nitride layer is subject to parabolic dependence, with the decrease of the semi cycle duration the total thickness of the layer increases, due to the internal nitriding area (ZNI, figure 3). As the duration of the half-cycles increases, the thickness of the layer decreases, which is related to the process of dissociation of the nitride layer at the second stage of the cycle [15, 16].

When comparing thermo-gas-cyclic nitriding and ordinary nitriding, a sudden increase of the total layer is observed, it can be seen on microstructures (figure 4) according to those characteristics of the phases - γ'. At the same time, the total thickness increases especially on the account of ZNI, which increases approximately 6 times, that of nitrides - about 2 times compared to normal nitriding.

The increase in the intensity of the process is also shown by the radiation distribution curves from the nitrogen atoms (figure 5) - as it moves away from the surface, the nitrogen content in the layer increases and reaches its maximum value, especially in its depth. Obviously, the distribution of nitrogen on the thickness of the nitride layer in the investigated samples has the same character and are characterized by the fact that the maximum nitrogen content is very slow and is not on the surface of the samples,

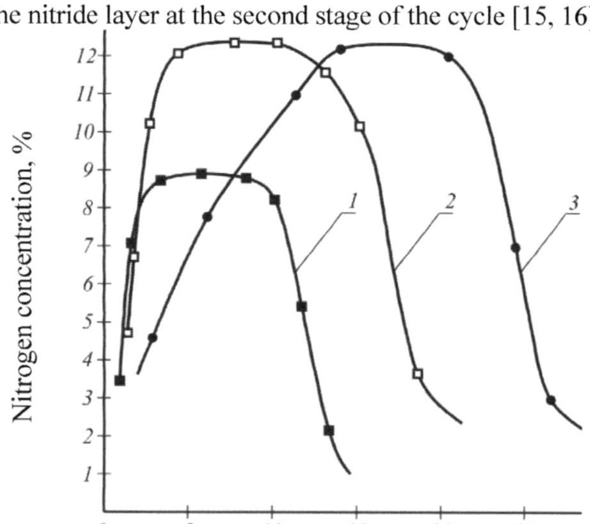

Figure 6. Distribution of nitrogen on the thickness of the nitrogen layer on the ARMCO iron: 1. Ordinary nitriding; 2. Gas - cyclic nitriding; 3. Thermo-gas-cyclic nitriding (enrichment).

which is in contact with the enrichment medium. In the internal nitride zone, at a distance from the sample surface.At the same time, taking into account the value of the intensity of the reflection of the lines, we can notice that, firstly, the thickness of the layer with high nitrogen content increases with the transition from ordinary to thermo-gas-cyclic nitriding, secondly, the intensity of radiation, and correspondingly the nitrogen content in the nitride area after thermo-gas-cyclic nitriding is obviously higher, compared to ordinary nitriding.

Figure 6 shows the results of the calculation of nitrogen consumption on the scanned curves. It can be seen that on the surface of the samples, which is in contact with the enrichment atmosphere, the nitrogen consumption is not high and constitutes about 3.5-4.8% (after meals). As the surface is removed, the nitrogen content in the layer increases and reaches a maximum value, for ordinary nitriding the nitrogen content increases maximum to 8.5%, and in the thermo-gas-cyclic - up to - 12%. This indicates that nitriding under the conditions of the thermo-gas-cyclic process is more intensive. At the same time, the maximum nitrogen content at its distribution curve on the layer thickness is very slow and moves in the depth of the layer when switching from ordinary to thermo-gas-cyclic nitriding, which can be clarified by intensifying the process of dissociation of the nitride layer in the material depth. The microhardness measurement showed: for the thermo-gas-cyclic nitriding which takes place 6.5 hours, we can obtain the same surface hardness as for the normal nitriding, only after 25-30 hours, the nitriding time was reduced by about 4-5 times.

Thus, by modifying only one parameter - the duration of the half-cycle, on the surface of the hardened product we can obtain layers with different phase structures at a corresponding correlation of the processes when nitrogen diffuses in iron and dissociation. Thus, we can use this method of nitriding to regulate the process, obtaining necessary phases on the surface of the product depending on the necessary operating conditions. As an example, in the thermo-gas-cyclic nitriding of 105MnCrW11 steel on the surface it is rational to obtain simultaneously the phase - γ'hard and plastic yes not the phase - ε fragile. In this case the nitriding is necessary to complete it at the dissociation stage, nor how at the enrichment stage. The wear nomograms of 105MnCrW11 steel samples after conventional gas-cyclic nitriding and thermo-gas-cyclic nitriding are shown in Figure 7. We point out that, following the research, it was established that the minimum wear corresponds to the thermo-gas-cyclic nitriding, which ended with the dissociation stage, at which the duration of the half-cycle was 0.5 hours. The wear of the samples is about 2.5 times lower than the usual nitriding. This fact can be clarified that on the surface formed at dissociation the phase - γ 'yes not the phase - ε was formed [17 - 18].

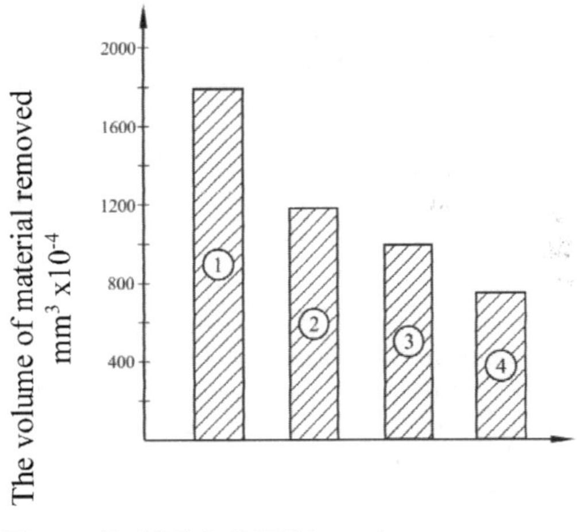

Figure 7. 105MnCrW11 steel wear nomograms STAS 3611-88 after:
1. Ordinary nitration; 2. Gas - cyclic nitriding;
3. Thermo-gas-cyclic nitriding (enrichment);
4. Thermo-gas-cyclic nitriding (dissociation)

5. Conclusions and results

A new nitriding method has been developed, which consists in periodic alternation of saturation cycles during flow nitriding and resorption of the nitride layer with the maximum possible decrease in the saturating capacity of the atmosphere. The method of thermo-gas-cyclic nitriding is an effective new method of strengthening parts that allows to reduce the consumption of ammonia from 2 to 10 times, the nitriding time by 4-6, 5 times with an increase in the thickness of the diffusion layer by 2-6 times without reducing the physical and mechanical characteristics, and also improve the environmental aspects of the nitriding process in flowing ammonia. A new technological parameter, the duration of half-cycles, has been investigated, which allows simply from the point of view of implementation and effectively regulates the phase composition and structure of the layer in order to obtain the required physical and mechanical characteristics.

References

[1] Demin D, Kostik V and Kostik V 2015 Modelirovaniye glubiny diffuzionnogo sloya poverkhnostnoy tverdosti stali pri azotirovanii *Vostochno-Yevropeyskiy zhurnal peredovykh tekhnologiy* **2** (5) pp 45-49

[2] Gherasimov S, Krukovich M, Baderko E and Klyuchkov N 2013 Modelirovaniye protsessa azotirovaniya *Nauka i obrozovanie* pp 313-331

[3] Belous V A and Nosov G I 2004 Ionnoye azotirovaniye staley v dugovom razryade nizkogo davleniya *Tekhnologiya mashinostroyeniya* **3** 35-39.

[4] Anishchenko O, Kukhar V, Grushko A, Vishtak I, Prysiazhnyi A and Balalayeva E Yu 2019 *Mater. Sci. Forum* **945** 531-537

[5] Kukhar V V, Grushko A V and Vishtak I V 2018 *Solid State Phenom.* **284** pp 408–415

[6] Kukhar V, Balalayeva E, Prysiazhnyi A, Vasylevskyi O and Marchenko I 2018 *MATEC Web of Conferences* **178** 1-6

[7] Fossati A 2006 Glow-discharge nitriding of AISI 316L austenitic stainless steel: influence of treatment time *Surf. and Coat. Technol.* **200**(11) 3511-3517

[8] Tsujikawa M 2005 Behavior of carbon in low temperature plasma nitriding case of austenitic stainless-steel *Surf. and Coat. Technol.* **193**(1-3) 309-313

[9] Belousov V A 2009 Nanoindentirovanie poverkhnostnykh sloyev materialov *Voprosy atomnoi nauki i energetiki* 146-148

[10] Ponomarenko I V 2010 Rol' izmeneniya tverdosti v povyshenii konstruktsionnoy prochnosti taley posle ionno-plazmennoy obrabotki *Vestnik KHNADU* **51** 135-138

[11] Belashova I S and Shashkov D P 2003 Vliyaniye termogazotsiklicheskikh vozdeystviy na protsess azotirovaniya *Tekhnologiya metallov* **11** 8-56

[12] Sun Y. 1999 The response of austenitic stainless Steels to low temperature plasma nitriding *Heat Treat. Met.* **26** 9-16

[13] Lyakishev N P 2007 *Polucheniye i fiziko-mekhanicheskiye svoystva ob"yemnykh nanokristallicheskikh materialov* Moskow ELIZ 148p

[14] Aleksandrov V A 2005 Azotirovaniye instrumenta iz vysokokhromistykh i bystrorezhushchikh stalei *Uprochnyayushchiye tekhnologii i pokrytiya* 14-20

[15] Zinchenko V M 2001 *Upravleniye protsessom gazovogo azotirovaniya spomoshch'yu kislorodnogo zonda* MiTOM **8** M Mashinostroyeniye 28 – 31

[16] Shcherbedinskiy G V 2001 Fizicheskiye aspekty formirovaniya mnogofaznykhpokrytiy na metallicheskikh materialakh *Temperaturo ustoychivye funktsional'nyye pokrytiya* Tula: Izd. TGPU 29-33

[17] Topala P, Besliu V and Cosovschi P 2012 *The process for hardening of metal surfaces* Patent: **4184** MD

[18] Trifan N, Mazuru A and Mazuru S 2020 Termogazocyclic nitruration Process *Conference European Exibition of Creativity and Innovation* "Euroinvent" Iasi p 106

IOP Conf. Series: Materials Science and Engineering **1018** (2021) 012012 doi:10.1088/1757-899X/1018/1/012012

FEM simulation of C45 steel and Cu-ETP billet shaping at hot upsetting between convex conical dies

V V Kukhar[1,*], E Yu Balalayeva[1], M G Korenko[2], A H Prysiazhnyi[1] and O S Anishchenko[1]

[1] Pryazovskyi State Technical University, Universytetska str., 7, Mariupol, 87555, Ukraine
[2] Kryvyi Rih State University of Economics and Technology, Technological Institute, Stephana Tilhy str., 5, Kryvyi Rih, 50006, Ukraine

Email: kvv.mariupol@gmail.com

Abstract. On the basis of finite element analysis (FEA), geometric differences in the shaping during hot upsetting of cylindrical billets with a height-to-diameter ratio of 1.0 and 2.0 made of C45 (1.0503) steel and copper (Cu-ETP) between pointed convex conical dies with a cone slant angle of 12.5°, 15.0° and 17.5° are considered. The stroke velocity of the upper die is 0.5 m/s; process temperature (t) and accepted surface friction coefficient (μ): for steel – t = 1100 °C, μ = 0.32, for copper – t = 850 °C, μ = 0.34. An equation is obtained for the relation between values of reduction in height ratio and engineering strain on height of workpiece during upsetting between conical dies, taking into account the initial billet dimensions and unevenness of forming due to the geometry of tool. The presence of qualitative and quantitative differences in shaping along the height of upset steel and copper billets was revealed, i.e. influence on unevenness of forming the rheological properties of materials under process conditions. It is advisable to use the results obtained to design and improve of technological processes of open die and closed die forging with preforming of billets.

1. Introduction

The application in metal-forming the upsetting of billets or ingots between convex dies usually pursues several goals: reducing of shrinkage and porosity defects, void closing in the axial zone of the ingot [1–3], providing a technologically advanced form of a semi-finished product (preform) for optimization of forming at subsequent stages of open die or closed die forging [4–6]. The pressure on the tool at finishing forming is reduced, and increasing the tool life of which is always relevant and requires the use of additional hardening processes [7]. Herewith, descaling after heating of the workpiece occurs, which excludes its pressing into the forging body, reduces the likelihood of defect formation and improves the acceptance indices of shop products [1, 3, 8–10]. At the same time, among the processes of upsetting with convex working tool, the preforming of workpieces by pointed convex conical dies is the least studied [8].

Upsetting between convex conical dies under certain their shape characteristics and conditions of surface friction leads to a decrease in the uneven deformation of the workpiece, which is used in methods for determining the coefficient of friction when choosing an effective lubricant [11, 12]. It is known that surface friction coefficients for cases of plastic upsetting of heated workpieces of steel and copper between dies made of tool steel are $\mu = 0.32$ and $\mu = 0.34$ respectively [11, 12]. Usually, at the open die forging and pre-stages of closed die forging technological lubricants are not used, and the effect of friction conditions on the contact of dies and workpieces with a height H_0 to diameter D_0 ratio more than 1.0 is insignificant [1, 8, 12]. Therefore, a change of the upper die inclination affects the workpiece shaping not only at the face zones, but also on the forming of the lateral surface of the upsetted preform. The ongoing processes of material volume displacement when the upper conical die is wedge in to the billet face are accompanied by severe plastic deformation of the surface layer, which is the object of separate studies [13, 14].

To identify the conditions for obtaining the most technologically advanced form for preformed workpiece and to ensure the optimal shaping at further stages, it is necessary to establish the patterns of workpiece forming at the upsetting under study. Information in this area is fragmentary (for example, there is limited information about the shaping at the initial stage of the penetration of conical rotating die into the end face of the workpiece during orbital forging [15–17]), which unreasonably narrows the application of workpieces upsetting by conical dies as a preforming. The use of a conical tool reduces the workpiece unevenness of deformation in rotary forming methods [15–17], requiring the operation of complex kinematic units of equipment with several degrees of freedom and complex gears of special profile [18, 19], which complicates the technology.

Also, the effect on the shaping of the workpieces material (steel and copper) properties under certain thermo-mechanical and physical-geometric conditions of forming is not reliably known, which is objectively due to the difference in patterns of materials strain hardening [8, 10, 20]. Analytical methods are used to research of workpiece forming allow to take into account the properties of the material and moving from shape indices to the analysis of the stress-strain state in different zones [21, 22]. The most common numerical method is finite element analysis (FEA), which allow to research the shaping, stress-strain state, force modes, accuracy of equipment and assembles [23, 24], and the rheological model of the material can be included in the calculations [25]. It should be noted that the hot press forming process is applicable for low and medium carbon steels and copper [4, 8, 9, 27–30]. The influence on the preform shape of the differences between the strain hardening models of hot steel C45 and copper Cu-ETP under certain thermo-mechanical conditions is obvious. Comparison of the corresponding work hardening curves for the flow stress σ_s given in the sources [8, 26–30] confirms this; however, this effect is practically not taken into account when studying the upsetting processes.

The aim of the work is to reveal the differences in the shaping of solid cross-section cylindrical billets made of C45 steel (1.0503 EN 10083-2) and Cu-ETP copper (EN 1652) during hot upsetting between conical convex pointed dies.

2. Materials, method and assumptions

The most powerful means of researching metalforming processes are computer-aided FEA tools, one of which (QFORM) was used to solve the set goal. The visualization of the forming in QForm occurs simultaneously with the calculation, which allows to immediately interpret the obtained results and make the necessary changes to the research process. QForm provides automatic generation of computational models (splitting into finite elements) and, accordingly, high calculation accuracy. When implementing finite element method (FEM), the forming during upsetting is divided into a number of successive and time-identical stages, depending on the movement of the upper die. The final criterion is the smallest change in the strain rate field within the stage. The values of stresses, strains and displacements are calculated step by step, taking into account the totality of kinematic, dynamic and temperature boundary conditions passing from stage to stage. The final values of the accumulated shear strain, obtained at each stage, are used as initial values when calculating at the next stage. Accordingly, the method of FEM for upsetting included: creationb of a conformal mapping of the flow area on the canonical region; finding the fields of strain and stress rates; calculation of the accumulated strains and assessment of residual material plasticity.

The upsetting of billets by convex conical dies with a slant angle at the base of the cone (Fig. 1) $\alpha = 12.5°; 15°; 17.5°$ was simulated. The initial diameter of the billet $D_0 = 25$ mm, and the height –

$H_0 = 25$ mm and $H_0 = 50$ mm, i.e. ratio $H_0/D_0 = 1.0$ and $H_0/D_0 = 2.0$ respectively. Splitting the solid-state computer model of the billet into a mesh of finite elements with the determination of their types was provided by means of FAM tools, while fulfilling the condition of impermeability of the finite elements of the working dies model to the finite elements of the billet model. Billet materials: (i) C45 steel (forming temperature $t = 1100\,°C$) and (ii) Cu-ETP copper (forming temperature $t = 850\,°C$). The taken temperature environment was to isothermal. In accordance with the review, the surface friction coefficients during billet upsetting are taken for steel $\mu = 0.32$ and copper $\mu = 0.34$. As a technological process parameter characterizing the reduction in height ratio of billets during upsetting, $\varepsilon_h = \Delta h / H_0$ was taken, where $\Delta h = (H_0 - h_1)$ is the upsetting stroke, mm, and h_1 is the distance between opposite points of the upper and lower conical dies after deformation, mm, see Fig. 1(a). Forming at steady state of upsetting only is for practical importance, i.e. for $\varepsilon_h \geq 0.25 \ldots 0.3$. The maximum reducing ratio was limited to $\varepsilon_h = 0.5$.

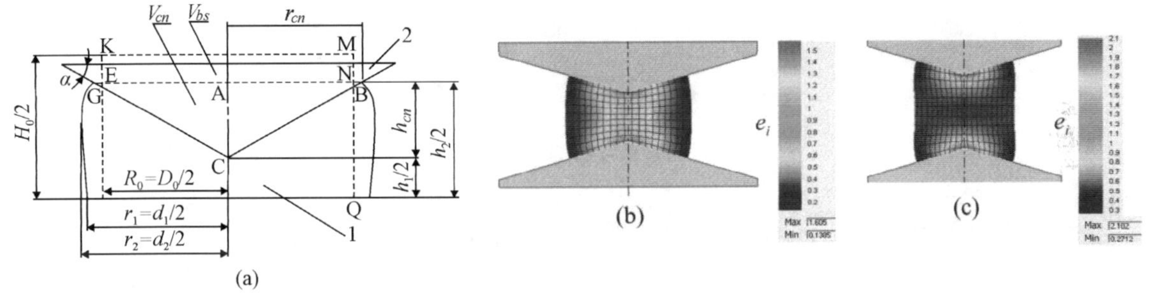

Figure 1. Upsetting steady state of the billets by conical dies: upper half of the workpiece (symmetry along the horizontal axis) (a); models of the billets upset up to $\varepsilon_h = 0.5$ (Cu-ETP, $t = 850\,°C$) with $H_0/D_0 = 1.0$ (b) and $H_0/D_0 = 2.0$ (c) respectively at $\alpha = 17.5°$; 1 – workpiece with sizes h_1, h_2, r_1, r_2 after upsetting; 2 – upper cone upsetting dies; V_{cn} – metal volume GBC, displaced by the cone (increases along the upsetting due to an increase in the values r_{cn} and h_{cn}); V_{bs} – metal volume KMNE, displaced with respect to the vertical axis; e_i – equivalent strain (scale).

The stroke velocity of the upper die $\upsilon = 0.5$ m/s was considered, which corresponds to the displacement velocity of the sliders of mechanical presses and provides a strain rate of the workpiece of about $\xi = 5 \ldots 10$ s^{-1}. The changes in the properties of the workpiece materials during forming under the given conditions were taken into account by choosing models of their strain hardening from the data list of the used package for finite element modeling (FEM) as functions $\sigma_s = f(e_i, \xi, t)$.

In the process of simulation, the values of displaced volumes (displaced mass) were determined as $V_{dsv} = V_{cn2} + V_{bs2}$, where $V_{cn2} = 2 \cdot V_{cn}$ and $V_{bs2} = 2 \cdot V_{bs}$ are doubled volumes according to Fig. 1 referring to the entire workpiece. Accordingly, the true engineering strain e_h in the vertical direction was found as $e_h = V_{dsv} / V_{wp}$, where $V_{wp} = \pi D_0^2 H_0 / 4$ is the volume of the billet. The maximum difference in computer calculation of displaced volumes for steel and copper billets did not exceed 0.8%. This error is negligible and is associated with differences in finite element mesh generation for materials with different properties. Thus, the graphs of the relation between the technological parameter ε_h and the values e_h (Fig. 2) for steel and copper billets will not differ.

3. Results and discussion

Considering that the nature of the dependences shown in Fig. 2, is close to linear, then based on the simulation results, a planning matrix for a full factorial experiment design (2^3) was compiled taking into account the interaction of influencing factors (H_0 / D_0, ε_h and α). The variable factors and their levels, as well as the simulation plan (2^3), are given in table 1.

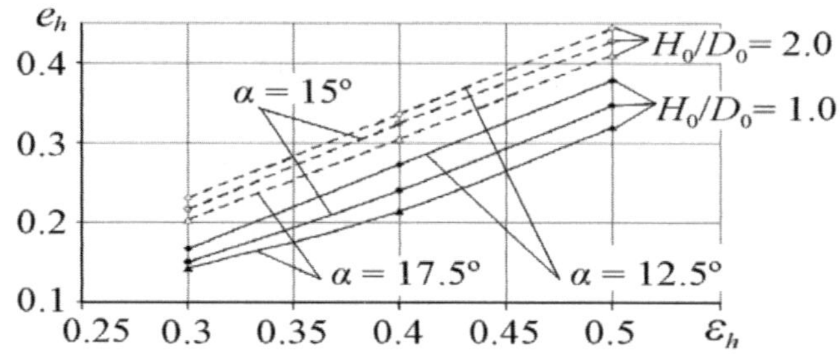

Figure 2. The relation between the reduction in height ratio of billets ε_h and the true engineering strain e_h, calculated through displaced volumes for upsetting of billets ($H_0/D_0 = 1.0$ and $H_0/D_0 = 2.0$) by convex conical dies with $\alpha = 12.5°, 15°, 17.5°$.

Table 1. Variable factors, their levels and simulation plan.

	H_0 / D_0 (x1)	ε_h (x2)	α ° (x3)	No.	H_0 / D_0 (x1)	ε_h (x2)	α ° (x3)	e_h (y)
Basic level	1.5	0.4	15	1	1.0	0.3	12.5	0.164
				2	2.0	0.3	12.5	0.233
Variation interval	0.5	0.1	2.5	3	1.0	0.3	17.5	0.142
				4	2.0	0.3	17.5	0.202
Top level	2.0	0.5	17.5	5	1.0	0.5	12.5	0.377
				6	2.0	0.5	12.5	0.444
Lower level	1.0	0.3	12.5	7	1.0	0.5	17.5	0.319
				8	2.0	0.5	17.5	0.408

The processing of the plan made possible to obtain the regression equation:

$$e_h = 0.287 + 0.035X_1 + 0.098X_2 - 0.017X_3 + 0.004X_1X_2 - 0.006X_2X_3 + 0.005X_1X_2X_3 \quad (1)$$

where $X_1 = (H_0 / D_0 - 1.5)/0.5$, $X_2 = (\varepsilon_h - 0.4)/0.1$ and $X_3 = (\alpha - 15°)/2.5°$ are coded values of specified influencing factors.

As the main indices of the billet shaping during hot upsetting between conical dies, the following ratios were chosen h_2 / H_0; d_1 / D_0 and d_2 / d_1, Fig. 1(a), and graphs of their dependences on influencing factors for steel and copper billets are plotted (Fig. 3–5). Obviously, $d_2 / d_1 = r_2 / r_1$. At the same time, the material flow of the workpiece to lateral surface is forming the double barrel profile with the presence of r_1 and r_2 values occurs when upsetting relatively high billets with $H_0/D_0 = 2.0$ ratio. When forming a lateral profile in the shape of a single barrel, which is typical for billets with $H_0/D_0 = 1.0$ ratio, it was assumed that $r_1 = r_{cn}$, see Fig. 1(a), where r_{cn} is the radius of the contact circle of the workpiece with the convex conical dies.

Graph $h_2 / H_0 = f(\varepsilon_h; \alpha; H_0 / D_0)$, drawn in Fig. 3, shows the change in the vertical dimension of the workpiece depending on the reduction in height ratio ε_h, the slant angle at the base of the cone α of the upsetting dies and H_0 / D_0 ratio of the initial billet. The obtained dependences of the change in the height dimension on the conditions of upsetting by conical dies of steel and copper billets are qualitatively identical, and the quantitative differences are insignificant (no more than 2%). This is due to the insignificant difference in the surface friction coefficients, the difference in the generated finite element mesh, and allows using the same graph for different materials. Consequently, the influence of the rheological properties of materials on the shaping on vertical of billets during upsetting by convex conical dies can be neglected.

Graph $d_1 / D_0 = f(\varepsilon_h;\alpha;H_0 / D_0)$, drawn in Fig. 4, shows that an increase in H_0 / D_0 ratio increases the relative diameter after upsetting in the middle of the height of the workpiece at the same reductions, which is relate with a larger metal volume involved in plastic forming. The obtained dependences of the change in the relative median diameter are characterized by its rising with an increase in the reduction in height ratio and a decrease with an increase in the slant angle at the base of the cone. These dependences for steel and copper billets are qualitatively identical, and the quantitative differences are very insignificant (up to 1%). This is explained in the same way as in the previous case and allows using the same graph for different materials also. Consequently, the influence of the rheological properties of materials on the change in the relative median diameter of the workpiece during upsetting by conical dies is insignificant.

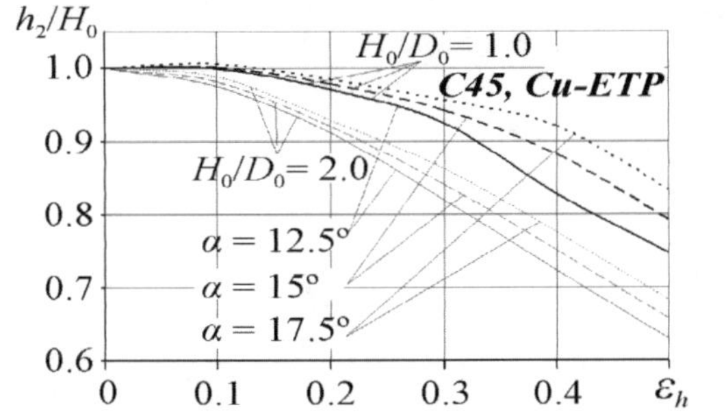

Figure 3. Dependency graph $h_2 / H_0 = f(\varepsilon_h;\alpha;H_0 / D_0)$ for steel and copper billets while upsetting by convex conical dies.

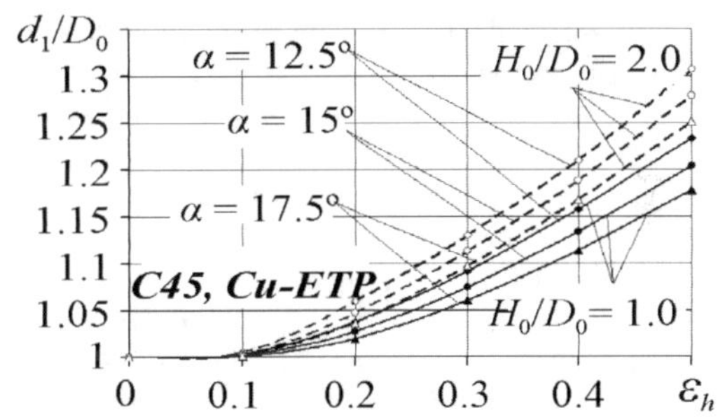

Figure 4. Dependency graph $d_1 / D_0 = f(\varepsilon_h;\alpha;H_0 / D_0)$ for steel and copper billets while upsetting by convex conical dies.

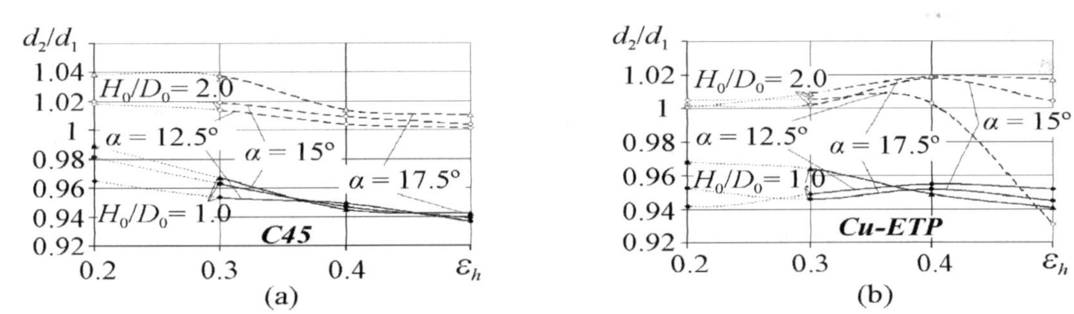

Figure 5. Dependency graph $d_2 / d_1 = f(\varepsilon_h;\alpha;H_0 / D_0)$ for steel (a) and copper (b) billets while upsetting by convex conical dies.

Graph $d_2 / d_1 = f(\varepsilon_h;\alpha;H_0 / D_0)$, drawn in Fig. 5, describes the change in the unevenness of workpiece deformation along the height. In general, for billets with $H_0/D_0 = 1.0$ ratio, values d_2 / d_1 are less than 1.0, and for billets with $H_0/D_0 = 2.0$, values d_2 / d_1 are greater than 1.0. At the same time, for copper workpiece, the presence of maxima is observed on the graphs related to the upsetting of higher billets. The obtained dependences of the change in the value of the ratio of the front end (d_2) diameter to the median diameter (d_1) for steel and copper billets have qualitative and quantitative differences. This testifies to the significant influence of the rheological properties of materials on the change in the unevenness of the workpiece deformation along the height during upsetting by convex conical dies, which cannot be neglected in the design of forging processes.

4. Conclusions

An equation is obtained for the relation between the upseting reduction ratio (adjustment index) and engineering strain on height of a cylindrical workpiece during upsettng between conical dies, taking into account initial billet dimensions and upsetting dies geometry (cone slant angle to the base). The patterns of the workpiece shaping during upsetting with conical dies have been established, which make it possible to determine the dimensions of upsetted preform, depending on its initial height-to-diameter ratio, reduction in height ratio and taper of convex cone dies. It was revealed that the unevenness of deformation along the height of the upsetted preforms is significantly influenced by the rheological properties of the billet material. It is advisable to use the results obtained to improve the technological processes of open die and closed die forging on the basis of preforming of workpiece by upsetting with profiled dies.

References

[1] Tyurin V A and Sapunov A L 2014 *Steel in Translation* **44(3)** 190-195
[2] Markov O E, Aliiev I S, Aliieva L I and Hrudkina N S 2020 *J. Chem. Technol. Metall.* **55(3)** 640-648
[3] Aliev I, Zhbankov Y and Martynov S 2015 *J. Chem. Technol. Metall.* **51(4)** 393-400
[4] Kukhar V, Balalayeva E, Hurkovska S, Sahirov Y, Markov O, Prysiazhnyi A and Anishchenko O 2020 *Advances in Intelligent Systems and Computing* **989** 325-331
[5] Artiukh V, Kukhar V and Balalayeva E 2018 *MATEC Web of Conferences* **224** 01036
[6] Vlasenko K, Hrudkina N, Reutova I and Chumak O 2018 *Eastern-European Journal of Enterprise Technologies* **3(1)(93)** 51-59
[7] Zurnadzhy V I, Efremenko V G, Wu K M, Azarkhov A Yu, Chabak Yu G, Greshta V L, Isayev O B, Pomazkov M V 2019 *Materials Science and Engineering: A* **745** 307-318
[8] Bhaduri A 2018 *Mechanical Working of Metals and Alloys* (Singapore: Springer, SSMATERIALS, 264)
[9] Kukhar V, Balalayeva E, Prysiazhnyi A, Vasylevskyi O and Marchenko I 2018 *MATEC Web of Conferences* **178** 02003
[10] Kukhar V, Kurpe O, Klimov E, Balalayeva E and Dragobetskii V 2018 *International Journal of Engineering and Technology(UAE)* **7(4)** 35-39
[11] Kramer P and Groche P 2019 Lubricants **7(2)** 12
[12] Dalmaz G, Lubrecht A A, Dowson D and Priest M 2004 *Transient Processes in Tribology* (Boston: MA: Elsevier, 43)
[13] Alexandrov S, Šidjanin L, Vilotic D, Movrin D and Lang L 2018 *Metals* **8** 71
[14] Botnari V and Mazuru S 2014 *Applied Mechanics and Materials* **657** 147-152
[15] Aksenov L B and Kunkin S N 2018 *Lecture Notes in Mechanical Engineering* **F5** 1-8
[16] Baglyuk G A, Kurikhin V G and Gozhii S P 2015 *Powder Metall. Met. Ceram.* **53(9-10)** 522-528
[17] Nowak J, Madej L, Ziolkiewicz S, Plewinski A, Grosman F and Pietrzyk M 2008 *International Journal of Material Forming* **1** 387-390
[18] Mazuru S, Casian M and Scaticailov S 2017 *MATEC Web of Conferences* **112** 01026
[19] Bostan I, Mazuru S and Casian M 2018 *MATEC Web of Conferences* **178** 06024
[20] Razali M K, Irani M and Joun M 2019 *Journal of Materials Research and Technology* **8(3)** 2710-2720
[21] Anishchenko O S, Kukhar V V, Grushko A V, Vishtak I V, Prysiazhnyi A H and Balalayeva E Yu 2019 *Materials Science Forum* 945 531-537
[22] Kukhar V V, Grushko A V and Vishtak I V 2018 *Solid State Phenomena* **284** 408-415
[23] Li Y H and Zhang J 2012 *Advanced Materials Research* **472-475** 1870-1873
[24] Markov O E, Gerasimenko O V, Shapoval A A, Abdulov O R and Zhytnikov R U 2019 *International Journal of Advanced Manufacturing Technology* **103(5-8)** 3057-3065
[25] Markov O, Gerasimenko O, Aliieva L and Shapoval A 2019 *EUREKA: Physics and Engineering* **2** 52-60
[26] Spittel M and Spittel T 2009 *Metal forming data of ferrous alloys – deformation behaviour* (Berlin, Heidelberg: Springer-Verlag)
[27] Hojny M and Głowacki M 2011 *Procedia Engineering* **10** 2353-2362
[28] Vorel I, Vančura F, Pileček V, Jirková H and Mašek B 2015 *Procedia Engineering* **100** 714-721
[29] Zasadzińska M, Knych T, Strzępek P, Jurkiewicz B and Franczak K 2019 *Archives of Civil and Mechanical Engineering* **19(1)** 186-193
[30] Ssemakula H, Ståhlberg U and Öberg K 2006 *J. Mater. Process. Technol* **178(1-3)** 119–127

Numerical modeling of pipe parts of agricultural machinery expansion by stepped punches

R H Puzyr[1], V T Shchetynin[2], R H Arhat[2], Yu B Sira[1], V V Muravlov[3] and S I Kravchenko[3]

[1] Kremenchuk Mykhailo Ostrohradskyi National University College, Chumatskyi Shliakh str. 7, Kremenchuk 39621, Ukraine
[2] Kremenchuk Mykhailo Ostrohradskyi National University, Pershotravneva str. 20, Kremenchuk 39600, Ukraine
[3]Poltava State Agrarian Academy, Skovorody str. 1/3, Poltava, 36003, Ukraine

E-mail: puzyruslan@gmail.com

Abstract. The paper presents the research of plastic expansion of a pipe workpiece by punches of various configurations. It is shown that interest in this process is caused by the development of oil production, where there is a need to increase the diameter of well pipes, the automobile industry, where the protection of the driver and passengers in case of accidents comes first. Here, stable plastic deformation plays an important role as a good absorber of kinetic energy, etc. Data on the problems of expanding pipe ends are given, where special attention is paid to the destruction of the end in the form of localization of deformations and appearance of a longitudinal crack. To eliminate this phenomen onvarious technological methods are used. Here the influence of the design and shape of the punch on the localization of deformations along the thickness of the workpiece is researched. Several spatial outlines of the tool are proposed to create an effective reloading of the pipe workpiece in the deformation zone. Based on numerical modeling, the distribution of stresses according to Mises, the displacement of the end of the semi-finished product and logarithmic deformations in the radial direction is shown for each standard size of the punch. Comparative results reveal an increase in the thickness of the workpiece in the dangerous section for all presented punches. The comparison was carried out with a reference tool that has a straight generatrix of a tapered surface. The best results are shown by a punch with an enlarged generatrix at the end of the working stroke. Attention is focused on conducting further research in order to study the possibility of overcritical deformation by the proposed tool.

1. Introduction

Tubular metal parts are widely used in the structure so mechanism so ships, automobiles, airplanes, spacecraft, electrical appliances, household and agricultural devices. Normal functioning of the oil and gas production and oil refining, chemical, food, engineering and agricultural industries of any country is unthinkable without pipelines. Pipe parts are also used in the construction industry as various types of connecting elements and power welded structures.

To obtain a strong pipe connection from pipes of different diameters, adapters or expansion of the ends of pipes of a smaller diameter are used. This operation in cold forming is called expansion. And although the first scientific substantiation of this process was given more than 70 years ago by R. Hill

[1], where the influence of the main parameters of the expansion on the distribution of stresses and deformations in the selected section of the workpiece was considered, the task of further studying and improving the expansion remains relevant at the moment. This is facilitated by the steady development of oil and gas production, where expansion is used to increase the passage of well pipes using the Solid Expandable Tubular (SET) technology [2, 3] or for the conservation of oil wells [4, 5]. An intensive subject of research is the improvement of the passive protection of the automobile in a collision with an obstacle, where plastic deformation of the tubular body elements by expanding and bending acts as an effective absorber of kinetic energy [6, 7]. Similar tubular stabilizers are used for soft landing of spacecraft and aircraft [8, 9]. Heat exchangers are widely used in the world, and their designs are constantly being improved. They are used into hold and industrial ventilation and air conditioning systems [10], in the food and oil refining industries as heaters and evaporators [13, 14], in refrigeration systems for ships, railway transport and agricultural machinery [11,12]. The expansion operation is used here when connecting the tube soft the tube bundle with the tube plate to seal the intertubular space [15].

In the above examples, researchers solve various problems that are inherent in the chosen research subject. Reducing the expansion force when pushing the ferrule through the well pipe. The solution is achieved by reducing the coefficient friction between the tool and the pipe by using modern lubricants, as well as by varying the geometry of the tool. The influence of the geometry of the tool and the coefficient of friction on the expansion force was considered in papers [16-18]. In research on the protection of automobile occupants in a collision, expansion is analyzed in the context of the stability of the plastic deformation process. A number of papers are devoted to this problem [19-21]. When coupling the tubes with the tube plate of the heat exchanger, it is important to take into account the joint deformation of the tube and the plate in order to obtain a tight connection. This is achieved by tool geometry and deformation rate control. [22, 23].

It should be especially noted that with an increase in the degree of deformation of a pipe workpiece, in some cases, processes of localization of deformations at the end of a semi-finished product can occur with the subsequent appearance of longitudinal cracks and destruction, which is not permissible from the point of view of quality.[24]. Therefore, the efforts of researchers are aimed at studying various factors of expansion that affect the increase in the degree of deformation without loss of stability and destruction of the workpiece. Analytical dependences and numerical results of research are proposed, which make it possible to assess the influence of the initial and acquired anisotropy on the stability of the plastic deformation process [25, 26], the geometric dimensions of the workpiece and the taper angle of the punch [27-29] on crack formation and the formation of a longitudinal fold on the non-deformable part of the workpiece. Numerous methods of supporting the end of the workpiece during expansion and changing the rectilinear shape of the generatrix of the conical surface of the tool to increase the degree of deformation are also presented.[30-32]. Therefore, the purpose of this paper is to numerically model the process of expansion by conical punches of various spatial configurations, compare the modeling results, develop recommendations and substantiate further areas of research.

2. Research Methodology

The desire of researchers from various scientific schools to increase the degree of deformation of a cylindrical workpiece in one pass of the punch results in a complication of the design of the latter, excessive consumption of additional materials or an increase in the duration of the part manufacturing cycle. The creation of triangular or toroidal surfaces on the straight generatrix of the punch, which reduce the contact area of the workpiece and the tool and facilitate the lubrication conditions, leads to a rather laborious process of manufacturing such a tool [31, 32]. The additional support of the end face in the form of retaining rings complicates the design, and the support in the form of rings made of easily deformable alloys leads to the consumption of this additional material. This increases the technological cost of the operation [30, 33].

To study the process, three types of punches were proposed. They differed in the location of the transition step along the generatrix of the tapered part. The results of the numerical experiment were compared with each other, as well as with the reference version of expansion, where a punch with a straight generatrix of a cone acted as a tool. For all variants, the material and dimensions of the work piece were identical. A material such as aluminum 5005 with the following mechanical characteristics was modeled: yield point $\sigma_{0.2}$=50 MPa; ultimate strength σ_v=110 MPa; Young's modulus E=70 GPa; Poisson's coefficient ν=0.33, density ρ=2700 kg/m^3. An elastoplastic isotropic model of the material with isotropic hardening was taken, the diagram of the true hardening curve was approximate in the form of $\sigma_{0.2}=50+0.64\varepsilon^{0.62}$ [34, 35]. That is, the following assumptions were made: the initial and acquired anisotropy in different planes was absent; during deformation the material was strengthened equally in all coordinate directions. The workpiece external diameter D_0= 25 mm, wall thickness s_0= 2 mm, height h_0= 50 mm. The geometrical dimensions of the reference punch: larger taper diameter D_p= 29 mm, the height fo the taper part H_p= 15 mm, the angle of the generatrix of the cone to its axis of symmetry $\alpha\approx15°$.The expansion ratio was k = 1.38. This parameter was chosen in accordance with the recommendations [36, 37] and ensured stable expansion without loss of stability and destruction of the workpiece end. The geometry of the stepped punches was varied so that the broken generatrix coincided with the points of the beginning and end of the generator of the reference punch and α_1=45° for all stepped punches (figure 1).

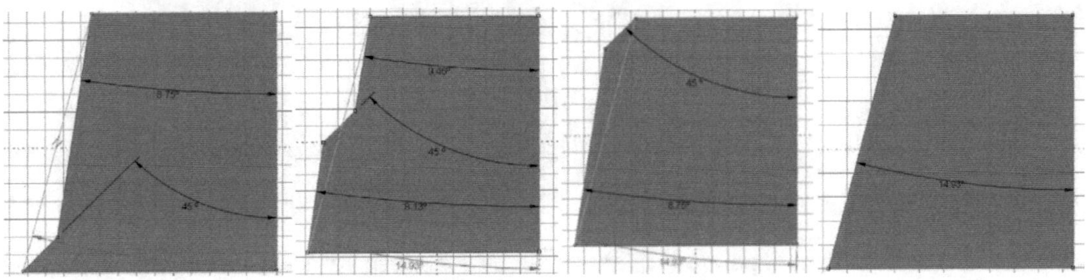

a) b) c) d)

Figure 1. Sections of 1 \ 2 part of punches for expansion: a) a punch with a protrusion at the end of the tool stroke α_1=45°, $\alpha_2\approx$ 9°; b) a punch with a protrusion in the middle part of the generatrix of the toolα_1=45°, $\alpha_2\approx$ 9.5°, $\alpha_3\approx$ 8°; c) a punch with a protrusion at the beginning of the tool working strokeα_1=45°, $\alpha_2\approx$ 9°; d) a reference punch with an inclined taper generatrix α_1=15°.

Finite element modeling was carried out in the applied complex Simullia Abaqus - student edition, which is supplied by "TESIS" in the free access and only limits the number of finite elements in the assembly of the model [38].

The tool and workpiece were modeled in a 3D setting by the rotation method. Moreover, since the problem is axisymmetric, ¾ share of the part and the punch participated in the assembly. To imitate the deformation process in each problem, a velocity equal to v=2.5 mm/sec directed along the axis of the cylinder was applied to the punch, which was represented as an undeformable solid [39, 40]. Also, the loading conditions were modeled by setting boundary conditions in displacements, while the lower end of the workpiece was fixed during the entire loading period, which modeled a rest against a rigid base, the ends of the workpiece were limited in displacements along two axes and rotations also along two axes, and on the side surfaces and the upper end were restricted in rotation around the axis, which coincides with the axis of symmetry of the workpiece..One of the main assumptions of the problem was the friction between the workpiece and the tool, which was taken in mechanical constants in a kinematic formulation with the final formulation of the contact and the coefficient of friction equal to k_f= 0.1. The coefficient of friction was chosen on the basis of the data of works [15, 18, 36], as well as production experience when distributing cylindrical billets made of aluminum. It takes this value when lubricating tools with industrial oil.

When generating a finite-element grid for the tool, the free method of its construction was determined and the element type R3D4 was selected, setting the bilinear order of the element that is contained in the standard library from the category of solid quadrangular ones. A hexahedral form was chosen for the workpiece, expansion was taken as the method of its creation, element C3D8R with the linear order from the standard library, category 3D Stress, i.e. operating in all three directions across the whole volume was selected.

3. Results

The results of modeling the expansion process are shown in figures 2, 3, 4, 5. The figures contain diagrams of stress fields according to Mises, displacements of the upper edge of the workpiece and logarithmic deformations of the end of the cylinder at the end of the working stroke of the tool for each standard size of the punch.

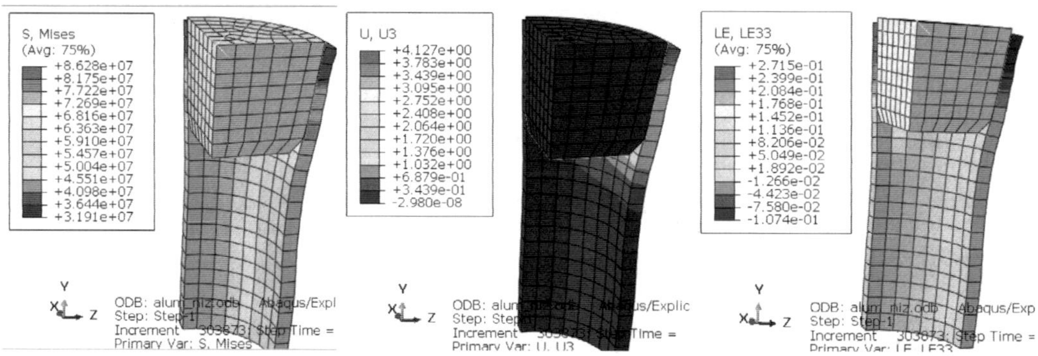

a) b) c)

Figure 2. Diagrams of deformation parameters at the end of the punch working stroke with alow tooth arrangement: a) stress distribution according to Mises; b) displacement distribution; c) distribution of logarithmic deformations.

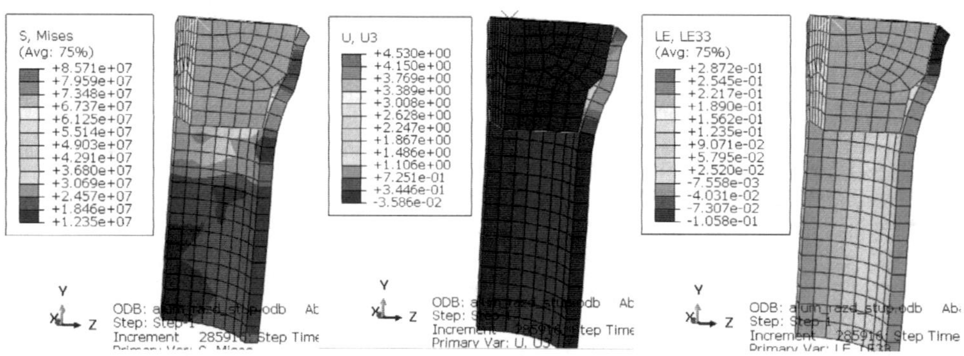

a) b) c)

Figure 3. Diagrams of deformation parameters at the end of the punch working stroke with a middle tooth arrangement: a) stress distribution according to Mises; b) displacement distribution; c) distribution of logarithmic deformations.

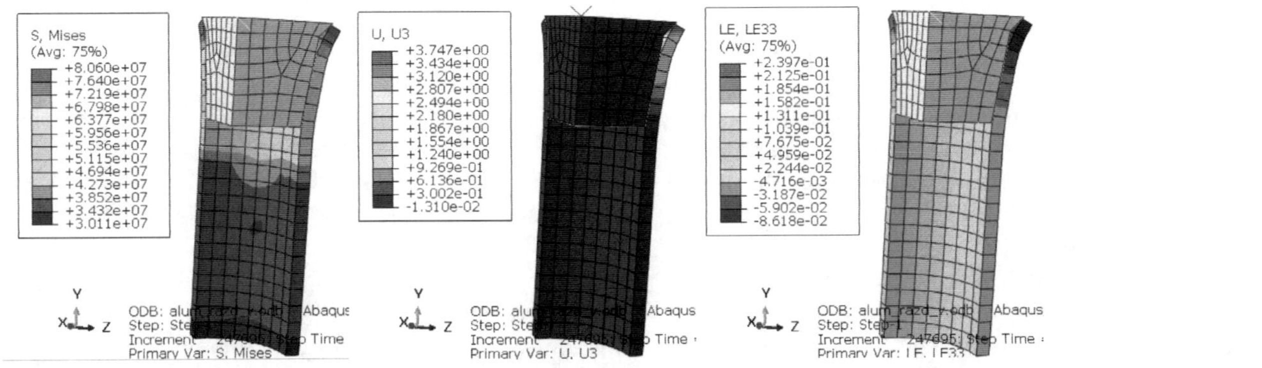

a) b) c)

Figure 4. Diagrams of deformation parameters at the end of the punch working stroke with a middle tooth arrangement: a) stress distribution according to Mises; b) displacement distribution; c) distribution of logarithmic deformations.

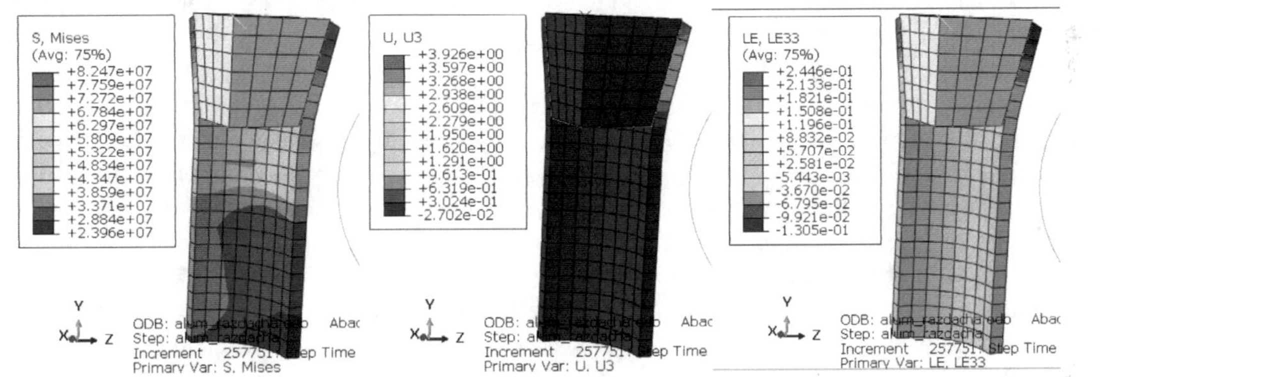

а) б) в)

Figure 5. Diagrams of deformation parameters at the end of the punch working stroke of the reference punch: a) stress distribution according to Mises; b) displacement distribution; c) distribution of logarithmic deformations.

Comparison of the given diagrams reveals that the greatest Mises stresses affect the edge part of the tubular workpiece in all variants, except for expansion by a punch with the tooth located in the center of the tool. Here the middle part of the semi-finished product is subjected to the greatest stresses. The largest value of equivalent stresses according to Mises σ_i=86.2 MPa belong to the workpiece, which is deformed by a punch with a low tooth arrangement, the workpiece that is deformed by a punch with an upper tooth arrangement has the smallest value – σ_i=80.6 MPa. The movement of the end of the workpiece in the direction perpendicular to the axis is the greatest for the workpiece, which is deformed by a punch with a middle tooth location U = 4.5 mm, the smallest - U = 3.92 mm is for the workpiece, which is expanded by the reference punch. The largest logarithmic deformations, which indicate a decrease in the thickness of the workpiece, occur, as expected, at the end. The workpiece end loaded by the reference punch is subject to the greatest deformations ε = 0.13, the end of the workpiece which is deformed by a punch with an upper tooth location is subject to the smallest deformations ε = 0.086.

Also, an interesting result of the research consists in the fact that during expansion by a punch with a step in the middle of the generatrix, the zone of the middle part of the workpiece is subject to the highest equivalent stresses according to Mises, where σ_i=85.7 MPa, while at the end, where the greatest stresses are to be expected, they are less than σ_i=67.4 MPa. This indicates that the semi-finished product receives the main deformation of the expansion in the middle zone, passing the step of a sharp increase in the diameter, and when approaching the upper end of the punch practically does not experience plastic deformations. As figure 3 demonstrates, the edge section of the semi-finished product does not come in to contact with the punch wall, having received a greater expansion at the punch step. This is evidenced by the largest displacement of the end face U = 4.5 mm.

4. Conclusions

The research results presented in the previous section lead to certain conclusions. The nature of the Mises stress distribution, where the highest stresses act on the end of the semi-finished product, is similar for all loading options, except for deformation by a punch with a step in the middle part of the conical surface generatrix. However, their value is not the same. For a punch with a step at the top, it is 2% less, and for a punch with a step at the bottom, it is 4% more in comparison with the results of deformation by the reference tool. The displacements of the end surfaces of the workpiece, which indicate its increase in diameter, are practically equal to U≈4.0 mm, except for deformation by a punch with a step in the middle part. This suggests that the workpieces receive the planned degree of expansion with the coefficient k = 1.38 and the semi-finished product end bears against the tool at the end of the working stroke. This result does not apply to the above case. However, the logarithmic deformations acting in the radial direction and indicating a decrease in the thickness of the workpiece differ significantly from each other. For a punch with a step at the top of the generatrix, they are 32% less, and for a punch with a step at the bottom, they are 18% less compared with identical deformations obtained with the reference punch. This fact indicates the positive effect of the step on the amount of thinning in the dangerous section of the workpiece. The step of a sharp change in the diameter of the tool creates a sharp jump in stresses during deformation, which imitates the support of the end of the workpiece, which leads to an increase in the thickness of the end part of the semi-finished product. This fact will make it possible to use the stepped punches for expansion within the established intervals of a steady expansion process. The next stage of research can include the modeling of the expansion with a stepped tool beyond the boundaries of the process stability, so the thickness margin of the workpiece metal makes it possible to predict such a possibility.

As for the results of the research on loading with a punch with a step in the middle part, a decrease in stresses at the end of the workpiece at the end of the working stroke is of interest. In this case, the logarithmic deformation in the radial direction at the end is 19% less than in the reference version. This indicates presence of a rest on the side of the step, but it is not as effective as a punch with a step at the upper part.

The following recommendations could be given for production: use punches with a step at the top for distributing the ends of pipe blanks. This tool design gives the best result in terms of increasing the thickness of the finished part at the end. This, in turn, can reduce the weight of the structure, since the initial blank will be thinner.

References

[1] Hill R 1950 *The Mathematical Theory of Plasticity. Oxford University Press* (New York)
[2] Al-Abri O S and Pervez T 2013 Structural behavior of solid expandable tubular undergoes radial expansion process – Analytical, numerical, and experimental approaches *International Journal of Solids and Structures* **50** 2980-94
[3] Pervez T 2010 Experimental and numerical investigations of expandable tubular structural integrity for well applications *J. Achiev. Mater. Manuf.* **41** 147-154
[4] Pervez T and Qamar S 2011 Finite element analysis of tubular ovality in oil well *Advanced Materials Research* **264-265** 1654-59

[5] Sanchez F and Al-Abri O 2013 Tube expansion under various down-hole end conditions *The Journal of Engineering Research* **10** 25-40

[6] Drusina N, Mahapatra R, Abdul-Latif A, Baleh R, Wilhelm C, Stoyanov P and Es-Said O 2008 Microstructure Analysis of Aluminum Alloy and Copper Alloy Circular Shells After Multiaxial Plastic Buckling *Mechanical Engineering Faculty Works* **17** 755-766

[7] Baleh R and Abdul-Latif A 2016 Quasi-Static Biaxial Plastic Buckling of Tubular Structures Used as an Energy Absorber *Journal of Applied Mechanics* **74** 628-635

[8] Jialing Yang, Min Luo, Yunlong Hua and Guoxing Lu Energy absorption of expansion tubes using a conical–cylindrical die: Experiments and numerical simulation *International Journal of Mechanical Sciences* **52** 716–725

[9] Alves M, Almeida B, Rosa P and Martin P 2006 End forming of thin-walled tubes *Journal of Materials Processing Technology* **1** 177–183

[10] Xiaowen M, Dian-ce G and Dong L 2017 Improved Control Strategy of Variable Speed Pumps in Complex Chilled Water Systems Involving Plate Heat Exchangers *Procedia Engineering* **205** 2800–06

[11] Palomba V, Dino G, Ghirlando R, Micallef C and Frazzica A 2019 Decarbonising the Shipping Sector: A Critical Analysis on the Application of Waste Heat for Refrigeration in Fishing Vessels Appl. Sci. **9** 5143

[12] Xu X, Li Y, Yang S and Chen G 2017 A review of fishing vessel refrigeration systems driven by exhaust heat from engines *Appl. Energy* **203** 657–676

[13] Haikova T, Puzyr R. and Levchenko R. 2020 Experimental Studies on the Stress-Strain State under Drawing Aluminum–Copper Bimetal Parts Rectangular in Plan. *Russian Journal of Non-Ferrous Metals*, **61** (4) 404–412

[14] Prafulla D, Kozminski M, Peterson J and Kumar S 2019 Fouling Diagnosis of Pennsylvania Grade Crude Blended with Opportunity Crude Oils in a Refinery Crude Unit's Hot Heat Exchanger Train *Industrial & Engineering Chemistry Research* 17918–927

[15] Tynchenko V S, Petrovsky E A, Bukhtoyarov V V, Bashmur K A, Tynchenko V V and Kukartsev V V 2019 *Similarity criteria for modeling the process of pipes pulse pressing into the tube sheet* (J. Phys.: Conf. Ser)

[16] Omar S AL -Abri and Pervez T 2013 Structural Behavior of Solid Expandable Tubular Undergoes Radial Expansion Process- Analytical, Numerical, And Experimental *Approaches International Journal Solids and Structures* **50** 2980- 94

[17] Puzyr R, Savelov D, Shchetynin V, Levchenko R, Haikova T, Kravchenko S, Yasko S, Argat R, Sira Y and Shchipkovakyi Y 2018 Development of a method to determine deformations in the manufacture of a vehicle wheel rim *Eastern-European Journal of Enterprise Technologies* **4** 55–60

[18] Sosenushkin E, Yanovskaya A, Sosenushkin A and Emel'yanov V 2015 Mechanics of nonmonotonic plastic deformation *Russian Engineering Research* **35** 902–06

[19] Yang C, Li P and Fan L 2014 Blank Shape Design for Sheet Metal Forming based on Geometrical Resemblance *Procedia Engineering* **81**1487–92

[20] Wang X, Ouyang K and Xia J 2009 FEM analysis of drawing-thickening technology in stamping-forging hybrid process *Forging & Stamping Technology* **34** 73–78

[21] Houliara S and Karamanos S 2010 Stability of long transversely-isotropic elastic cylindrical shells under bending *International Journal of Solids and Structures* **47** 10–24

[22] Puzyr R, Savelov D, Argat R and Chernish A 2015 Distribution analysis of stresses across the stretching edge of die body and bending radius of deforming roll during profiling and drawing of cylindrical workpiece *Metallurgical and Mining Industry* **1** 27–32

[23] Kukhar V, Kurpe O, Klimov E, Balalayeva E and Dragobetskii V 2018 Improvement of the method for calculating the metal temperature loss on a Coilbox unit at the rolling on hot strip mills *International Journal of Engineering and Technology* **7** 35–39

IOP Conf. Series: Materials Science and Engineering **1018**(2021) 012013 doi:10.1088/1757-899X/1018/1/012013

[24] Fung-Huei Yeh 2007 Study of tube flaring forming limit in the tube flaring process *The Journal of Strain Analysis for Engineering Design* **42** 315-324

[25] Puzyr R, Kukhar V, Maslov A and Shchipkovskyi Y 2018 The Development of the Method for the Calculation of the Shaping Force in the Production of Vehicle Wheel Rims *International Journal of Engineering and Technology* **7** 30–34

[26] Razooqi A 2018 Experimental and Numerical Analysis of Expanded Pipe using Rigid Conical Shape *Journal of Engineering* **24** 106-119

[27] Omar S, AL -Abri and Pervez T 2013 Structural Behavior of Solid Expandable Tubular Undergoes Radial Expansion Process- Analytical, Numerical, And Experimental Approaches *International Journal Solids and Structures* **50** 2980-94

[28] Puzyr R, Haikova T, Majerník J, Karkova M and Kmec J 2018 Experimental Study of the Process of Radial Rotation Profiling of Wheel Rims Resulting in Formation and Technological Flattening of the Corrugations Manufacturing Technology **18** 106–111

[29] Karrecha A and Seibi A 2010 Analytical model for the expansion of tubes under tension *Journal of Materials Processing Technology* **210** 356–362

[30] Yeh F 2007 Study of flaring forming limit in the tube flaring process *Journal of Strain Analysis* **42** 315–342

[31] Patent of Polska WO 2006/135266. M. Burda, Z. Sova, K. Drzal, A. Masiolek. Method and apparatus for shaping various cross-sections on metal pipes

[32] Kalyughniyi V, Zaporozchenko A and Pimanov V 2012 Intensification of technologies of making of good is «Bulb 180x184» *Treatment of materials pressure: collection of scientific labour* **2** 136–140

[33] Puzyr R, Haikova T, Trotsko O and Argat R 2016 Determining experimentally the stress-strained state in the radial rotary method of obtaining wheels rims *Eastern-European Journal of Enterprise Technologies* **4** 52–60

[34] Geselbracht, Margaret J, Ellis Arthur B, Penn Rona L, Lisensky George C, Stone and Donald 1994 Mechanical Properties of Metals: Experiments with Steel, Copper, Tin, Zinc, and Soap Bubbles *J. Chem. Educ* **71** 254

[35] Markov O, Gerasimenko O, Khvashchynskyi A, Zhytnikov R and Puzyr R 2019 Modeling the techological process of pipe forging without a mandrel *Eastern-European Journal of Enterprise Technologies* **3** 42-48

[36] Popov E 1977 *Fundamentals of the theory of sheet stamping* (Mechanical engineering, Moscow)

[37] Fischer F D, Rammerstorfer F G and Daxner T 2006 *Flaring-an analytical approach International Journal of Mechanical Sciences* **48** 1246–55

[38] Maslov A, Batsaikhan J, Puzyr R and Salenko Y 2018 The determination of the parameters of a vibration machinef the internal compaction of concrete mixtures *International Journal of Engineering & Technology* **7** 12-19

[39] Savelov D, Dragobetsky V, Puzyr R and Markevych A 2015 Peculiarities of vibrational press dynamics with hard-elastic restraints in the working regime of metal powders molding *Metallurgical and Mining Industry* **2** 67–74

[40] Haikova T, Puzyr R, Dragobetsky V, Symonova A and Vakylenko R 2019 *Finite-Element Model of Bimetal Billet Strain Obtaining Box-Shaped Parts by Means of Drawin (DSMIE 2019, LNME)* ed V Ivanov et al (Lutsk, Ukraine: Springer) pp 85 – 94

[41] Salenko Y, Puzyr R, Shevchenko O, Kulynych V and Pedun O 2020 *Numerical Simulation of Local Plastic Deformations of a Cylindrical Workpiece of a Steel Wheel Rim (DSMIE 2020, LNME)* ed V Ivanov et al (Kharkiv, Ukraine: Springer) pp 442 – 451

IOP Conf. Series: Materials Science and Engineering **1018**(2021) 012014 doi:10.1088/1757-899X/1018/1/012014

Study on grinding mode effect on external conical thread quality

L Ya Ropyak, V S Vytvytskyi, A S Velychkovych, T O Pryhorovska and M V Shovkoplias

Ivano-Frankivsk National Technical University of Oil and Gas, Department of Mechanical Engineering, 15 Karpatska Str., Ivano-Frankivsk, Ukraine, 76019

E-mail: pryhorovska@gmail.com

Abstract. The article has grounded the necessity to develop an empirical power dependence of cutting modes influence on part surface roughness obtained during thread grinding. The calculations were carried out for grinding of parts made of steel and hard alloy. There have been developed a relation of part material properties, grinding wheel options, lubricating and cooling liquid (constant for concrete conditions of thread grinding), depth of cutting (thickness of a grinded layer per path), revolutions of a part and a wheel, longitudinal feed refer to obtained roughness. It has been stated that components of the grinding mode contribute in roughness values in different way, in particular they are arranged in the ascending order as follows: grinding depth, rotation speed, and detail's diameter.

1. Introduction

Modern hydrocarbon production is associated with technologies of opening of productive layers, increasing of well's depth for both vertical and inclined drilling. Reliable and quality drilling equipment provides productive performance in conditions of high temperature, pressure, and abrasive and corrosion environment. Design of drill and oil/gas field equipment envisage conical thread as the main type of connection, in particular a pipe thread according to State Standard GOST 6357–81 and API, and a lock thread according to State Standard GOST 28487–90. Threaded surface accuracy and quality ensure tightness of connections, rapid screwing, and increases the uniformity of load distribution between the turns of the thread, which, in turn, increases the reliability of equipment in a whole. This way, modern drilling technologies need improved drilling equipment designs and technologies of machining.

Helical surfaces are widely used in conversion of rotational motion into translational motion, transmissions of rolling and sliding, which requires a rational choice of grinding mode to obtain the necessary parameters of surface roughness.

Reducing the roughness of threaded surfaces increases reliability and corrosion resistance of connections, as far as local stresses are reduced. This way stress concentrators do not cause destruction of parts under alternating loads and corrosive environments. Rolling and/or grinding are the most common methods used to improve thread surface quality in particular to reduce roughness, decrease surface micro-cracks, etc. Study of thread grinding and substantiation of rational cutting modes are especially relevant to provide necessary surface roughness of steel and hard alloys.

2. Literature review

Well drilling envisages improvement of drilling and oil/gas equipment's design and technologies of its manufacturing. Drilling tools include pipes, weighted pipes, adapters, calibrators, bits and others.

Design [1-3], technological [4-6], and operational methods [7-9] are used to increase drilling tool reliability and durability. Besides, there have been simulated threads [10, 11] and machining of contact interaction between parts of threaded connection [12, 13], rock-drilling pipe [14–16] and downhole-drill bit contact interaction [17-19], well washing process [20]. The researchers have tested static load [21] and alternating load effect on fatigue [22], substantiated drilling regimes, including taking into account thermal effects in the contact zone [23, 24] and effect of corrosive environments [25-27]. Much attention is paid to the environmental safety of equipment during drilling [28] and transportation of hydrocarbons [29].

The works [10, 11, 30] have studied accuracy of conical threads within turning. The work [31] studied effect of machining errors on stresses of thread connection. Design and technology simulation of assembling with system analysis is promising [32]. It should be noted that the most significant methods among the aforementioned ones are technological ones aimed at increasing of reliability and durability of drilling tools.

The problem of improving the quality of smooth surfaces [33–35] and surfaces of gears by milling [36] and grinding [37] is well covered in the scientific and technical literature.

Researchers pay considerable attention to grinding simulation [38, 39] and determining the cutting forces during grinding [40, 41], as well as to substantiation of grinding wheel profile [42]. But only few works consider the problem of ensuring the roughness of threaded surfaces [43, 44]. In particular, it is the problem to choose the abrasive material for tools and specify grinding wheel profile, schemes of grinding and rational cutting modes [44–46].

Therefore, it is relevant to develop empirical formulas to determine roughness refer to thread grinding modes in order to obtain the desired surface quality of parts.

3. Statement of the problem

Actual Ukraine regulations for oil and gas equipment, in particular State Standard GOST 28487-90, specify low roughness of threaded surfaces. In general, roughness of threaded surfaces should not exceed *Ra* 3.2 μm. Despite this fact, the problem of roughness obtained within conical thread grinding is not as elaborated in scientific and reference literature, as it is elaborated for other types of machining (turning, drilling, milling, etc.) or for grinding of smooth surfaces. This fact can be explained by technological complexity of grinding, such as the complexity of the selection of equipment and tools, and limited sizes of threads and steps.

In contrast to thread turning, grinding is a more complex process in terms of quality and accuracy of the surface. Accuracy parameters of the thread can be controlled by existing measuring tools, but the quality control of the thread surface layer is a rather time-consuming procedure and is often performed visually, but non-destructive and / or destructive control methods or not performed at all.

The method of copying of grinding wheel profile is used for thread grinding providing the coincidence of the formed surface and wheel profile. Grinding of threads can be carried out according to the schemes of single-threaded, multithreaded grinding or according to the combined scheme – rough grinding by a multithreaded wheel, or pure – by a single-threaded wheel. This imposes certain difficulties when specify grinding wheels: grain size, degree of hardness, bond material, which is determined by the radius of rounding, and the pitch of the thread.

Due to the lack of analytical relations between roughness and grinding modes, this study proposed herein aims to develop and analyze of an empirical power relation between roughness options and grinding mode, in particular the influence of a grinding mode and material properties on roughness of external conical threaded surfaces. Therefore, development of power relations for to thread grinding is an important scientific and practical problem, because this relation will simplify specification of appropriate grinding modes to provide necessary roughness of threaded surfaces and make possible establishing each factor influence on roughness of threaded surfaces.

Mainly drilling equipment is made of steel, but some details (in particular bodies of PDC drill bits are made of steel or hard alloy by sintering), so there have been developed relations for steel and hard alloy.

4. Results of study

Traditionally, an empirical power relation between roughness and cutting (grinding) modes is developed as

$$R_a = f(C_a, t, V_d, V_к, s) \tag{1}$$

where R_a – roughness of the machined surface depends on machining more. For thread grinding, R_a depends on C_a – material, grinding wheel characteristics and type of coolant (constant value for specific processing conditions), t – depth of grinding (thickness of removed layer of metal by grinding per path of a machine table), detail rotation speed V_d, grinding wheel rotation speed $V_к$, and longitudinal feed s.

The options of part rotation speed V_d and a machine table longitudinal path are interrelated; in particular, one path of the machine table is performed during one rotation of the machined part, so these options cannot be independent variables in the regression relation. Therefore, the following options were selected:

- physical and mechanical properties of the machined material, the grinding wheel, and lubricating and cooling fluid (constant value for specific grinding conditions);
- depth of grinding (thickness of the removed layer of metal by grinding per path of the machine table);
- rotation speed of the part; and
- rotation speed of the grinding wheel.

Grinding is assumed to be performed on a universal grinding machine 5822M, so the range of grinding modes was specified according to its certificate, in particular grinding wheel rotation speed and grinding wheel diameter were specified with combination of the number of spindle revolutions.

Grinding wheels are made of white electrocorundum and monocorundum, less often of green silicon carbide. They have a ceramic bond with a grain size of 12 to 20, on a bakelite bond with a grain size of 16 to 8 and on a bond SKN (V2) with a grain size of 4. At present, grinding wheels made of Elbor are widely used for grinding of single-start and multi-start threads. Single-start grinding with Elbor wheels is carried out with the following grinding modes: low detail rotation speed (0.16–0.30 m/min) and large longitudinal feed (0.4–0.5 mm/path). This grinding mode provides high accuracy of thread grinding, in particular high accuracy of thread profile and its step, with complete absence of surface structural transformations of hardened steel.

For grinding, there have been used T20 Д300 grinding wheels according to State Standard 2424-83 made of Elbor for steel details (25Г2С2Н2МА State Standard 5632-72) and diamond ones for hard alloy details. The rotation speed of the wheel was $V_к = 35$ m/s.

In order to simplify the mathematical model that describes profile roughness R_a (arithmetic mean deviation of the surface roughness R_a, μm, i.e. the average deviation of the profile points from the midline of the profile within the base length) refer to grinding mode options, there have been taken the power function:

$$R_a = C_a \cdot t^m \cdot V_d^{\,n} \cdot d^p \tag{2}$$

where C_a – physical and mechanical properties of the machined material, grinding wheel, and lubricating and cooling fluid (constant value for specific grinding conditions); t – depth of grinding (thickness of removed layer of metal by grinding per path of a machine table); V_d –rotation speed of the part; d – diameter of the processed thread and m, n, p – unknown exponents.

The rotation speed of the wheel has not been not introduced into this function, and has been taken as a constant for the studied grinding – $V_к$ =35 m/s. Roughness ranged from 1.6 to 3.2 μm.

This relation is nonlinear, so it has been reduced it to a linear form by logarithming:

$$\ln R_a = \ln C_a + m \ln t + n \ln V_d + p \ln d. \quad (3)$$

To simplify the calculations, there have been introduced the following notation:

$$\ln R_a = y,\ \ln C_a = b_0,\ b_1 = m,\ X_1 = t,\ b_2 = n,\ X_2 = V_d,\ b_3 = p \text{ and } X_3 = d$$

substituting them in (3) there have been obtained:

$$y = b_0 + b_1 \ln X_1 + b_2 \ln X_2 + b_3 \ln X_3 \quad (4)$$

Equation (3) is an empirical relation of roughness *Ra* refers to grinding mode for circular diamond grinding with longitudinal feed.

There have been used the full-factor experiment 2^k, where $k = 3$ is the number of variables (total number of experiments N = 8) to determine the unknown coefficients of equation (3). Table 1 presents the planning matrix of this experiment.

Table 1. Planning matrix of the full-factor experiment 2^3

Experiment	Value of coded factors				Interaction of coded factors				Output parameter, *Y* repeatability of experiments		Average value of the output parameter
	x_0	x_1	x_2	x_3	x_1x_2	x_1x_3	x_2x_3	$x_1x_2x_3$	Y_{u1}	Y_{u2}	$\bar{Y}_v$
1	+1	−1	−1	−1	+1	+1	+1	−1	Y_{11}	Y_{12}	Y_{1c}
2	+1	+1	−1	−1	−1	−1	+1	+1	Y_{21}	Y_{22}	Y_{2c}
3	+1	−1	+1	−1	−1	+1	−1	+1	Y_{31}	Y_{32}	Y_{3c}
4	+1	+1	+1	−1	+1	−1	−1	−1	Y_{41}	Y_{42}	Y_{4c}
5	+1	−1	−1	+1	+1	−1	−1	+1	Y_{51}	Y_{52}	Y_{5c}
6	+1	+1	−1	+1	−1	+1	−1	−1	Y_{61}	Y_{62}	Y_{6c}
7	+1	-1	+1	+1	−1	−1	+1	−1	Y_{71}	Y_{72}	Y_{7c}
8	+1	+1	+1	+1	+1	+1	+1	+1	Y_{81}	Y_{82}	Y_{8c}

Grinding mode options were transformed into dimensionless coded variables for exponential-degree polynomial by the formula:

$$x_i = \frac{2(\ln X_i - \ln X_{i\max})}{\ln X_{i\max} - \ln X_{i\min}} \quad (5)$$

where X_i – values of the grinding mode options for diamond grinding with longitudinal feed; $X_{i\min}, X_{i\max}$ – minimum and maximum values of the factors and i – respectively the serial number of the independent factor.

The numerical values of the grinding mode options for diamond grinding were specified according to grinding machine 5822M certificate.

Initially, studies have been carried out for heat-resistant steel 25Г2С2Н2МА State Standard GOST 5632-72. Table 2 summarizes factor variation ranges.

Table 2. Ranges of grinding factor variation for diamond grinding with longitudinal feed.

Interval of variation and factor ranges	Feed per path s_t, mm/path		Grinding speed v, m/min		Average diameter of the processed thread d, mm	
	X_1	$\ln X_1$	X_2	$\ln X_2$	X_3	$\ln X_3$
Basic level, $x_i = 0$	0.045	–	40	–	125	–
Variation range, $\Delta x = 1$	0.035	–	15	–	75	–
Upper level, $x_i = +1$	0.08	−2.52573	55	4.007333	200	5.298317
Lower level, $x_i = -1$	0.01	−4.60517	25	3.218876	50	3.912023

Substituting the numerical values from table 2 in formula (5) there have been obtained expressions for coded factors:

$$x_1 = \frac{2(\ln s_t + 4.605)}{-4.605 + 5.298} = 0.961797 \ln s_t + 3.429237 \quad (6)$$

$$x_2 = \frac{2(\ln V_d - 3.912)}{3.912 - 2.303} = 2.536599 \ln V_d - 9.165 \quad (7)$$

$$x_3 = \frac{2(\ln d - 5.3132)}{5.3132 - 3.912} = 1.442695 \ln d - 6.64386 \quad (8)$$

Equations with the introduction of variables taking into account factor interaction have been written in the coded variables:

$$Y = b_0 + b_1x_1 + b_2x_2 + b_3x_3 + b_{12}x_1x_2 + b_{13}x_1x_3 + b_{23}x_2x_3 + b_{123}x_1x_2x_3 \quad (9)$$

where b_0, b_1, b_2, b_3, b_{12}, b_{13}, b_{23}, b_{123} – regression coefficients that characterize the free term of the equation, linear effects and effects of factor interaction, respectively;

x_1, x_2, x_3 – experimental factors (parameters of the grinding mode).

The order of the experiments according to the matrix has been chosen using a table of random numbers to eliminate systematic errors. To ensure good reproduction of the results, each experiment has been repeated twice at the same level of factors and there have been calculated the average value of the initial parameter in each experiment.

$$\overline{Y_v} = \frac{1}{r}\sum_{u=1}^{r} Y_{ui} \quad (10)$$

Table 3 presents experimental data of thread roughness after grinding.

Table 3. The experimental data of thread roughness after grinding.

Experiment number	Value of coded factors				Output parameter, Y repeatability of experiments reproducibility				Average value of the output parameter		Dispersion of the experimental data
	x_0	x_1	x_2	x_3	Y_{u1}	$\ln Y_{u1}$	Y_{u2}	$\ln Y_{u2}$	$\overline{Y}_v$	$\ln \overline{Y}_v$	s_v^2
1	+1	–1	–1	–1	1.4	0.336	1.6	0.470	1.5	0.405	0.0090
2	+1	+1	–1	–1	1.7	0.531	1.5	0.405	1.6	0.470	0.0079
3	+1	–1	+1	–1	1.2	0.182	1.4	0.336	1.3	0.262	0.0119
4	+1	+1	+1	–1	1.3	0.262	1.5	0.405	1.4	0.336	0.0102
5	+1	–1	–1	+1	3.0	1.099	3.2	1.163	3.1	1.131	0.0020
6	+1	+1	–1	+1	4.1	1.411	3.5	1.253	3.8	1.335	0.0125
7	+1	–1	+1	+1	2.4	0.875	2.8	1.030	2.6	0.956	0.0120
8	+1	+1	+1	+1	3.1	1.131	3.1	1.131	3.1	1.131	0

The obtained experimental results have been subjected to statistical analysis. There have been determined:

– ordinal variances

$$S_u^2 = \frac{1}{r-1}\sum_{v=1}^{N}\left(Y_{uv} - \overline{Y}\right)^2 \quad (11)$$

– variance of reproduction experiments

$$S^2(Y) = \frac{1}{N}\sum_{v=1}^{N} S_v^{\ 2} \quad (12)$$

Variance homogeneity was checked by Cochran's test

$$G = \frac{S_{\max}^2}{\sum_{v=1}^{N} S_v^{\ 2}} \quad (13)$$

Calculated value of Cochran's test have been compared with the referenced one for $f_{ad} = N - \lambda = 8 - 4 = 4$ and $f = N(r - 1) = 8(2 - 1) = 8$ degrees of freedom and the level of significance of $\alpha = 0{,}05$. As far as $G_{table} = 0.6798 > G = 0.1908$ the hypothesis of regression variance homogeneity has been accepted.

The full-factor experiment is an orthogonal plan and it can greatly simplify the calculation of regression coefficients:

$$b_0 = \frac{\sum_{v=1}^{N} \overline{Y_v}}{N} \quad (14)$$

$$b_i = \frac{\sum_{v=1}^{N} x_{iv} \cdot \overline{Y}_v}{N} \quad (15)$$

$$b_{i,j} = \frac{\sum_{v=1}^{N} x_{iv} \cdot x_{jv} \cdot \overline{Y}_v}{N}, i \neq j \quad (16)$$

$$b_{i,j,w} = \frac{\sum_{v=1}^{N} x_{iv} \cdot x_{jv} \cdot x_{wv} \cdot \overline{Y}_v}{N}, i \neq j \neq w \quad (17)$$

where $i = 0, 1, 2, 3$ – factor number.

The coefficients of the regression equation determined by formulas (11) – (14) have been summarized in table 4.

Table 4. Regression coefficients.

Coefficient	b_0	b_1	b_2	b_3	b_{12}	b_{13}	b_{23}	b_{123}
Value	0.7533	0.3850	–0.0820	0.0648	–0.0128	0.0300	–0.0025	–0.0048

The obtained relation has been subjected to statistical analysis. To specify coefficient significance by Student's t-test, there have been calculated regression coefficient variance:

$$S^2(b_i) = \frac{1}{Nr}\sum_{v=1}^{N} S^2(Y) \tag{18}$$

The critical value of the Student's t_{kr} criterion for $f_E = N(r-1) = 8(2-1) = 8$ degrees of freedom and significance level $\alpha = 0{,}05$ is t_{kr}=2.306. This way, confidence interval is:

$$\Delta b_i = t_{kr} S(b_i) \tag{19}$$

where $S(b_i)$ – coefficient errors.

The variance of the regression coefficients have been calculated by formula (14), and it was 0.000512. Then according to (17) $\Delta b_i = 2.306 \cdot 0.02262 = 0.05216$. The regression coefficient is considered to be significant if $|b_1| \geq \Delta b_i$ (here b_i means $b_0, b_i, b_{ij}, b_{ijw}$).

Statistically insignificant coefficients are discarded, and the regression equation (4) has been specified as:

$$Y = 0.75325 + 0.385x_1 - 0.082x_2 + 0.06475x_3 \tag{20}$$

Fisher's F-test was used to test adequacy of the developed relation:

$$F = \frac{S_{ad}^2}{S^2(Y)} \tag{21}$$

where S_{ad}^2 – between group variability, $S^2(Y)$ – within-group variability, which was calculated as follows

$$S_{ad}^2 = \frac{r\sum_{v=1}^{N}\left(\overline{Y}_v - \widehat{Y}_v\right)^2}{N-\lambda} \tag{22}$$

where λ – the number of significant coefficients in the regression equation; $\widehat{Y}_v$ – calculated values of the initial parameter obtained after substitution into the refined regression equation (20) of factor levels (–1 and +1) according to the planning matrix of the full-factor experiment.

To check the adequacy of the obtained relation (20), there have calculate the values of the thread roughness (Table 5).

Table 5. The calculated data of thread roughness after grinding

Experiment number		1	2	3	4	5	6	7	8
Calculated thread roughness after grinding	$\ln R_a$	0.3855	0.5150	0.2215	0.3510	1.1555	1.2850	0.9915	1.1210
	R_a, μm	1.4703	1.6736	1.2479	1.4205	3.1756	3.6147	2.6953	3.0679

Within-group variability according to (22) is $S_{ad}^2 = 0.00873$.

Calculated Fisher's criterion has been compared with the referenced one for $f_{ad} = N - \lambda = 8 - 4 = 4$ and $f_{\tilde{A}} = N(r-1) = 8(2-1) = 8$ degrees of freedom and $\alpha = 0{,}05$ level of significance. As far as $F_{tabl} = 0.53 < F = 3.84$, the regression relation has been accepted.

Besides, the multiple correlation coefficient has been calculated as:

$$R = \sqrt{1 - \sum_{v=1}^{N}\left(\overline{Y}_v - \hat{Y}v\right)^2 / \sum_{v=1}^{N}\left(\overline{Y}_v - \overline{Y}\right)^2} \qquad (23)$$

where $\overline{Y}$ – average value $\overline{Y}_v$.

The initial data for calculating the multiple correlation coefficients are summarized in table 6.

When $R = \sqrt{1 - 0.00873/1.28186}$.

Thus, the regression equation (20) completely met the experimental data.

After re-coding, the variables in the regression equation (20) there have been obtained:

$$\ln Ra = 2.39485 + 0.37029 \ln s_t - 0.208 \ln V_d + 0.093242 \ln d \qquad (24)$$

Potentiating the regression equation (24) there have been obtained:

$$R_a = 10.967 \frac{s_t^{0.37} d^{0.093}}{V_d^{0.208}}. \qquad (25)$$

Table 6. Data for multiple correlation coefficient calculation.

Experiment number	The average value of the output parameter $\overline{Y}_v$	Calculated value of the output parameter $\hat{Y}_v$	$(\overline{Y}_v - \hat{Y}_v)^2$	$(\overline{Y}_v - \overline{Y})^2$
1	0.4050	0.3855	0.00038	0.12128
2	0.4700	0.5150	0.00203	0.08023
3	0.2620	0.2215	0.00164	0.24133
4	0.3360	0.3510	0.00022	0.17410
5	1.1310	1.1555	0.00060	0.14270
6	1.3350	1.2850	0.00250	0.33843
7	0.9560	0.9915	0.00126	0.04111
8	1.1310	1.1210	0.00010	0.14270
Total	$\overline{Y} = 0.75325$	–	0.00873	1.28186

According to the obtained power dependence (25), the response surfaces there have developed (Fig. 1).

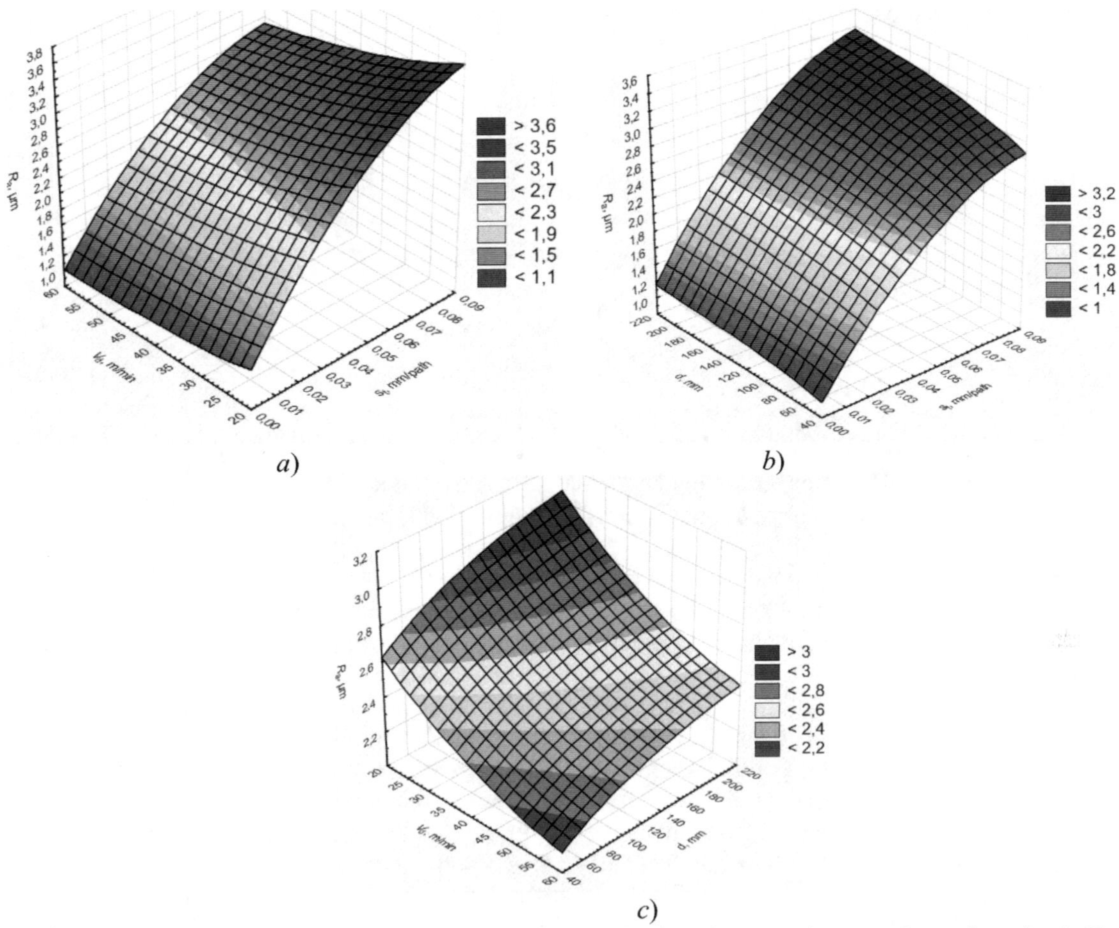

Figure 1. Response surfaces of the developed relation of surface roughness refer to thread grinding modes (for heat-resistant steel 25Г2С2Н2МА State Standard GOST 5632-72)
a – s_t, V_d, d = 125 mm; *b* – s_t, d, V_d = 40 m/min; *c* – V_d, d, s_t = 0.045 mm/path

Similarly, there has been obtained the relation for hard alloy ВК 8 GOST 3882 –74

$$R_a = 9.606\frac{s_t^{0.331} d^{0.086}}{V_d^{0.173}}. \tag{26}$$

According to the obtained power dependence (26), the response surfaces there have been developed (Fig. 2).

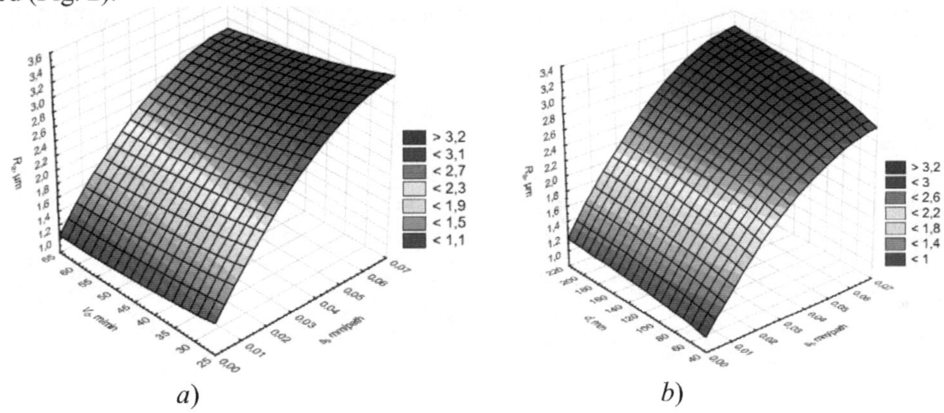

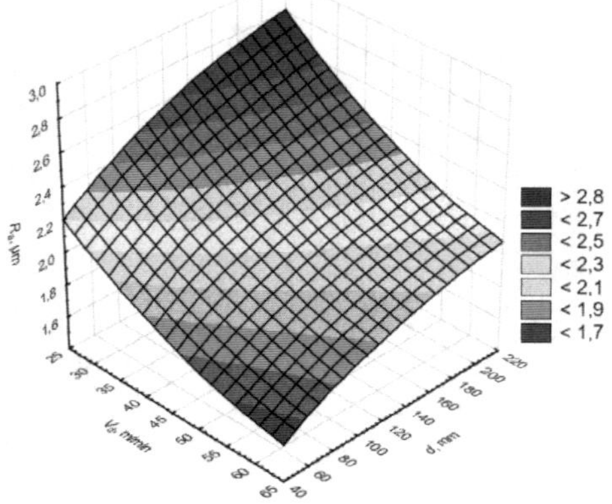

c)

Figure 2. Response surfaces of the developed relation of surface roughness refer to thread grinding modes (for hard alloy BK 8 State Standard GOST 3882-74)
a – s_t, V_d, d = 125 mm; *b* – s_t, d, V_d = 45 m/min; *c* – V_d, d, s_t = 0.035 mm/path

The response surfaces developed for steel and hard alloy show that roughness *Ra* of thread surface obtained by diamond grinding depends mostly on longitudinal feed per path s_t and part rotation speed (Fig. 1 and Fig.2). Much less roughness depends on the part's diameter *d*. Roughness increases more significant if rotation speed of the part is variable and longitudinal feed is constant than if rotation speed of the part is constant and longitudinal feed is variable (Fig. 1, a and Fig. 2, a). Similarly, considering the response surfaces in Fig. 1, c and 2, c, it should be noted that longitudinal feed influences more significantly on roughness *Ra* if grinding depth is constant than longitudinal feed is constant. Similarly, considering the response surfaces in Fig. 1, c and 2, c, it should be noted that variant revolution speed of the part and constant longitudinal feed make nonsignificant greater effect on the surface roughness then variant longitudinal feed and constant revolution speed of the part.

The developed relations show that roughness of threaded parts made of hard alloy (Fig. 2) is 3–12% less then roughness of threaded parts made of steel (Fig. 1) for the same input data. This improvement in the quality of the machined surface can be explained by the fact that the abrasive grains blunt faster during the processing of harder material – hard alloy BK 8 for this case.

The results of this study, in contrast to thread grinding studies presented in [43], allow to rationally specify the technological modes of thread grinding to obtain the required surface roughness, both for steel and hard alloy parts.

5. Conclusions

1. There has been developed the relation between part surface roughness obtained during thread grinding refer to grinding modes (physical and mechanical properties of the machined material, grinding wheel, and lubricating and cooling fluid; depth of grinding; rotation speed of the part; and rotation speed of the grinding wheel). The calculations were carried out for parts made of heat-resistance steel 25Г2С2Н2МА and hard alloy BK 8.
2. It was stated that components of the grinding modes contribute in roughness values in different way, in particular they are arranged in the ascending order as follows: grinding depth, rotation speed, and detail's diameter.
3. The developed relations show that roughness of threaded parts made of hard alloy is 3–12% less then roughness of threaded parts made of steel for the same input data.

In further research it is planned to optimize the modes of thread grinding of parts made of steel and hard alloy by linear programming taking into account kinematic, force and temperature limitations during conical thread machining.

References

[1] Vlasiy O, Mazurenko V, Ropyak L and Rogal O 2017 *Eastern-European Journal of Enterprise Technologies* **1**(7-85) 25–31
[2] Shats'kyi I P, Lyskanych O M and Kornuta V A 2016 *Strength of Materials* **48**(3) 469–472
[3] Onysko O, Borushchak L, Kopei V, Lukan T, Medvid I and Vryukalo V 2020 *Lecture Notes in Networks and Systems* 128 LNNS 720–729
[4] Ropyak L, Schuliar I and Bohachenko O 2016 *Eastern-European Journal of Enterprise Technologies* **1**(5) 53–62.
[5] Ropyak L Ya, Pryhorovska T O and Levchuk K H 2020 *Progress in Physics of Metals* **21**(2) 274–301
[6] Kopei V, Onysko O and Panchuk V 2020 *Lecture Notes in Mechanical Engineering* 149–158
[7] Velychkovych A, Petryk I and Ropyak L 2020 *Shock and Vibration* 3292713
[8] Dutkiewicz M, Gołębiowska I, Shatskyi I, Shopa V and Velychkovych A 2018 *MATEC Web of Conferences* **178** 06010
[9] Velichkovich A S 2007 *Chemical and Petroleum Engineering* **43**(7-8) 458–461
[10] Onysko O R, Kopey V B and Panchuk V G 2020 *IOP Conference Series: Materials Science and Engineering* **749**(1) 012007
[11] Onysko O, Kopei V, Medvid I, Pituley L and Lukan T 2020 *Lecture Notes in Mechanical Engineering* 432–441
[12] Shatskyi I, Ropyak L and Velychkovych A 2020 *Engineering Solid Mechanics* **8**(4) 301–312
[13] Xu H, Shi T, Zhang Z and Shi B 2014 *Mathematical Problems in Engineering* **2014** 287076
[14] Levchuk K G 2018 *Metallofizika i Noveishie Tekhnologii* **40**(5) 701–712
[15] Grydzhuk J, Chudyk I, Velychkovych A and Andrusyak A 2019 *Eastern-European Journal of Enterprise Technologies* **1**(7-97) 6–14
[16] Levchuk K G 2017 *SOCAR Proceedings*, (2) 23–33
[17] Pryhorovska T 2018 *Engineering Solid Mechanics* **6**(4) 315–330
[18] Pryhorovska T O, Chaplinskiy S S and Kudriavtsev I O 2015 *Shiyou Kantan Yu Kaifa/Petroleum Exploration and Development* **42**(6) 812–816
[19] Pryhorovska T A and Chaplinskiy S S 2018 *Neftyanoe Khozyaystvo – Oil Industry* (1) 38–41
[20] Vytyaz O, Chudyk I and Mykhailiuk V 2015 *New Developments in Mining Engineering 2015: Theoretical and Practical Solutions of Mineral Resources Mining* 591–595
[21] Spanu P, Amza C G, Abaza B F 2018 *Materiale Plastice* 55 580-583
[22] Belkacem L, Abdelbaki N, Gaceb M, Bouali E and Bettayeb M 2017 *Chemical Engineering Transactions* **61** 1099–1104
[23] Tatsiy R M, Pazen O Y, Vovk S Y, Ropyak L Y and Pryhorovska T O 2019 *Journal of the Serbian Society for Computational Mechanics* **13**(2) 36–55
[24] Volchenko N, Volchenko A, Volchenko D, Poliakov P, Malyk V, Zhuravliov D, Vytvytskyi V and Krasin P 2019 *Eastern-European Journal of Enterprise Technologies* **1**(5-97) 47–53
[25] Striletskyi Y, Rovinskyi V and Yevchuk O 2016 *Eastern-European Journal of Enterprise Technologies* 3(9) 24-29
[26] Saakiyan L S, Efremov A P, Ropyak L Ya and Gorbatskii A V 1987 *A method of microelectrochemical investigations Soviet Materials Science* **23**(3) 267–269
[27] Saakiyan L S, Efremov A P and Ropyak L Ya 1989 *Effect of stress on the microelectrochemical heterogeneity of steel Protection of Metals (English translation of Zaschita Metallov)* 25(2) 185–189
[28] Skitsa L, Yatsyshyn T, Liakh M and Sydorenko O 2018 *Mining of Mineral Deposits* **12**(3) 71–79
[29] Mandryk O, Artym, V, Shtohry M and Zaytsev V 2020 *Management Systems in Production Engineering* **28**(3) 168–177
[30] Onysko O, Kopei V, Panchuk V, Medvid I and Lukan T 2020 *Lecture Notes in Mechanical Engineering* 236–245
[31] Pryhorovska T and Ropyak L 2019 *Proceedings of the International Conference on Advanced Optoelectronics and Lasers, CAOL*, 2019-September 9019544 493–497
[32] Pasichnyk V and Khmurenko O 2015 *Eastern-European Journal of Enterprise Technologies*, **4**(7)

15–20
[33] Zdenko L, Pokorny P, Vaclav S and Buransky I 2010 *Research papers Faculty of materials science and technology in trnava Slovak University of technology in Bratislava* **28** 27–37
[34] Klymenko S A, Kopejkyna M Yu, Lavrynenko V Y, Majboroda V S, Akulovych L M, Levyn M L, Xejfecz M L, Xudolej A L and Chyzhyk S A 2017 Finishnaya obrabotka poverxnostej pry proyzvodstve detalej (Minsk: Belaruskaya navuka) p 378
[35] Mladenovic G, Bojanic P, Tanovic Lj and Klimenko S 2015 *Journal of Manufacturing Science and Engineering, Transactions of the ASME* **137**(3) 034502
[36] Heisel U, Pasternak S, Storchak M, Schaal M and Danilchenko S Y 2010 *ZWF Zeitschrift fuer Wirtschaftlichen Fabrikbetrieb* **105**(7-8) 649–654
[37] Larshin V P and Lishchenko N V 2018 *Russian Engineering Research* **38**(9) 712–713
[38] Guo M, Li B, Ding Z, Liang S Y 2016 *International Journal of Advanced Manufacturing Technology* **86**(9-12) 3395–3405
[39] Wang W, Feng X 2013 *Advances in Mechanical Engineering* 2013 827831
[40] Akyildiz H K, Livatyali H 2011 *Machining Science and Technology* **15**(4) 485-506
[41] Fang C, Yang C, Cai L, Zhao Y, Liu Z 2019 *International Journal of Advanced Manufacturing Technology* **104**(1-4) 943-956
[42] Abu Shreehah T A, Abdullah R A 2006 *Machining Science and Technology* **10**(4) 539-547
[43] Fang C, Zhao Y, Yang C and Liu Z 2020 *IOP Conference Series: Materials Science and Engineering* **784**(1) 012018
[44] Lavrinenko V I, Il'nyts'ka G D and Smokvyna V V 2011 *Journal of Superhard Materials* **33**(4) 261–267
[45] Medvedev V and Volkov A 2018 *Mechanisms and Machine Science* **51** 305–325
[46] Grechishnikov V A 2020 I *Russian Engineering Research* **40**(1) 51–54

IOP Conf. Series: Materials Science and Engineering **1018** (2021) 012015 doi:10.1088/1757-899X/1018/1/012015

Automatic positioning methodology and algorithm for modular jigs and fixtures components

T Savu, S Nanu and I C Ene

University POLITEHNICA of Bucharest, Manufacturing Engineering (TCM) Dept., 313 Spl. Independentei, 060042 sector 6, Bucharest, Romania

tom.savu@upb.ro

Abstract. One of the steps in designing a modular jig or fixture assembly is that of selecting the appropriate components. Among these components, those forming the jig's body must be chosen for closing the dimensional chains between the other locating, clamping, or indexing components. The methodology starts by defining the components' virtual connectors, specifying the rules for establishing their coordinates systems, and then is defining the rules to be followed for aligning and assembling two adjacent components. Different cases are described and only some of them are detailed, for clarity reasons. The algorithm is looking to determine a series of components from which the first is in a fixed position, related to other locating or clamping jig's component. A hierarchical tree is generated, using available components, having in each node a component which is assembled with a previous one using the above-mentioned connectors. Interference criteria is checked for selecting possible solutions only. A stop criterion related to the number of components in a branch is used, Solutions from the different branches are compared using the distances to an end component.

1. Introduction

When designing a modular jig or fixture assembly, having a set of available modular components, one has to solve the problem of identifying an optimal subset of modular components which must satisfy, between others, some geometrical conditions, established in most cases for closing dimensional chains between other jig or fixture components.

The paper is describing an approach for T-slot-based systems [1], in which the different possible combinations of components are built by using components connectors, defined in the specific points where which component can be assembled with another component. These connectors are entities like the ones used in different CAD software packages for describing the assembly of two parts.

To apply and implement an algorithm for the automated selection of the optimal subset, data representation structures have to be defined for describing the content of the set of available modular components, the components' dimensions, the position and the orientation of the components' connectors and also for representing the different possible solutions.

The algorithm has to be able to identify possible new solutions, to check for their compatibility with various criteria, to add the compatible solutions to a set, to check for an end of searching condition and then to apply some criteria for selecting the optimal solution.

An implementation was developed using LabVIEW [2], which will later allow storing the data in databases, presenting the solutions in 3D formats, and accessing the application as a Web service.

IOP Conf. Series: Materials Science and Engineering **1018** (2021) 012015 doi:10.1088/1757-899X/1018/1/012015

2. Notation conventions for describing the components

It is considered that a component Ci has two types of virtual connectors (figure 1):

- a main connector, named Ci_ConP, placed on the component's face on which a T-slot nut is mounted on the component:
    - the Z axis of the main connector Ci_ConP is parallel with the axis of the screw which is assembled with the T-slot nut, with the positive sense toward the nut;
    - the X-axis of the main connector Ci_ConP is parallel with the direction on which the T-slot nut can slide;
- none or more secondary connectors, placed on the faces where T-slots are present:
    - the Z-axis of a secondary connector Ci_ConSj is parallel with the axis of the screw which is assembled with the T-slot nut, with the positive sense toward the neighbouring component (outside the parent component);
    - the X-axis of a secondary connector Ci_ConSj is parallel with the direction on which the T-slot nut can slide

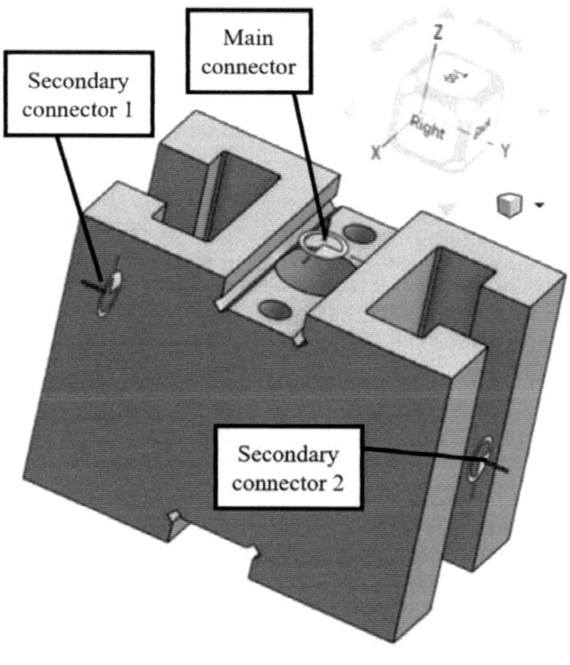

Figure 1. Main and secondary connectors for a CDC 048 type component.

The X(Ci_ConP), Y(Ci_ConP) and Z(Ci_ConP) coordinates of a main connector are expressed using the equipment's coordinate system, while the coordinates of a secondary connector Ci_ConSj are expressed using the main connector's coordinate system.

The rotation angles alfa(Ci_ConSj), beta(Ci_ConSj) and gamma(Ci_ConSj) of a coordinate system belonging to a secondary connector, relative to the main connector's coordinate system, are expressed considering the situation in which the rotations are performed in the X, Y, Z order. The positive sense of a rotation is the counterclockwise one.

One component's dimensions are described using its faces coordinates, expressed in the coordinates system of the component's main connector. Using as an example the CDC 048 type component from the DISROM set [3], its dimensions are Xmin = -19, Xmax = 19, Ymin = -32, Ymax = 32, Zmin = -48, Zmax = 0 (figure 2).

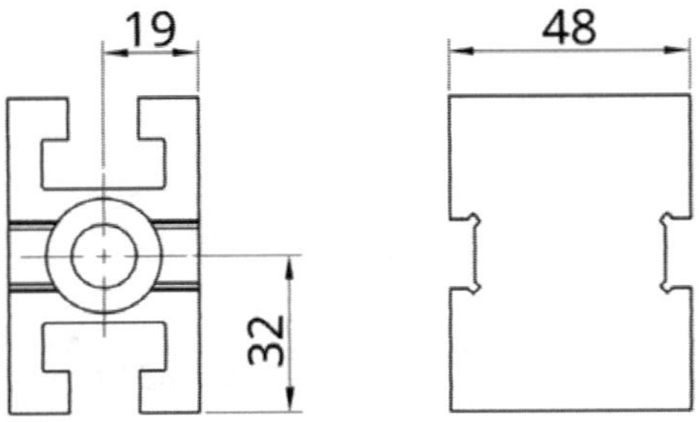

Figure 2. CDC 048 type component dimensions.

Also, the coordinates of the first secondary connector are X(Ci_ConS1) = 0, Y(Ci_ConS1) = -32, Z(Ci_ConS1) = -24, while the coordinates of the second secondary connector are X(Ci_ConS2) = 0, Y(Ci_ConS2) = 32, Z(Ci_ConS2) = -24. The rotation angles of the first secondary connector are alfa(Ci_ConS1) = 90°, beta(Ci_ConS1) = -90°, gamma(Ci_ConS1) = 0°, while the rotation angles of the second secondary connector are alfa(Ci_ConS2) = -90°, beta(Ci_ConS2) = 90°, gamma(Ci_ConS2) = 0°.

3. Assembly rules

When two components C1 and C2 must be assembled, one main connector of one component must be aligned with one secondary connector of the other component. There are two possible cases for aligning the connectors:

- the main connector C1_ConP of the first component is aligned with a secondary connector C2_ConSj of the second component;
- the main connector C2_ConP of the second component is aligned with a secondary connector C1_ConSj of the first component.

For the first case, the following conditions must be met for aligning the two connectors, all expressed in the coordinate system of the C1_ConP connector:

- Z(C2_ConSj) = 0, the origins of the two connectors have to stay in the same X-Y plane;
- (alfa(C2_ConSj) = π and beta(C2_ConSj) = 0) or (alfa(C2_ConSj) = 0 and beta(C2_ConSj) = π), the Z-axis of the two connectors can be made parallel and in opposite directions either using a rotation around the X-axis or using a rotation around the Y-axis;
- Y(C2_ConSj) = 0, no translation on the Y-axis;
- gamma(C2_ConSj) = 0, no rotation around the Z-axis;
- X(C2_ConSj) $\in [\delta 1, \delta 2]$, limited translation on the X-axis, where $\delta 1$ and $\delta 2$ depend on the dimensions of the second component.

Further discussing only for the simplified case when X(C2_ConSj) = 0, there are two possible cases for the C2_ConSj relative coordinates in the C1_ConP system:

- X(C2_ConSj) = Y(C2_ConSj) = Z(C2_ConSj) = 0, alfa(C2_ConSj) = π, beta(C2_ConSj) = 0, gamma(C2_ConSj) = 0;
- X(C2_ConSj) = Y(C2_ConSj) = Z(C2_ConSj) = 0, alfa(C2_ConSj) = 0, beta(C2_ConSj) = π, gamma(C2_ConSj) = 0.

Discussing about the first of the two cases above, when the C1_ConP absolute coordinates are known, the absolute coordinates of C2_ConSj can be computed by rotating with π around the X-axis the absolute coordinates of C1_ConP [4].

Similarly, for the second case, the following conditions must be met for aligning the two connectors, all expressed in the coordinate system of the C1_ConSj connector:

- Z(C2_ConP) = 0, the origins of the two connectors have to stay in the same X-Y plane;
- (alfa(C2_ConP) = π and beta(C2_ConP) = 0) or (alfa(C2_ConP) = 0 and beta(C2_ConP) = π), the Z-axis of the two connectors can be made parallel and in opposite directions either using a rotation around the X-axis or using a rotation around the Y-axis;
- Y(C2_ConP) = 0, no translation on the Y-axis;
- gamma(C2_ConP) = 0, no rotation around the Z-axis;
- X(C2_ConP) $\in [\delta 1, \delta 2]$, limited translation on the X-axis.

Further discussing only for the simplified case when X(C2_ConP) = 0, there are two possible cases for the C2_ConP relative coordinates in the C1_ConSj system:

- X(C2_ConP) = Y(C2_ConP) = Z(C2_ConP) = 0, alfa(C2_ConP) = π, beta(C2_ConP) = 0, gamma(C2_ConP) = 0;
- X(C2_ConP) = Y(C2_ConP) = Z(C2_ConP) = 0, alfa(C2_ConP) = 0, beta(C2_ConP) = π, gamma(C2_ConP) = 0.

Discussing about the first of the two cases above, when the C1_ConSj absolute coordinates are known, the absolute coordinates of C2_ConP can be computed by rotating with π around the X-axis the absolute coordinates of C1_ConSj.

4. Algorithm description

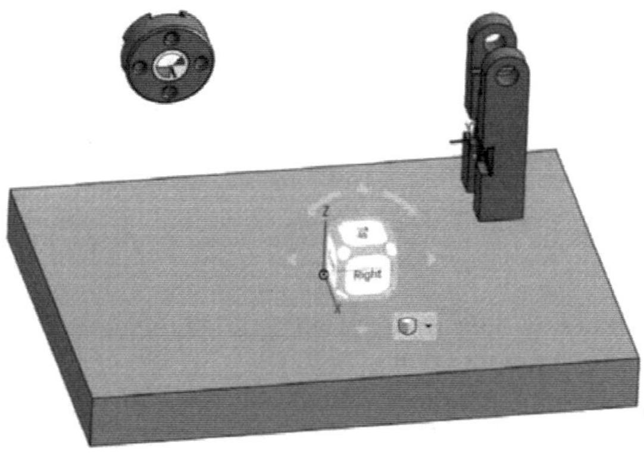

Figure 3. Initial components of the dimensional chain.

The algorithm is starting from the situation when the dimensional chain must be closed between two previously established components (figure 3) and is looking to build, for each initial component, a tree of modular components Ci.

It was assumed, for simplifying the initial cases for which the algorithm was implemented, that these initial components have only one free main connector each.

Each branch of one tree will contain a possible subset of modular components. The tree is initialized with one of the components between which the dimensional chain must be closed.

At a certain moment, there are a number of free connectors in a tree, belonging to the modular components in the subset (figure 4). For each free connector in the tree, the algorithm is looking in a list of available modular components ACk and is trying to extend the tree with more branches, by assembling available modular components on the free connectors.

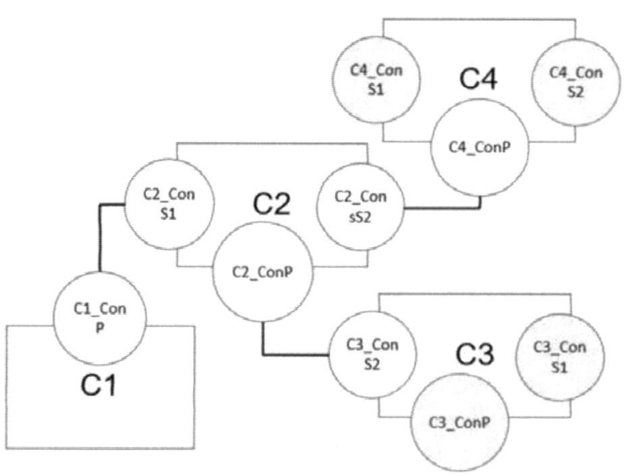

Figure 4. Tree structure, with free connectors marked in yellow.

IOP Conf. Series: Materials Science and Engineering 1018 (2021) 012015 doi:10.1088/1757-899X/1018/1/012015

For evaluating the possibility of assembling an available modular component on a free connector, interference conditions are checked between the available component and all the other components already on the tree branch.

The algorithm is not building branches with more than a specified number of components maxC, this being a limitation imposed by the rigidity of the jig or fixture assembly.

Because the two subsets of modular components are presumed to be assembled on the same base plate, the algorithm is computing the vertical distances between free connectors, belonging to terminal components of branches in the two separate trees, with vertical Z-axis. A stopping condition appears if the distance between a pair of such free connectors is smaller than a predefined value.

5. Data structures definitions

Data structures were defined for describing some of the conventions mentioned before. The data structure used for organizing a connector's linear and angular coordinates is presented in figure 5, while the data structure for describing a component's dimensions is presented in figure 6.

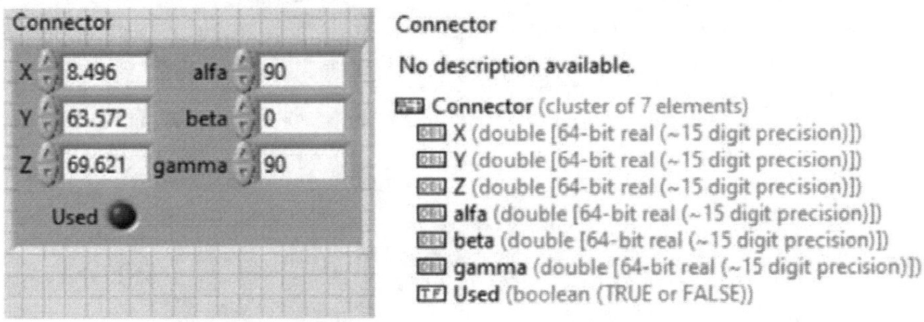

Figure 5. Data structure for connector's linear and angular coordinates.

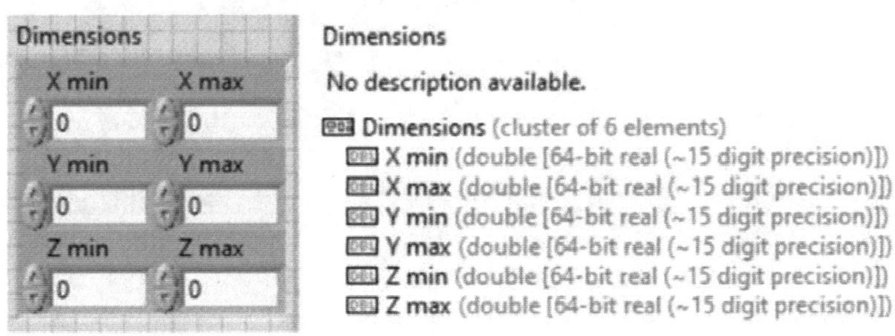

Figure 6. Data structure for describing a component's dimensions.

The data structure for completely describing a component contains substructures for: the component's main connector's coordinates, the component's dimensions, an array containing substructures for the secondary connectors' coordinates and a value specifying the component's type (figure 7).

IOP Conf. Series: Materials Science and Engineering **1018** (2021) 012015 doi:10.1088/1757-899X/1018/1/012015

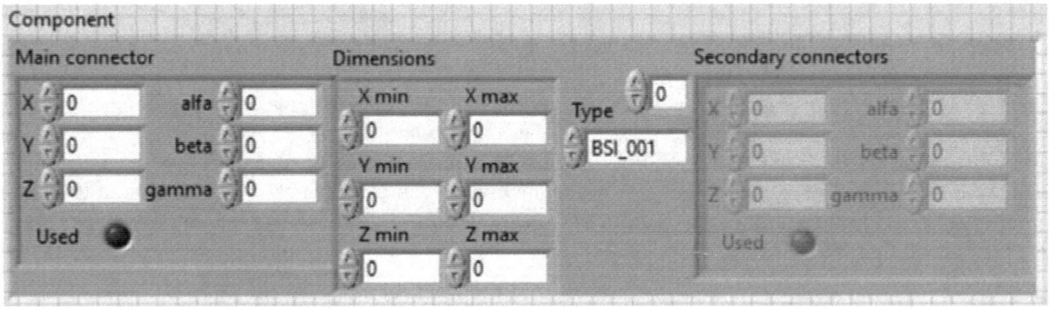

Figure 7. Data structure for complete component description.

When a component is described as part of the tree which is built by the algorithm, its data structure (figure 8) contains its type, the main connector's coordinates, a substructure containing the connection information, a rank value and an array describing the states of the secondary connectors (used or not). The rank value was not used for the moment but was placed for future developments.

The substructure containing the connection information has three unsigned byte values. If the substructure belongs to a component which has the index j in the tree branch and is connected to a component with index i, the three values are: the index i, the index of the used connector on the component with index i and the index of the used connector on the component with index j.

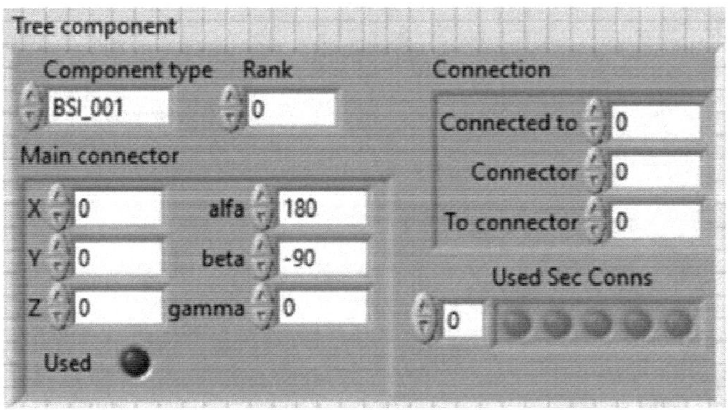

Figure 8. Data structure for a component part of a tree.

When a component is described as part of the set of available modular components, its data structure (figure 9) contains its type, the main connector's coordinates, an array with its secondary connectors' coordinates and an array of availability data.

Because each branch of the solutions tree built by the algorithm represents a possible solution, separate from the other branches, the number of instances of a component of a certain type in the whole tree may be greater than the number of real instances in the set of available modular components. Therefore the availability of a component of a certain type is described separately for each tree branch and the values are stored in an array structure.

The tree branches are stored in a two-dimensional array of components, where each line is representing a branch.

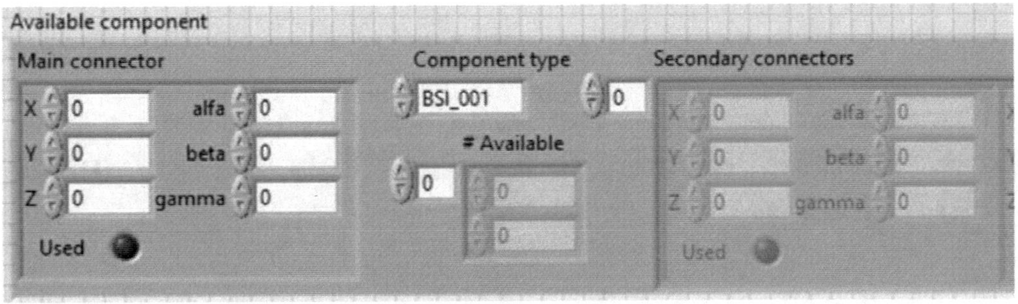

Figure 9. Data structure for an available modular component.

For checking the interference conditions [5] between an available component ACk and another component Cj already on the tree branch where ACk could be placed, with the modular components presumed to be shaped as simple boxes, for each face of ACk and for each edge of Cj, a subroutine is computing the coordinates of the intersection point Pjk between ACk face and the Cj edge. If Pjk belongs to the Cj edge, then it is checked if Pjk belongs also to the ACk face. If it exists at least one combination between an ACk face and a Cj edge for which the above conditions are met, it means that interference exists between the two components, so the tested solution is not acceptable and the ACk component is not placed in the tree.

6. Conclusions

The algorithm was tested for a simple implementation, which is taking into account only some of the assembly cases mentioned in the paper, with only a limited number of types of available components, but the implementation proved to offer correct solutions and is able to be further developed with minimum efforts. Further validation methods which will use 3d representations of the designed solutions will be also developed.

References

[1] Ghatpande P 2008 *Study of Fixturing Accessibilities in Computer-Aided Fixture Design – Master of Science Thesis* (Massachusetts: Worcester Polytechnic Institute)
Rong Y, Huang S and Hou Z 2005 Advanced Computer-Aided Fixture Design (Burlington: Elsevier Inc.)
Rong Y and Zhu Y 1999 *Computer-Aided Fixture Design* (New York: Marcel Dekker Inc.)
Bgoi Kok Ann B 1990 Computer Aided Design of Modular Fixture Assembly – Ph.D. Thesis (Cristchurch: University of Canterbury)

[2] Bitter R, Mohiuddin T and Nawrocki M 2017 *LabVIEW: Advanced Programming Techniques, Second Edition* (Boca Raton: CRC Press – Taylor & Francis Group)

[3] Bragaru A, Panus V, Dulgheru L and Armeanu A 1982 SEFA-DISROM Sistem si metoda, Volumul I - *Teoria si practica proiectarii dispozitivelor pentru prelucrari pe masini-unelte* (Bucharest: Editura Tehnica)

[4] Bona B 2015 *Reference Frames and Rotations* (Turin: Politecnico di Torino)

[5] Hughes J, van Dam A, McGuire M, Sklar D, Foley J, Feiner S and Akeley K 2013 *Computer Graphics: Principles and Practice* (USA: Addison-Wesley Professional)

Acknowledgments

The authors were inspired to start working on this subject during the support they offered to Mr. Marian Claudiu Ursei during the development of his graduation project. Mr. Ursei, who also developed and presented an own method, showed interest, and appreciated that the subject will be promising.

Optimal selection of milling cutting tools for machining of triangular pockets

A Belloufi, M Abdelkrim, I Rezgui, R Saci, M Mezoudj and M T Amira

Department of Mechanical Engineering, Faculty of Applied Sciences, University Kasdi Merbah Ouargla, 30000 Ouargla, Algeria

abelloufi@yahoo.fr

Abstract. Minimizing machining time is an objective that must be taken into account when choosing the cutting tool. The diameter of the cutting tool is one of the parameters which greatly influences the machining time of different geometric shapes, especially pockets. In this study, a methodology for optimizing the choice of milling cutting tools for machining triangular pockets was proposed. An analytical model expressing the length of cutting tool paths has been optimized using sequential quadratic programming. the study carried out in this work leads to the conclusion that the spiral contour strategy makes it possible to have a minimum length for the machining of triangular shaped pockets.

1. Introduction

Machining by chip removal was the subject of numerous studies since the beginning the twentieth century, which has allowed this process to remain one of the most used in industry mechanical, despite the arrival and development of competing processes such as EDM or, more recently, manufacturing processes by adding material. The machining techniques are constantly changing, in order to maintain their performance at the highest level and to be able to meet new industrial requirements in terms of quality and productivity, but also impact on the environment and health [1].

In industrial mechanics, the manufacture of a part from a quantity of material delivered in the form of semi-finished products (sheets, bars, etc.) requires the implementation of a set of techniques. One of them is machining, that is to say removal of material by a cutting tool. The machining of a part is broken down into a succession of operations, defined by the range of machining established by the methods office from the definition drawing from the design office [1]. Today the machining is carried out, by optimizing the cutting conditions in order to achieve several objectives. Several works are carried out in this direction with various objectives [2-5].

Pocket machining is one of the problems encountered when optimizing milling operations. Several works have tackled the problem of pocket hollowing [6-10].

In general, the recessing of the pockets requires the removal of a large volume of material in roughing or in roughing and finishing combined [1]. From a geometric point of view, a pocket consists of a bottom and a side surface resting on an open or closed contour. The problem of increasing productivity requires that the operation be carried out in minimal time, which in turn requires knowing the geometric shape of the path of the cutting tools used.

Knowledge of the path of cutting tools for pocketing is primarily based on execution methods as well as manufacturing resources.

Firstly, a modelling of the trajectory of cutting tools for machining triangular-shaped pockets using three machining strategies is presented in this study. Subsequently, an optimization strategy consists of minimizing the machining time of a triangular pocket, taking into account the technological constraints that have been proposed.

2. Analytical model of cutting tool path

The parameters of the pocket are: three sides, three angles, three connecting radii. The figure 1 shows the outline of a triangular shaped pocket with the different parameters.

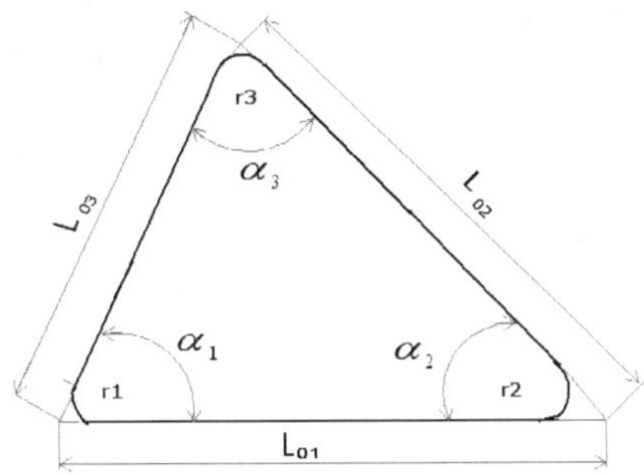

Figure 1. Parameters of a triangular contour pocket

2.1. Zig Zag Strategy

Calculation of Height (h)

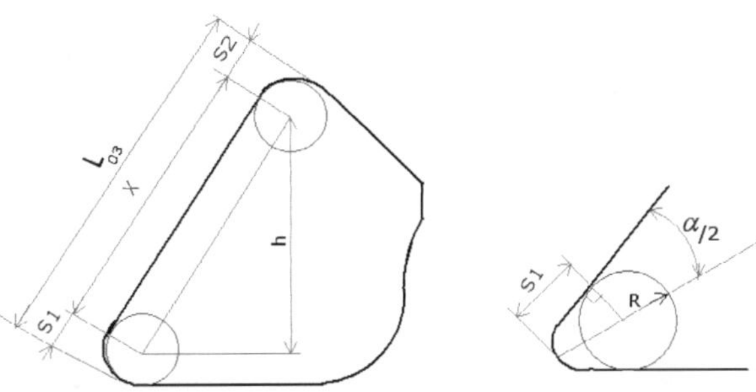

Figure 2. Geometric representation of h

From figure 2:

$$\sin \alpha_1 = \frac{h}{x} \Rightarrow h = x \sin \alpha_1$$

with :

$$x = L_{03} - s_1 - s_2$$

and

$$\operatorname{cotg}\left(\frac{\alpha_1}{2}\right) = \frac{s_1}{R} \Rightarrow s_1 = R \operatorname{cotg}\left(\frac{\alpha_1}{2}\right)$$

the same for s_2 :

$$s_2 = R \operatorname{cotg}\left(\frac{\alpha_3}{2}\right)$$

X becomes

$$x = L_{03} - R \operatorname{cotg}\left(\frac{\alpha_1}{2}\right) - R \operatorname{cotg}\left(\frac{\alpha_3}{2}\right)$$

IOP Conf. Series: Materials Science and Engineering **1018** (2021) 012016 doi:10.1088/1757-899X/1018/1/012016

$$h = \left[L_{03} - R \operatorname{cotg}\left(\frac{\alpha_1}{2}\right) - R \operatorname{cotg}\left(\frac{\alpha_3}{2}\right) \right] \sin\alpha_1 \qquad (1)$$

Calculation of the increment (p)

The number of increments is given by the following equation:

$$n = E\left(\frac{h}{2\rho R}\right) + 1 = E\left[\frac{\left[L_{03} - R \operatorname{cotg}\left(\frac{\alpha_1}{2}\right) - R \operatorname{cotg}\left(\frac{\alpha_3}{2}\right) \right] \sin\alpha_1}{2\rho R} \right] + 1 \qquad (2)$$

with: $\rho < 1$ Machining overlap coefficient.

For the calculation of E there are two cases:

1st case:

$$E\left(\frac{h}{2\rho R}\right) + 1 = \frac{h}{2\rho R} - 1 + 1 \Rightarrow E = 1 - \frac{2\rho R}{h}$$

2nd case:

$$E = \frac{\operatorname{integer}\left(\frac{h}{2\rho R}\right)}{\frac{h}{2\rho R}}$$

The increment "p" between two passes is calculated for an integer "n", it is expressed by the equation:

$$p = \frac{\left[L_{03} - R \operatorname{cotg}\left(\frac{\alpha_1}{2}\right) - R \operatorname{cotg}\left(\frac{\alpha_3}{2}\right) \right] \sin\alpha_1}{n} \qquad (3)$$

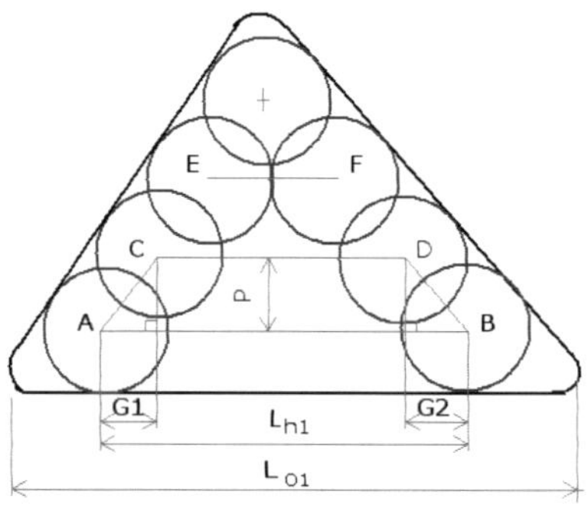

Figure 3. Horizontal segments.

Length of horizontal segments (figure 3)

The number of horizontal segments equals: $n+1$

The total length of the horizontal segments is: $L_h = \sum_{i=1}^{n+1} L_{h\,i}$

with: $L_{h\,1} = L_{01} - R \operatorname{cotg}\left(\frac{\alpha_1}{2}\right) - R \operatorname{cotg}\left(\frac{\alpha_2}{2}\right)$

$L_{h\,2} = L_{h\,1} - G_1 - G_2$

While : $\text{cotg}\ \alpha_1 = \frac{G_1}{\text{p}} \Rightarrow G_1 = \text{p cotg}\ \alpha_1$ et $G_2 = \text{p cotg}\ \alpha_2$

$$\text{L}_{\text{h}\,2} = \left[\text{L}_{01} - R\ \text{cotg}\left(\frac{\alpha_1}{2}\right) - R\ \text{cotg}\left(\frac{\alpha_2}{2}\right) - \text{p}\left(\text{cotg}\ \alpha_1 + \text{cotg}\ \alpha_2\right)\right]$$

$\vdots$

$$\text{L}_{\text{h}\,i} = \left[\text{L}_{01} - R\ \text{cotg}\left(\frac{\alpha_1}{2}\right) - R\ \text{cotg}\left(\frac{\alpha_2}{2}\right) - \text{p}(i-1)\left(\text{cotg}\ \alpha_1 + \text{cotg}\ \alpha_2\right)\right]$$

The total length of the horizontal segments becomes:

$$\text{L}_\text{h} = \sum_{i=1}^{n+1}\left[\text{L}_{01} - R\ \text{cotg}\left(\frac{\alpha_1}{2}\right) - R\ \text{cotg}\left(\frac{\alpha_2}{2}\right) - \text{p}(i-1)\left(\text{cotg}\ \alpha_1 + \text{cotg}\ \alpha_2\right)\right] \quad (4)$$

with : $n = \text{integer}\left(\frac{lv * \sin\alpha_2}{p}\right)$

$$lv = \text{L}_{02} - R * \cot\frac{\alpha_2}{2} - R * \cot\frac{\alpha_3}{2} \quad (5)$$

Calculating the path between increments

$$L_m = \frac{L_{12}}{2} + \frac{L_{13}}{2}$$

$$L_{12} = L_{02} - R\left(\cot g\left(\frac{\alpha_2}{2}\right) + \cot g\left(\frac{\alpha_3}{2}\right)\right)$$

$$L_{13} = L_{03} - R\left(\cot g\left(\frac{\alpha_1}{2}\right) + \cot g\left(\frac{\alpha_3}{2}\right)\right)$$

$$L_m = \frac{1}{2}\left[L_{02} + L_{03} - R\left(\cot g\left(\frac{\alpha_1}{2}\right) + \cot g\left(\frac{\alpha_2}{2}\right) + 2\cot g\left(\frac{\alpha_3}{2}\right)\right)\right] \quad (6)$$

Calculating the contouring path

$$L_c = L_{11} + L_{12} + L_{13}$$

$$L_c = L_{01} + L_{02} + L_{03} - 2R\left(\cot g\left(\frac{\alpha_1}{2}\right) + \cot g\left(\frac{\alpha_2}{2}\right) + \cot g\left(\frac{\alpha_3}{2}\right)\right)$$

The total length of the path of the cutting tool in the inner part " L_{pi} " is given by:

2.2. Parallel contour strategy

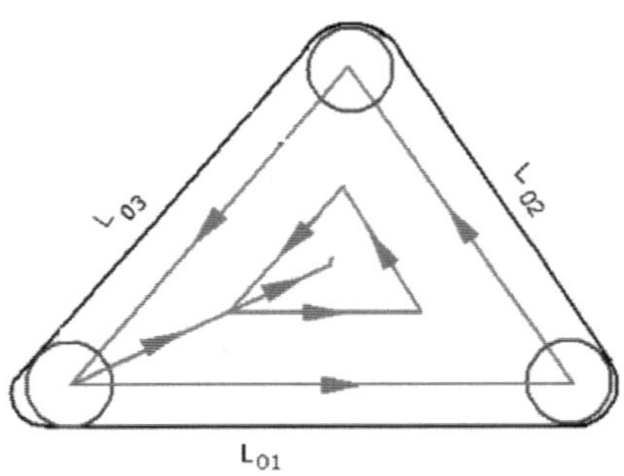

Figure 4. Representation of parallel contours.

From figure 4: $X_t = X \times n$

$\sin\frac{\alpha_i}{2} = \frac{2\rho R}{X}$ hence $X = \frac{2\rho R}{\sin\frac{\alpha_i}{2}}$

X_t : segment of passage from contour to another
n : number of contours minus one

$$L_{pi} = L_{11} + L_{12} + L_{13} + L_{21} + L_{22} + L_{23} + L_{31} + L_{32} + L_{33} + ... + L_{n1} + L_{n2} + L_{n3} + X_t$$

$$L_{ij} = L_{(i-1)j} - Y\left(\cot g\left(\frac{\alpha_j}{2}\right) + \cot g\left(\frac{\alpha_{j+1}}{2}\right)\right) \qquad (7)$$

$$L_{pi} = \sum_{i=1}^{n}\sum_{j=1}^{3} L_{ij} + X_t \qquad (8)$$

To determine the length of the first contour, $Y = R$ is set besides $Y = 2\rho R$

2.3. Spiral outline strategy

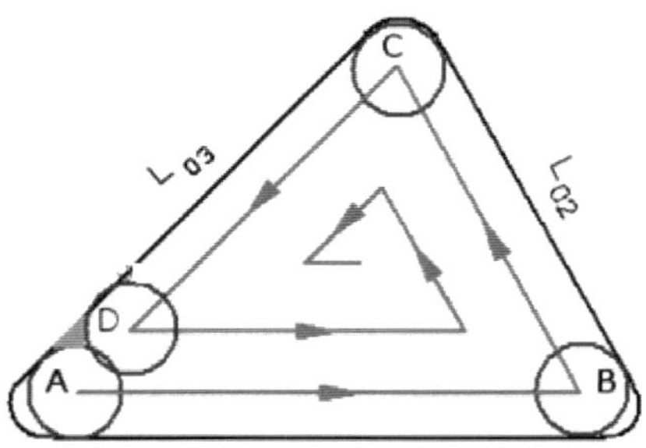

Figure 5. Representation of the spiral trajectory.

From figure 5:

$$L_{pi} = L_{11} + L_{12} + L_{13} + L_{21} + L_{22} + L_{23} + L_{31} + L_{32} + L_{33} + ... + L_{n1} + L_{n2} + L_{n3}$$

$$L_{ij} = L_{(i-1)j} - Y\left(\cot g\left(\frac{\alpha_j}{2}\right) + \cot g\left(\frac{\alpha_{j+1}}{2}\right)\right) + \overline{AD} \qquad (9)$$

$$\overline{AD} = \frac{2\rho R}{\sin\alpha_1}$$

3. Formulation of the optimization model for the choice of cutting tools

The problem of choosing the cutting tool is to determine the tool diameter values for each area, taking into account the limits due to the geometric shape of the pocket to be made, as well as the tool magazines available.

3.1. Choice of model variables

The variables used in the models for optimizing the choice of cutting tools are: the radii of the available cutters.

3.2. Formulation of the objective function

The objective function considered in this study is the cutting time.
The cutting time is calculated by the following relation:

$$t_{pi} = \frac{L_{pi}}{V_f} \qquad (10)$$

IOP Conf. Series: Materials Science and Engineering 1018 (2021) 012016 doi:10.1088/1757-899X/1018/1/012016

3.3. Formulation of constraints

The limits due to the geometric shape of the pocket result in constraints on the determination of the radii.

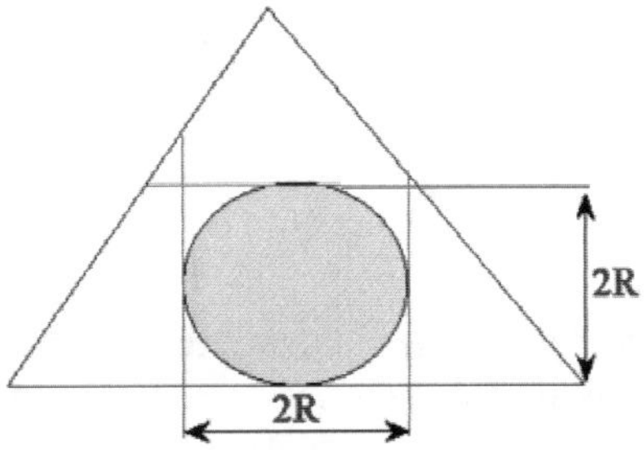

Figure 6. Representation of the limits due to the geometric shape of the triangular pocket.

As it is illustrated in figure 6 the constraints due to the geometrical form are:

$$2R \leq a - 2\left[\cos\alpha_1\left(c - \left(\frac{a}{2}\right)\cos\alpha_1\right)\right] \quad (11)$$

$$2R \leq b - \left(\frac{b}{2}\right)\sin\alpha_2 \quad (12)$$

$$R_1 \leq r_{\min} \quad (13)$$

The available sets of tools impose the following constraint:

$$R_{J\min} \leq R \leq R_{J\max} \quad (14)$$

4. The resolution method

The formulated optimization problem can be solved by sequential quadratic programming (SQP), this method is used for solving general problems of nonlinear optimization with constraints.

$$\begin{cases} \min\limits_{d \in \mathbf{R}^n} t_{pi} \\ sous \quad 2R \leq a - 2\left[\cos\alpha_1\left(c - \left(\frac{a}{2}\right)\cos\alpha_1\right)\right] \\ 2R \leq b - \left(\frac{b}{2}\right)\sin\alpha_2 \\ R_1 \leq r_{\min}; \qquad R_{J\min} \leq R \leq R_{J\max} \end{cases} \quad (15)$$

5. Case study

A mechanical part has a triangular pocket of any contour and identical connection radii. The geometric parameters defining the pocket are presented in Table 1.

Table 1. Geometric parameters of the triangular pocket

Sides (mm)			Angles °			Connections (mm)		
a	b	c	α_1	α_2	α_3	r_1	r_2	r_3
94	85	80	57.82	52.8	69.38	4	4	4

In this example a tool set consisting of 14 tools is considered. Table 2 illustrates the different diameters of the tools as well as the associated numbering (The tools are numbered from 1 to 14 in ascending order of diameter).

Table 2. Cutting tool diameters

N°	1	2	3	4	5	6	7	8	9	10	11	12	13	14
d (mm)	8	10	12	14	16	18	20	22	25	28	30	32	36	40

Table 3 is taken from the Mémotech book [11], where "d" represents the diameter of the tool in mm, "z" the number of teeth of the tool, "f_z" the feed rate per tooth in mm / $tooth$. These values are given for the machining of non-alloy steels: for general use, for heat treatment and for a tool life of 60 min, cutting speed $V_c = 28\ m/\text{min}$. The tool is made of uncoated high speed ARS steel.

Table 3. Cutting parameters

d	8	10	12	14	16	18	20	22	25	28	30	32	36	40
z	4	4	4	4	4	4	4	4	4	5	5	5	6	6
$f_z(10^{-3}\,mm/dent)$	13	20	25	30	35	40	45	50	55	50	50	55	55	55

6. Results and discussion

Figure 7 shows the variation of the total path length of the cutting tool depending on the diameter of the tools used. This length tends towards infinity for tools with diameters close to zero, on the other hand it becomes short while taking tools of large diameters but the most interesting is that the spiral strategy allows to have a minimum length compared to the Zig Zag and parallel contour strategy.

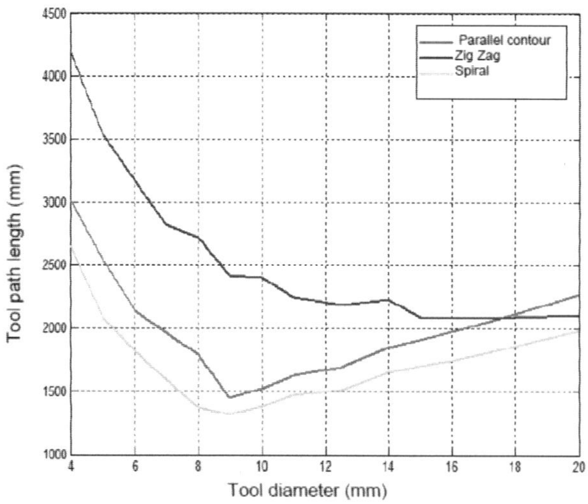

Figure 7. Tool path lengths

As for the tool path lengths, the pocket machining times by the three strategies are very similar, with an advantage for the spiral outline strategy as shown in Figure 8.

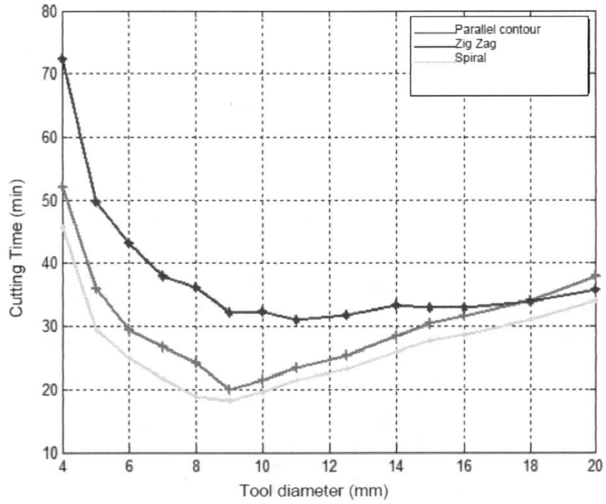

Figure 8. Machining time of the triangular pocket

The results of the optimization are shown in the three tables 4.

Table 4. Minimum machining time and optimum tool radius

Strategy	Minimum machining time (min)	The optimum tool radius (mm)
Zig Zag	19.3664	11
Parallel contour	13.8124	9
Spiral	12.2018	9

7. Conclusion

From the results obtained it is clear that the minimum machining time does not correspond to a large diameter tool but to a tool of optimal diameter which takes into account the technological machining constraints.

The spiral contour strategy is more efficient than the parallel contour strategy in the pocket machining; it makes it possible to reduce the pocket hollowing time, as well as the number of discontinuities in the path of the tool, which means a increased production.

References

[1] Al-Ahmad M 2008 *Industrialisation de Procède : Contribution a La maîtrise de L'opération de tréfilage ou fraisage vertical - Approches analytique et expérimentale*, Thèse doctorat, ENSAM, CER de Metz 2008.

[2] Abdelkrim M, BRIOUA M and Belloufi A 2017 Measurement and numerical simulation of the cutting temperature in cutting tool during turning operation *J. of Eng. Sci. Technol.* **12**(5) 1307 – 1317.

[3] Belloufi A, Abdelkrim M, Bouakba M and Rezgui I 2017 Minimization of energy and surface roughness of the products machined by milling *IOP conf. ser., Mater. sci. eng.* **227**.

[4] Abdelkrim M, Belloufi A, Rezgui and Brabie G 2018 Approach by the response surface method to determine the cutting temperature of AISI 1060 steel *IOP conf. ser., Mater. sci. eng.* **400**(2) 022001.

[5] Xie S and Guo Y 2011 Intelligent selection of machining parameters in multi-pass turnings using a GA-based approach *J. of Computational Information System* **7**(5) 1714-1721.

[6] Park S C and Chung Y C 2002 Offset tool-path linking for pocket machining *Comput. Aided Des.* **34**(4) 299–308.

IOP Conf. Series: Materials Science and Engineering **1018** (2021) 012016 doi:10.1088/1757-899X/1018/1/012016

[7] Park S C, Chung Y C and Choi B K 2003 Contour-parallel offset machining without tool-retractions *Comput. Aided Des.* **35**(9) 841–849.

[8] Zhao Z Y, Wang C Y, Zhou H M and Qin Z 2007 Pocketing toolpath optimization for sharp corners *J. Mater. Process. Technol.* **192–193** 175–180.

[9] Göloðlu C and Arslan Y 2009 Zigzag machining surface roughness modelling using evolutionary approach, *J. Intell. Manuf.* **20**(2) 203–210.

[10] Park S C and Choi B K 2001 Uncut free pocketing tool-paths generation using pair-wise offset algorithm *Comput. Aided Des.* **33** (10) 239-746.

[11] Barlier C and Poulet B 1999 *Mémotech génie mécanique : Productique mécanique*, Deuxième édition – Editions CASTEILLA.

IOP Conf. Series: Materials Science and Engineering **1018**(2021) 012017 doi:10.1088/1757-899X/1018/1/012017

Smart prediction of surface micro-hardness after milling based on fuzzy inference model

M T Amira, A Belloufi, M Abdelkrim, F Abdelkrim and M Mezoudj

Dpartemnt of Mechanical Engineering, Facultu of Applied Sciences, University Kasdi Merbah Ouargla, 30000 Ouargla, Algeria.

E-mail: amiramedtoufik@gmail.com

Abstract. The lack of comprehending and control of the micro-hardness of the machined surface is an important obstacle to the use of the milling process. In order to optimize the machining process by milling, this work has focused on the problem of micro-hardness changing of machined surfaces by milling, which has been the subject of several scientific works. A fuzzy inference model was developed to study the influence of cutting conditions (cutting speed, feed per tooth and depth of cut) on the micro-hardness of machined surfaces by milling. The predicted values, obtained by fuzzy model, are compatible with the experimental values, with an average error percentage of 0.63%.

1. Introduction

During the last decades, many experiments have been carried out to investigate the milling conditions that affect the variation of microhardness of milled surfaces. These experiments are extremely difficult and require a lot of time and material. And here is a small bibliographic search for some scientific works that have studied the microhardness of milled surfaces:

A.L. Mantle et al. [1] studied the surface integrity of a high speed milled gamma titanium aluminide, they found that the microhardness evaluation of the subsurface indicated a hardened layer to a depth of 300 µm.

W. Bouzid Saı et al. [2] studied the influence of machining by finishing milling on surface characteristics, two different materials were used in their study, the carbon steel (CS) and the duplex stainless steel (DSS). For carbon steel samples the maximum value of microhardness is higher and penetrates deeper into the surface layer with the increasing of feed. This is due to an increase in both chip thickness and tool chip contact length. Hence, cutting temperature and cutting forces change in the same direction as the feed. In addition, they noticed also that, microhardness is high at high cutting speeds which confirms that temperatures reach higher levels. No major changes of surface layer depth are noted. They consider that to improve wear and fatigue material resistance, the cutting speed must be high.

Xiaoh Wang [3] did an intelligent prediction of surface micro-hardness after milling based on smooth support vector regression, but he didn't show the results of the predicted microhardness obtained by his model and he didn't specify the accuracy of his model compared to the experimental values of the microhardness.

A. Ginting et al. [4] in their study Surface integrity of dry machined titanium alloys, they found that the microhardness is altered down to 350 μm beneath the machined surface. The first 50 μm is the soft subsurface with microhardness value 8% less than the bulk material hardness caused by thermal softening in the ageing process. Down to 200 μm is the hard subsurface with microhardness value 8% harder than the bulk material hardness caused by the cyclic internal work hardening and then it gradually decreased to the bulk material hardness.

Du Jin et al. [5] investigated the damage of the machined surface and subsurface in orthogonal milling of FGH95 superalloy, they concluded that the machined surface microhardness is higher than the bulk material demonstrates the appearance of work hardening during hard machining of FGH95. And also, higher cutting speeds mean higher surface microhardness.

F. wang et al. [6] studied the effects of cutting conditions on microhardness and microstructure in high-speed milling of H13 tool steel, they concluded that the subsurface microhardness is less high than that on the machined surface. The predominant factor determining the machined surface hardness would be the severe plastic deformation induced by mechanical load. Furthermore, the hardness of machined surface decreases with the increase of cutting speed and feed per tooth due to thermal softening effects.

P. Muñoz-Escalona et al. [7] in their study influence of cutting environments on surface integrity and power consumption of austenitic stainless steel, they found that the specimen machined in a cryogenic environment achieved higher values of hardness in subsurface compared to specimen machined in a dry environment. they reported that this is probably due to strain hardness during the cutting process been more noticeable in the cryogenic environment due to the decrease of the cutting temperature. In the case of the dry environment the increase of temperature probably helped to soften the material easing the removal of material. And they noticed also when the cutting speed or the feed per tooth changed there were no noticeable changes on microhardness, however all the specimens' hardness started to stabilize at 100 μm from the machined surface.

Yang Houchuan et al. [8] investigated the influence of cutting speed and tool wear on the surface integrity of the titanium alloy Ti-1023 during milling, they concluded that the Hardening can occur when machining Ti-1023. The depth of hardening decreases with the increase in cutting speed. The effect of tool wear on hardening is significant. When VB ranges from 0 to 0.2 mm, the depth of hardening increases from 35 to 55 μm. Then, at VB=0.35 mm, the depth of hardening reaches 130 μm. And also softening regions appear on the subsurface after a certain degree of tool wear, and when tool wear increases, the softening regions are more pronounced.

H. Hassanpour et al. [9] In their research, hard milling of 4340 alloy steel using MQL system is investigated and the effects of this process on microhardness are determined by applying RSM design of experiment method (DOE). The analysis of variance showed that quadratic polynomial models estimate the surface microhardness correctly and the average estimated error was 1.18% and the maximum error of the model was 5%. And they found that all hard-milling parameters had an increasing effect on microhardness. Also, feed rate with 73.1%, cutting speed with 14.4% and axial depth of cut with only 5.1% affected microhardness.

W. Akhtar et al. [10] studied the effect of machining parameters on surface integrity in high speed milling of super alloy GH4169/Inconel 718, they used to type of cutting insert, the cemented carbide inserts and whisker reinforced coated ceramic inserts. They found that the work hardening of the work piece surface took place in all the cutting experiments whether for carbide inserts or ceramic inserts, resulting in increased hardness of the machined surface as compared to the bulk material. However, due to wet cutting conditions employed for carbide tools as compared to dry condition for ceramic tools, carbide tools always resulted in comparatively greater surface hardness. However, the difference in the maximum surface hardness value obtained for both types of the tools was not big.

H. Zheng et al. [11] studied the effect of ball milling treatment on the microstructure and microhardness of Nb-25Cr alloy, they found that the microstructure refinement becomes much more significant as ball milling duration is extended, and the microhardness is obviously improved after ball milling and that is related to decreased grain size. And an abrupt drop in hardness was observed in the 48h treated sample due to the emergence of cracks on the surface of the Nb-25Cr alloy.

J. C. Pereira et al. [12] in their study the surface integrity of AISI 1010 and AISI 4340 steels subjected to face milling, they reported that the cutting fluid possesses a beneficial influence on surface integrity, i.e. it slightly reduced residual stresses and microhardness values. And they concluded also that the microhardness distribution beneath the surface of AISI 1010 steel was not altered by changes in any of the investigated parameters, as well as by alterations in the axial depth of cut and the use cutting fluid. And also, they stated that higher microhardness values near the surface of AISI 4340 steel is generated due the increasing of cutting speed and feed per tooth.

Y. Zhenchao et al. [13] in their research the effect of milling parameters on surface integrity in high-speed milling of ultrahigh strength steel, they concluded that Surface microhardness decreases with milling speed and feed per tooth, but increases with milling depth.

W. huang et al. [14] in their comparison of surface integrity and fatigue performance for hardened steel ball-end milled with different milling speeds, they found that the fatigue life for milled samples raise at least 83% compared with that of polished samples. This is due to the high surface residual compressive stress and the enhanced microhardness. Moreover, excessive surface strain hardening will increase surface brittleness (which can accelerate the crack propagation ratio) which is detrimental to fatigue resistance. Different cutting speed causes a difference in degree of strain hardening because of the discrepancy in thermomechanical coupling field.

I. Mathoho et al. [15] investigated the effect of milling parameters on microhardness and microstructure during dry and flood milling of Ti-6Al-4V, they found that the use of flood milling generated a maximum value of microhardness compared to the use of dry milling.

This bibliographic study show that some prediction methods have been developed to solve the problem of the changing in the microhardness of machined surfaces by milling and this for their ability to model different phenomena.

However, the progress on the use of Numerical Approaches to predict the microhardness of machined surfaces still lagging behind the other advances in the industry.

This is why we propose in this work to use the fuzzy logic approach to predict the microhardness and to study the effect of cutting parameters on the microhardness of surfaces machined by milling.

2. Experimental database

The aim of our work is to make a smart prediction of the microhardness of surfaces after milling, so we will create our own model using one of the artificial intelligence methods which is fuzzy logic based on an experimental database, which includes the variation of the micro-hardness according to the cutting parameters (cutting speed, feed per tooth and depth of cut) on which we will develop our model on MATLAB.

The table 1 presents the experimental values of the variation of the microhardness (in HV) as a function of the cutting parameters (cutting speed V_c, feed per tooth f_z and depth of cut a_p).

The microhardness measurement of the surface of the milled steel AISI 1060 was done by the Vickers method with HV3 using the Ultrasonic Portable Hardness Tester 'Ultramatic 2 HV400'.

Table 1. Experimental data.

$a_p (mm)$		**0.25**	**0.5**	**0.75**
$f_z (mm / tooth)$	$V_c (m/\min)$	The Microhardness (HV)		
0.09	**100**	125.6	117.4	129.6
	150	133.4	132.8	136.4
	200	126.4	142.4	130
	250	120.2	130.6	128.6
	300	122	128.6	125.4
0.12	**100**	130.6	127.4	132.2
	150	134.6	137.4	141.2
	200	117.4	138.2	130.3
	250	114.8	131.4	126.6
	300	125.2	122.2	127.8
0.15	**100**	131.6	130.2	130.2
	150	134.6	147.4	136.4
	200	116.8	130.4	131.6
	250	128.4	121.6	125.6
	300	126.2	127.6	127.6
0.18	**100**	125	140.8	138.4
	150	137.2	123.4	132.8
	200	131.4	136.4	124.4
	250	123	125.4	123.8
	300	137.2	135.4	122

3. Fuzzy modelling

3.1. Fuzzy variables

The table 2 represents the limit values of the input and output parameters, for the fuzzy model used, it is to define the universe (Domain) of discourse associated with this study.

Table 2. Limit values for the inputs and the outputs.

Parameter	**MIN value**	**MAX value**
$a_p (mm)$	0.25	0.75
$f_z (mm / tooth)$	0.09	0.18
$V_c (m/\min)$	100	300
$H(HV)$	114.8	147.4

3.2. Fuzzy system

The input (cutting parameters) and output (microhardness) variables of the fuzzy system used for the prediction of the microhardness during milling of AISI 1060 steel are shown in the figure 1.

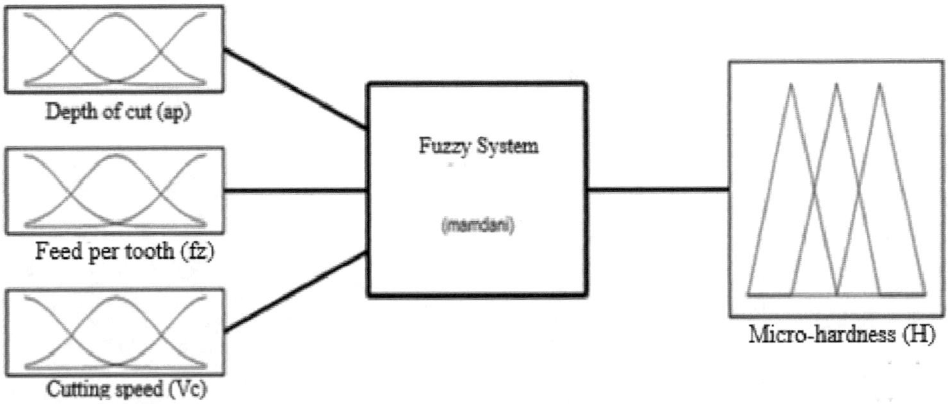

Figure 1. Input and output of the fuzzy system.

3.3. Definition of linguistic variables

In this part we will define all the linguistic variables associated with each parameter used from the experimental basis used as shown in the figures 2, 3 and 4.

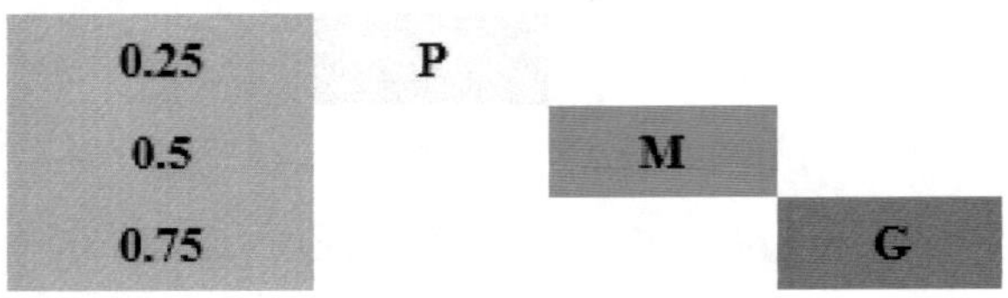

Figure 2. Linguistic variables for depth of cut.

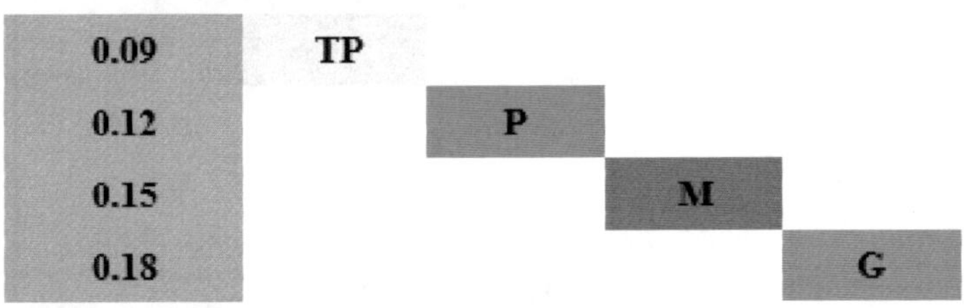

Figure 3. Linguistic variables for feed per tooth.

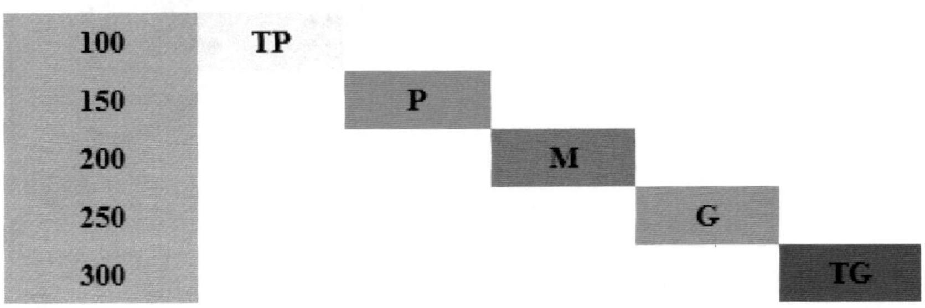

Figure 4. Linguistic variables for the cutting speed.

We will represent the experimental values of the output parameter which is the microhardness for all the experimental tests in the form of a scatter plot in order to define the linguistic variables associated with the microhardness. This representation allows us to divide the universe of discourse into a set of intervals, the purpose of which is to minimize the number of intervals as shown in the figures 5 and 6.

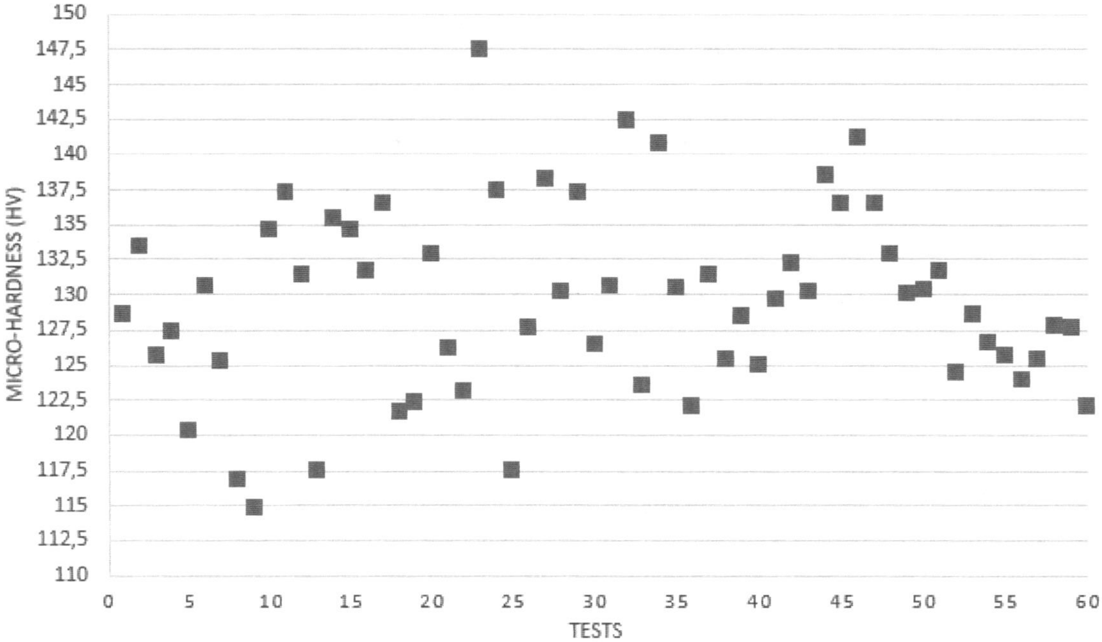

Figure 5. Scatter plot of the microhardness values for all tests.

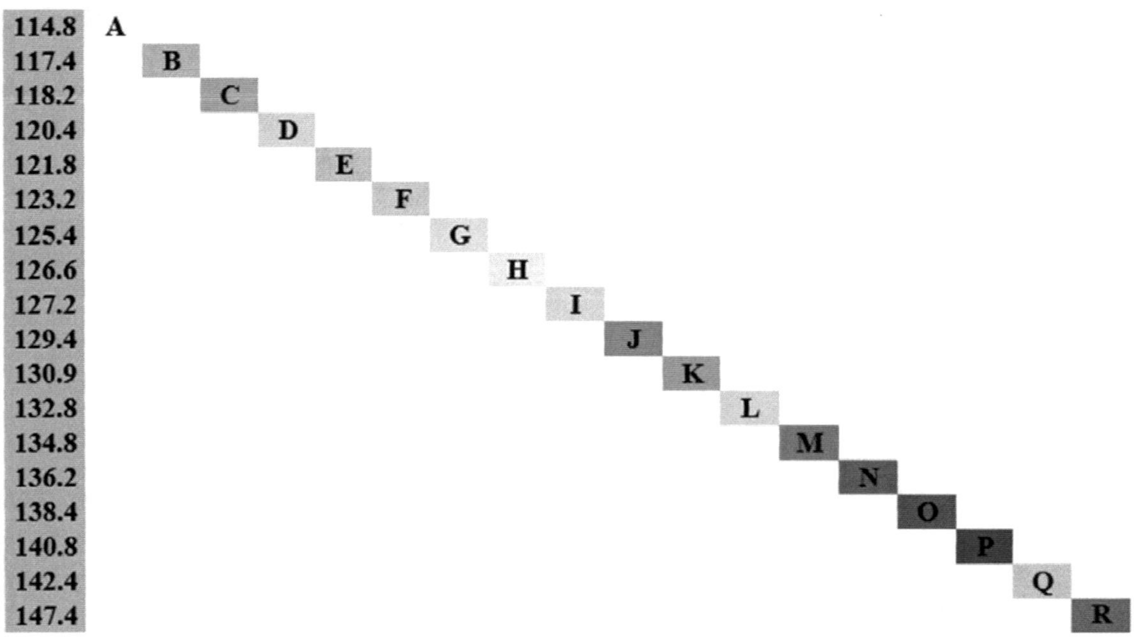

Figure 6. Linguistic variables for microhardness.

3.4. Membership functions

In this study we will use three types of membership function, the Triangular, Trapezoidal and Gaussian function as shown in the figures 7, 8 and 9.

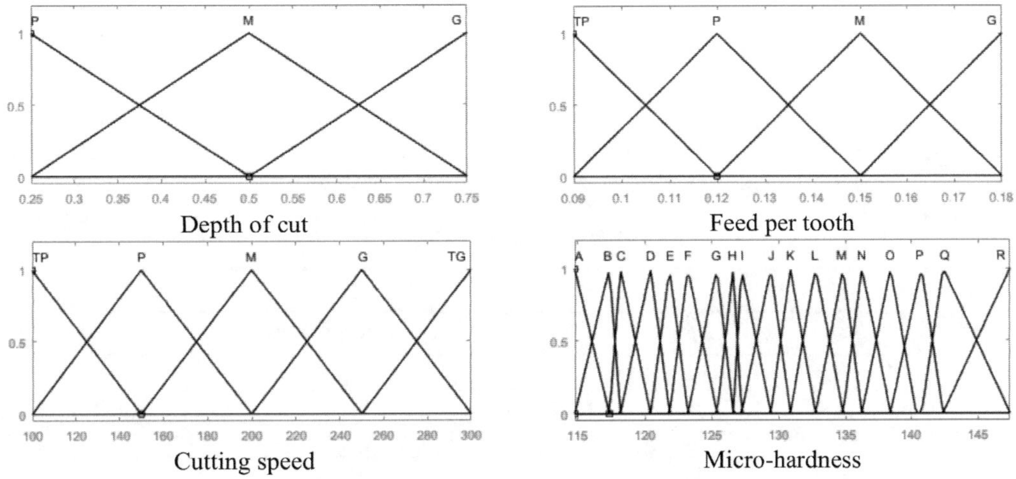

Figure 7. Triangular membership functions.

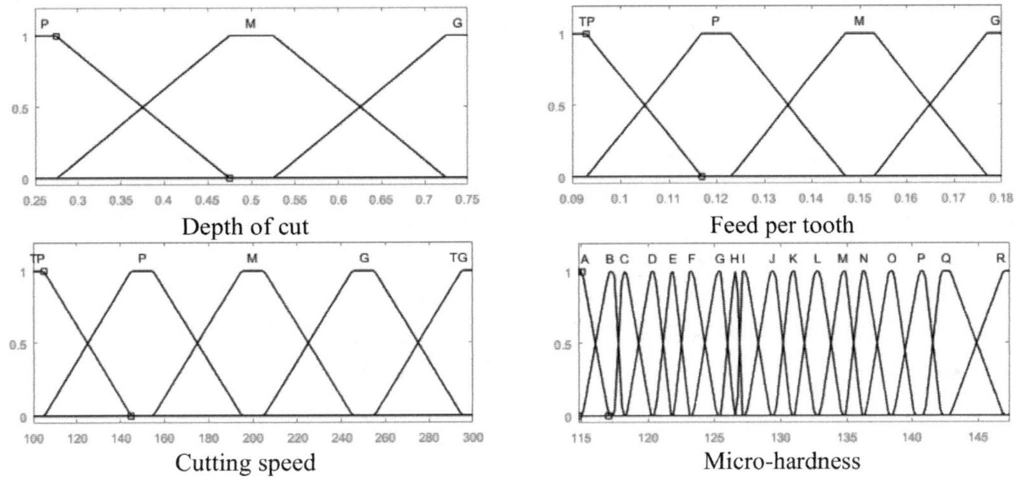

Figure 8. Trapezoidal membership functions.

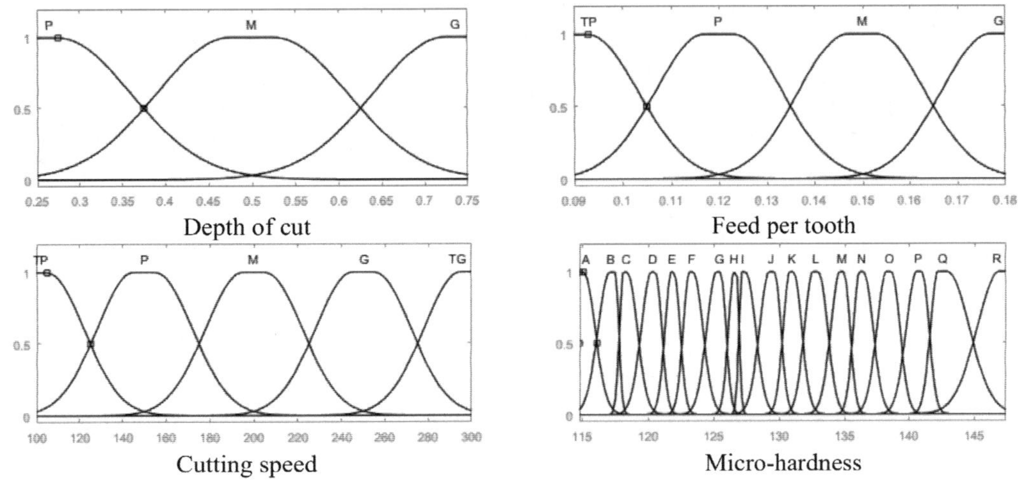

Figure 9. Gaussian membership functions.

3.5. *The fuzzy rules*

The sixty fuzzy rules, in table 3, were established according to the experimental conditions. Each rule takes the following form:

If a_p (linguistic variable) and f_z (linguistic variable) and V_c (linguistic variable) then H (linguistic variable).

The set of fuzzy rules developed are grouped together in the table 3.

Table 3. The fuzzy rules.

a_p		**P**	**M**	**G**
f_z	V_c	The Microhardness		
TP	**TP**	G	B	J
	P	L	L	N
	M	G	Q	J
	G	C	J	I
	TG	E	I	G
P	**TP**	J	I	K
	P	L	N	P
	M	B	N	J
	G	A	K	H
	TG	F	E	I
M	**TP**	K	J	J
	P	L	R	N
	M	A	J	K
	G	I	D	G
	TG	G	I	I
G	**TP**	F	P	O
	P	N	F	L
	M	K	N	F
	G	E	G	F
	TG	N	M	E

4. Results and discussion

4.1. *Results*

Now we are going to transform the linguistic values from the fuzzy model to numerical values, it is the defuzzification which represents the last step. The figure 10 represents the superposition of the values of the experimental microhardness and the values of the microhardness obtained with the three membership functions (Triangular, Trapezoidal and Gaussian):

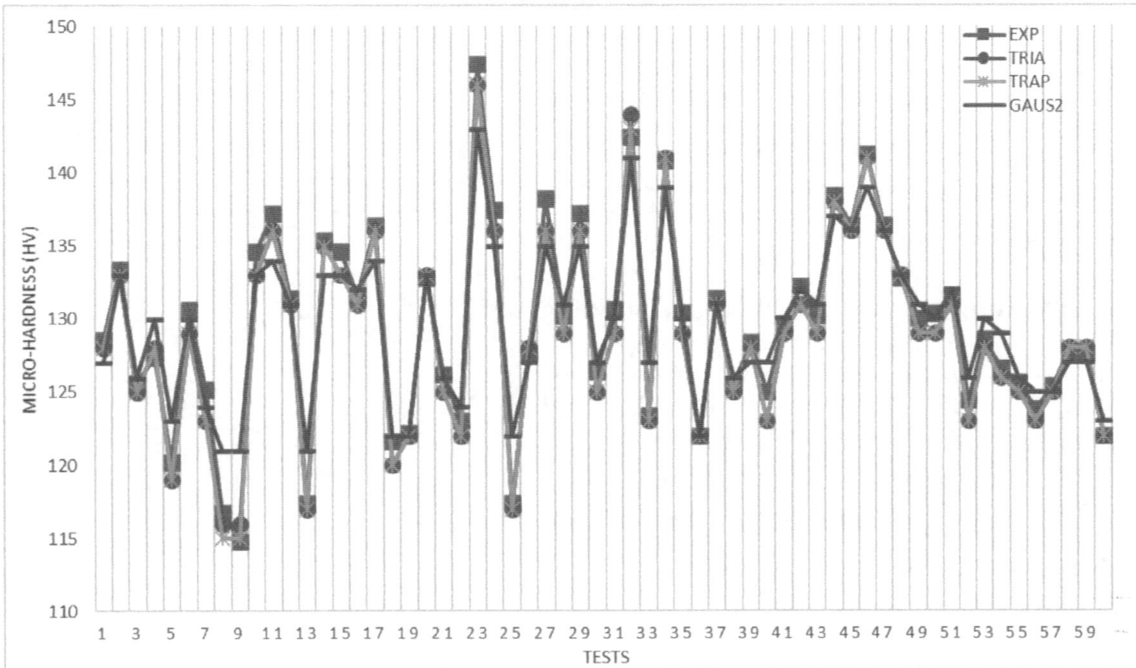

Figure 10. Variation of microhardness (Experimental, Trapezoidal, Triangular and Gaussian) according to the number of tests.

4.2. Validation of the developed model

Figure 10 shows the superposition between the predicted surface microhardness values and the experimental values.

We notice a similarity between the microhardness values using the three types of membership functions, and we notice that the Triangular and Trapezoidal membership functions are almost identical, so to make the choice between the three type of the membership functions: Triangular, Trapezoidal and Gaussian, we propose to use the standard deviation in order to define the values closest to the experimental results.

To calculate the standard deviation, the formula 1 is used:

$$\sigma = \left(\frac{1}{N} \sum_{i=1}^{N} (H_{pred} - H_{\exp})^2 \right)^{\frac{1}{2}} \tag{1}$$

$H_{\exp}$: The experimental microhardness.

H_{pred} : The predicted microhardness.

In our case $N = 60$ Tests.

- The standard deviation for the Triangular-type membership function is 1.0113.
- The standard deviation for the Trapezoidal-type membership function is 1.0030.
- The standard deviation for the Gaussian-type membership function is 1,9492.

We notice from the standard deviation that the results of the Trapezoidal membership function are closest to the experimental results compared to the results of the Triangular and Gaussian membership functions.

4.3. Study of the accuracy and error of the fuzzy system

To calculate the percentage error of the 60 tests, the formula 2 is used:

IOP Conf. Series: Materials Science and Engineering **1018**(2021) 012017 doi:10.1088/1757-899X/1018/1/012017

$$e_i = \frac{1}{N}\sum_{i=1}^{N}\left[\frac{\left|H_{\exp} - H_{pred}\right|}{H_{\exp}}\right]\times 100 \tag{2}$$

$H_{\exp}$: The experimental microhardness.

H_{pred} : The predicted microhardness (Trapezoidal).

In our case $N = 60$ Tests.

e_i : Error rate

To calculate the percentage of the accuracy of the 60 tests, the formula 3 is used:

$$A = \frac{1}{N}\sum_{i=1}^{N}\left[1 - \frac{\left|H_{\exp} - H_{pred}\right|}{H_{\exp}}\right]\times 100 \tag{3}$$

$H_{\exp}$: The experimental microhardness.

H_{pred} : The predicted microhardness (Trapezoidal).

In our case $N = 60$ Tests.

A : Accuracy

All results are grouped together in the table 4.

Table 4. The fuzzy system results.

Tests	Cutting parameters			Microhardness results			
	V_c	f_z	a_p	$H_{\exp}$	H_{trap}	**Error %**	**Accuracy %**
1	300	0.09	0.5	128.6	128	0.47	99.53
2	150	0.09	0.25	133.4	133	0.30	99.70
3	100	0.09	0.25	125.6	125	0.48	99.52
4	100	0.12	0.5	127.4	128	0.47	99.53
5	250	0.09	0.25	120.2	119	1.00	99.00
6	250	0.09	0.5	130.6	129	1.23	98.77
7	300	0.12	0.25	125.2	123	1.76	98.24
8	200	0.15	0.25	116.8	115	1.54	98.46
9	250	0.12	0.25	114.8	115	0.17	99.83
10	150	0.15	0.25	134.6	133	1.19	98.81
11	300	0.18	0.25	137.2	136	0.87	99.13
12	250	0.12	0.5	131.4	131	0.30	99.70
13	100	0.09	0.5	117.4	117	0.34	99.66
14	300	0.18	0.5	135.4	135	0.30	99.70
15	150	0.12	0.25	134.6	133	1.19	98.81
16	100	0.15	0.25	131.6	131	0.46	99.54
17	200	0.18	0.5	136.4	136	0.29	99.71
18	250	0.15	0.5	121.6	120	1.32	98.68
19	300	0.12	0.5	122.2	122	0.16	99.84
20	150	0.09	0.5	132.8	133	0.15	99.85
21	300	0.15	0.25	126.2	125	0.95	99.05
22	250	0.18	0.25	123	122	0.81	99.19
23	150	0.15	0.5	147.4	146	0.95	99.05
24	150	0.12	0.5	137.4	136	1.02	98.98
25	200	0.12	0.25	117.4	117	0.34	99.66

Table 4. Continued.

Tests	Cutting parameters			Microhardness results			
	V_c	f_z	a_p	$H_{\exp}$	H_{trap}	**Error %**	**Accuracy %**
26	300	0.15	0.5	127.6	128	0.31	99.69
27	200	0.12	0.5	138.2	136	1.59	98.41
28	100	0.15	0.5	130.2	129	0.92	99.08
29	150	0.18	0.25	137.2	136	0.87	99.13
30	200	0.09	0.25	126.4	125	1.11	98.89
31	100	0.12	0.25	130.6	129	1.23	98.77
32	200	0.09	0.5	142.4	143	0.42	99.58
33	150	0.18	0.5	123.4	123	0.32	99.68
34	100	0.18	0.5	140.8	141	0.14	99.86
35	200	0.15	0.5	130.4	129	1.07	98.93
36	300	0.09	0.25	122	122	0.00	100.00
37	200	0.18	0.25	131.4	131	0.30	99.70
38	250	0.18	0.5	125.4	125	0.32	99.68
39	250	0.15	0.25	128.4	128	0.31	99.69
40	100	0.18	0.25	125	123	1.60	98.40
41	100	0.09	0.75	129.6	129	0.46	99.54
42	100	0.12	0.75	132.2	131	0.91	99.09
43	100	0.15	0.75	130.2	129	0.92	99.08
44	100	0.18	0.75	138.4	138	0.29	99.71
45	150	0.09	0.75	136.4	136	0.29	99.71
46	150	0.12	0.75	141.2	141	0.14	99.86
47	150	0.15	0.75	136.4	136	0.29	99.71
48	150	0.18	0.75	132.8	133	0.15	99.85
49	200	0.09	0.75	130	129	0.77	99.23
50	200	0.12	0.75	130.3	129	1.00	99.00
51	200	0.15	0.75	131.6	131	0.46	99.54
52	200	0.18	0.75	124.4	123	1.13	98.87
53	250	0.09	0.75	128.6	128	0.47	99.53
54	250	0.12	0.75	126.6	126	0.47	99.53
55	250	0.15	0.75	125.6	125	0.48	99.52
56	250	0.18	0.75	123.8	123	0.65	99.35
57	300	0.09	0.75	125.4	125	0.32	99.68
58	300	0.12	0.75	127.8	128	0.16	99.84
59	300	0.15	0.75	127.6	128	0.31	99.69
60	300	0.18	0.75	122	122	0.00	100.00
The average						**0.63 %**	**99.37 %**

The average error rate is 0.63% and the average accuracy is 99.37%, that's mean that our prediction model based on fuzzy logic works correctly and with high accuracy and can be used as a solution to predict microhardness before starting milling provided that we respect a very specific range of parameters (defined by the universe of discourse) when using this model.

4.4. Graphical representation of results

Figure 11 shows the functions obtained using fuzzy logic simulation as follows:

- The surface (a) represents the variation of the surface microhardness as a function of the feed per tooth and the depth of cut for a cutting speed 100 m / min.
- The surface (b) represents the variation of the surface microhardness as a function of the depth of cut and the cutting speed for a feed per tooth 0.09 mm / tooth.
- The surface (c) represents the variation of the surface microhardness as a function of the feed per tooth and the cutting speed for a depth of cut 0.75 mm.

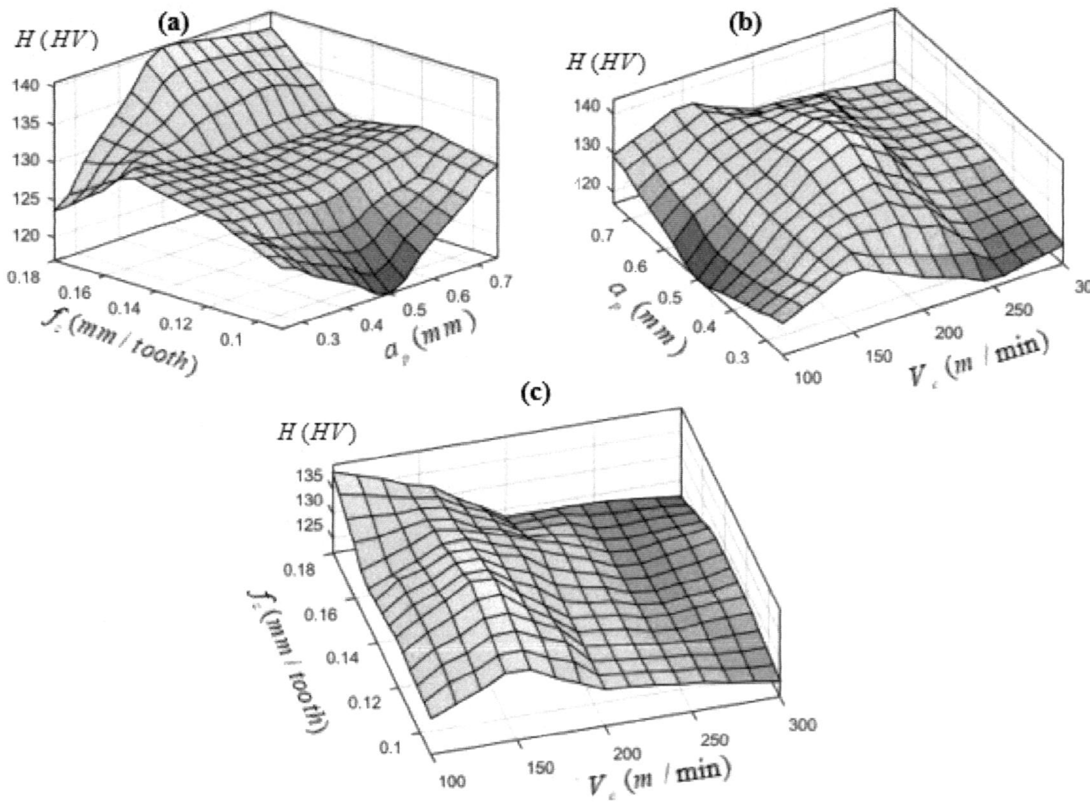

Figure 11. Variation of the predicted micro-hardness with the fuzzy logic model as a function of the cutting parameters.

Figure 11 illustrates the influence of cutting parameters on the microhardness of machined surfaces during milling operations.

From the figure 11 (a) it can be seen that the value of the microhardness becomes maximum for maximum values of the feed per tooth and average values of the depth of cut (0.5 mm). While the minimum values of microhardness are obtained for minimum values of feed per tooth and average values of depth of cut.

On the other hand, a higher feed per tooth leads to an increase in the microhardness of the machined surfaces.

From the figure 11 (b) it can be seen that the value of the microhardness becomes minimum for minimum values of cutting speed and average values of the depth of cut (0.5 mm) and also for maximum values of cutting speed and minimum values of the depth of cut. While the maximum values of microhardness are obtained for average values of cutting speed (200 m / min) and average values of depth of cut.

On the other hand, a higher cutting speed leads to a decrease in the microhardness of the machined surfaces regardless of the depth of cut.

From Figure 11 (c) it can be seen that the value of the microhardness becomes maximum for an average value of feed per tooth (0.12 mm / tooth) and a low value of cutting speed (150 m / min). While the minimum values of microhardness are obtained for maximum values of feed per tooth and cutting speed.

On the other hand, a higher cutting speed results in a decrease in the microhardness of the machined surfaces regardless of the feed per tooth.

5. Conclusion

In this work an approach based on fuzzy logic was carried out in order to predict the effects of cutting conditions on the micro-hardness of surfaces machined by milling. The main results of this work can be summarized as follows:

- The micro-hardness of the machined surface is much higher than that of the raw material.
- The severe plastic deformation induced by a mechanical load (the effect of the cutting tool), which varies according to the cutting parameters, is the predominant factor determining the microhardness of the machined surface by milling.
- The microhardness of the milled surface decreases with the increasing of cutting speed and Depth of cut due to the thermal softening effects.
- The change in micro-hardness results from the changes in microstructure due to the effects of the mechanical-thermal coupling.
- Work hardening induced by plastic deformation of the surface should be the predominant factor in the micro-hardness of the milled surface.
- The fuzzy model, developed during this study, makes it possible to predict the micro-hardness of surfaces machined by milling.
- The study of the fuzzy model efficiency was confirmed by the analysis of the error / accuracy ratio.
- The predicted values are in good agreement with the experimental values, with an average percentage error of 0.63% for the micro-hardness calculation of surfaces machined by milling.
- The approach based on fuzzy logic can be used also to predict other phenomena of milling process like cutting temperature and roughness.

References

[1] Mantle A L and Aspinwall D K 2001 Surface integrity of a high speed milled gamma titanium aluminide *J. Mater. Process. Technol.* **118** 143–50

[2] Bouzid Saï W, Ben Salah N and Lebrun J L 2001 Influence of machining by finishing milling on surface characteristics *Int. J. Mach. Tools Manuf.* **41** 443–50

[3] Wang X 2008 Intelligent prediction of surface micro-hardness after milling based on smooth support vector regression *Proc. - 2008 Int. Symp. Knowl. Acquis. Model. KAM 2008* 728–31

[4] Ginting A and Nouari M 2009 Surface integrity of dry machined titanium alloys *Int. J. Mach. Tools Manuf.* **49** 325–32

[5] Jin D and Liu Z 2013 Damage of the machined surface and subsurface in orthogonal milling of FGH95 superalloy *Int. J. Adv. Manuf. Technol.* **68** 1573–81

[6] Wang F, Zhao J, Li A and Zhang H 2014 Effects of cutting conditions on microhardness and microstructure in high-speed milling of h13 tool steel *Int. J. Adv. Manuf. Technol.* **73** 137–46

[7] Muñoz-Escalona P, Shokrani A and Newman S T 2015 Influence of cutting environments on surface integrity and power consumption of austenitic stainless steel *Robot. Comput. Integr. Manuf.* **36** 60–9

[8] Houchuan Y, Zhitong C and ZiTong Z 2015 Influence of cutting speed and tool wear on the surface integrity of the titanium alloy Ti-1023 during milling *Int. J. Adv. Manuf. Technol.* **78** 1113–26

[9] Hassanpour H, Sadeghi M H, Rasti A and Shajari S 2016 Investigation of surface roughness, microhardness and white layer thickness in hard milling of AISI 4340 using minimum

quantity lubrication *J. Clean. Prod.* **120** 124–34
[10] Akhtar W, Sun J and Chen W 2016 Effect of Machining Parameters on Surface Integrity in High Speed Milling of Super Alloy GH4169/Inconel 718 *Mater. Manuf. Process.* **31** 620–7
[11] Zheng H, Cheng S, Shu X and Li G 2017 Effect of ball milling treatment on the microstructure and microhardness of Nb-25Cr alloy *Int. J. Refract. Met. Hard Mater.* **66** 57–62
[12] Pereira J C C, Rodrigues P C M and Abrão A M 2017 The surface integrity of AISI 1010 and AISI 4340 steels subjected to face milling *J. Brazilian Soc. Mech. Sci. Eng.* **39** 4069–80
[13] Zhenchao Y, Yang X, Yan L, Jin X and Quandai W 2018 The effect of milling parameters on surface integrity in high-speed milling of ultrahigh strength steel *Procedia CIRP* **71** 83–8
[14] Huang W, Zhao J, Niu J, Wang G and Cheng R 2018 Comparison in surface integrity and fatigue performance for hardened steel ball-end milled with different milling speeds *Procedia CIRP* **71** 267–71
[15] Mathoho I, Titilayo Akinlabi E, Patrick Mubiayi M and Mbohwa C 2018 Effect of milling parameters on microhardness and microstructure during dry and flood milling of Ti-6Al-4V *IOP Conf. Ser. Mater. Sci. Eng.* **423**

IOP Conf. Series: Materials Science and Engineering **1018**(2021) 012018 doi:10.1088/1757-899X/1018/1/012018

Experimental study towards WIG welded joint. Case study: Repair by welding of X2CrNiMo17-12-2 pipes

F Ionescu, Ghe Solomon, I Voiculescu, G Gârleanu and D Gârleanu

University POLITEHNICA of Bucharest, Faculty of Industrial Engineering and Robotics, Quality Engineering and Industrial Technologies Department, 313 Splaiul Independentei, 060042 Bucharest, Romania

E-mail: florina.ionescu89@yahoo.com

Abstract. Nowadays, several technological problems were raised by industrial beneficiaries regarding spiral and pipe heat exchangers repairs and thus studies were done in this regard. X2CrNiMo17-12-2, austenitic stainless steel pipes were considered for the experiments on which artificial defects, simulating functioning defects, were done. In order to perform the repair activities on the designed defects, WIG welding process was used and an experimental stand was created. This paper presents some of the difficulties that can occur during the repair by welding of pipes and the defects identified using destructive and non-destructive testing on the probes and samples created for the experimental study scope. Significant conclusions can be drawn; up to a distance of about 100 - 150 mm from the access end, the repair has a good appearance with no serious imperfections.

1. Introduction

In engineering practice, the selection of construction materials is made according to the equipment for which they are designated, and the requirements imposed in operation. Nowadays, small pipes are increasingly used both in the construction of various equipment, such as heat exchangers, and for transporting fluids [1, 2].

The following main problems must be considered when performing welding repairs on an in-service pipe [1, 2, 3]:

a. Avoiding the burning through of the pipe wall by the electric arc used as thermic source while welding;
b. Avoiding the hydrogen induced cracking, also known as cold cracking or delayed cracking, which is strongly activated by the accelerated cooling of the pipe wall in the welding area, produced by the circulation with a certain speed of the fluid under pressure in the pipe.

When designing the welding technology of the pipes under pressure must first pay attention to the use of low hydrogen electrodes and to establish correctly the conditions for carrying out the welding processes with low levels of diffusible hydrogen content in the welded joints [3, 4]. Minimizing measures must also be applied when there exists the possibility of sensitive microstructures cracking forming in the welded joint [3,4].

This paper presents some of the problems that can occur when repairing by welding of pipes, what type of imperfection can appear, and the small differences in hardness measurements.

The reason why choosing the pipes for the experimental study is in regards to the spiral heat exchangers. At the moment, spiral heat exchangers repair involves stripping, cutting from the outside to the inside of the exchanger sheets until the defective sheet is reached. The current steps of performing the repair are: identifying the affected area; cutting the sheets from the outside to the inside to the affected area; repair by welding; the control of the repaired area; the welding of previously made cuts; the welding control of the cuts; and pressure testing. The method is easy to apply, but it has several disadvantages as the uncovering of the good sheets, which do not have a defect; the defects are at the spire / spirals of small diameter, so those inside and this means that we will uncover more than half of the sheets to reach the defective sheet; if the defects are disposed in several areas along the entire diameter of a turn and it is positioned inside the exchanger, it means that the cutting is executed several times in different areas and the repair costs are very high and not justified [8]. Performing low-access joints repair inside the pipes is similar to the spiral heat exchanger low-access repair and so, developing of new repair method for the equipment mentioned.

2. Methods and materials

The basic material used in the experimental plan was the X2CrNiMo17-12-2 austenitic stainless steel, which is a version of X5CrNiMo17-12-2 and is distinguished by a lower carbon content, but also by a lower mechanical property.

In Table 1 it is presented the chemical composition of X2CrNiMo17-12-2 austenitic stainless steel in accordance with the standard EN ISO 100088-3, and the mechanical properties of X2CrNiMo17-12-2 steel are presented in Table 2.

Table 1. Chemical composition of X2CrNiMo17-12-2 stainless steel [6]

EN ISO 10088-3	*C* *[%]*	*Mn* *[%]*	*P* *[%]*	*S* *[%]*	*Si* *[%]*	*Cr* *[%]*	*Ni* *[%]*	*Mo* *[%]*	*N* *[%]*
X2CrNiMo17-12-2	0,03	2	0,045	0,03	0,75	16 – 18	10 – 14	2 – 3	0,10

Table 2. Mechanical properties of X2CrNiMo17-12-2 stainless steel [6]

EN ISO 10088-3	Tensile Strength [MPa]	Yield strength [MPa]	Elongation [min, %]	Hardness	
				Brinell [HB]	Rockwell [HRB]
X2CrNiMo 17-12-2	485	170	40	217	95

The probes were repaired by welding using the WIG welding process with filler material. The filler material has the same chemical and mechanical properties as the basic material and the shielding gas was Ar 100%. The parameters of welding regime used for all the probes during the experiments were set according to the manufacturer's recommendations, monitored during the entire welding process and their values are presented in Table 3.

Table 3. Welding parameters

Parameter	Value
Welding amperage I_s [A]	137±5
Welding arc voltage U_a [V]	14.2±2
Feed rate of welding wire v_a [m/min]	1.8±1
Gas discharge, Ar 100%, D_g [l/min]	13±1
Welding speed v_s [cm/min]	12-15

3. Results and discussions

In order to carry out the experiments, three X2CrNiMo17-12-2 austenitic stainless-steel pipes were prepared with the following dimensions: 250 mm length, 3 mm thickness and 42 mm diameter. The cutting process of the pipes was done using a manual cooling cutting machine for avoiding structural changes of the austenitic stainless steel.

In order to validate the welding repair technology, an experimental program was developed for the realization of welded seam on the areas where the artificial defect was processed (holes with the diameters of 2, 3 and 4 mm, respectively) inside pipes. The holes were drilled at different distances from the access end, respectively at 50 mm, 100 mm, 150 mm and 200 mm. The probes prepared are presented in figure 1. The experimental stand used for the repair by welding process is presented in Figure 2.

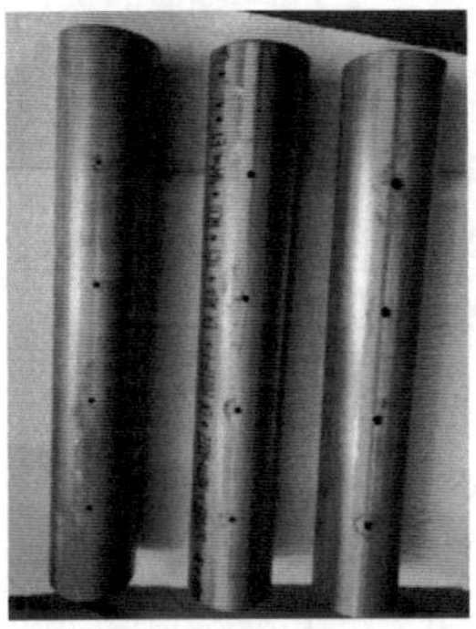

Figure 1. Prepared probes

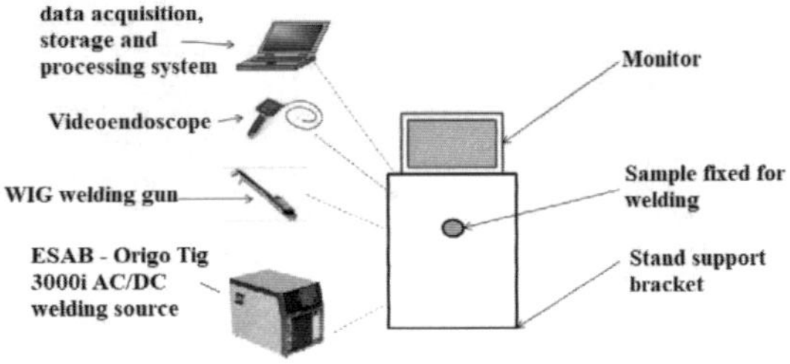

Figure 2. Stand for the repair by welding process

The probes were fixed for welding in the experimental stand realized and all the welded seams were executed by inserting the welding torch into the pipe and monitoring the molten metal bath using video-endoscopic cameras. The welding source used was ESAB Origo Tig 3000i AC/DC, the welding process was displayed on the monitor during the entire repair activity and images were stored using data acquisition, storage and processing system. Figure 3 shows images taken during welding on the pipe using the experimental stand and the welding obtained are presented in figure 4.

Figure 3. Welding process

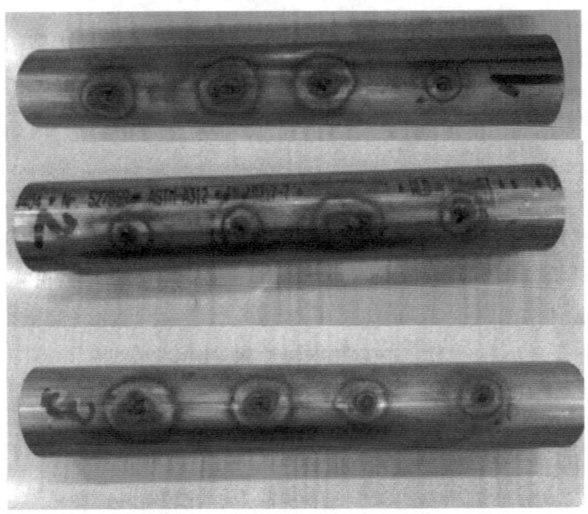

Figure 4. Obtained probes

After cooling, the probes were non-destructively examined using the following methods: visual testing penetrant testing and radiographic testing. The visual examination applied on the samples was made by direct visual testing and remote visual testing. For the detection of small imperfections on the surface of the samples (which cannot be visually detected) or in the immediate proximity of the surface, the penetrant testing was used. The examination of the samples with radiographic testing was performed using X-ray radiography installation and obtaining of the image on radiographic film [4].

After the non-destructive examination, there were found several imperfections. The imperfection found in the probes are presented in table 4.

Table 4. Imperfection found in the welded seam

Non-destructive method used	Code according to ISO 6520	Imperfection name	Probes
Visual testing	506	overlap	P1, P2, P3
	601	arc strike	P1, P2, P3
	602	spatter	P1, P2, P3
Penetrant testing	402	incomplete penetration (lack of penetration)	P1, P2, P3
	504	excessive penetration	P3
	5062	root overlap	P1, P2, P3
	509	sagging	P1, P2, P3
Radiographic testing	300	solid inclusion	P1, P2, P3
	2011	gas pore	P1
	2015	elongated cavity	P2, P3

During the inside visual examination, it could be observed that up to a distance of about 100 - 150 mm from the access end, the weld has a good appearance with no serious imperfections but at a distance of 200 mm, metal drops can be observed and also burns.

In order to assess the welding repairs quality and eventually inner defects that cannot be identified during previous nondestructive examination, probes P1 and P2 were taken to prepare samples for the destructive tests. Eight samples were prepared using mechanical cutting under cooling fluid as follows: Probe 1 – sample 1.1, 1.2, 1.3 and 1.4 and Probe 2 – sample 2.1, 2.2, 2.3 and 2.4 (2 mm and 3 mm holes placed at different distances from the access end, at 50 mm, 100 mm, 150 mm and 200 mm respectively). The sampled area was subjected to the grinding process and polishing in order to perform the hardness tests. The tests were performed on surfaces sanded with abrasive papers (Grit 400 - 600) and polished with alpha alumina suspension with a grain size of 0.25 μm, using the Shimadzu HMV 2T Micro hardness tester see figure 5, and the hardness values are presented in table 5.

Figure 5. Shimadzu HMV 2T Micro hardness tester

Table 5. Hardness values

Macroscopic image of the samples	Hardness values, HV0,5		Average value, HV0,5	Standard deviation	Coefficient of variation
	**MB 1.1**	**177, 177, 182, 189, 186**	**182**	**5,36**	**2,94**
	Welded seam	219, 223, 253, 236, 235	233	13,31	5,71
	HAZ	220, 208, 213, 193, 195	206	11,61	5,64
	MB 1.2	**188, 192, 191, 187, 191**	**190**	**2,17**	**1,14**
	Welded seam- first layer	187, 183, 177, 172, 171	178	6,93	3,89
Samples 1.1 and 1.2	Welded seam- last layer	228, 216, 197, 198, 224	213	14,45	6,80
	HAZ	165, 158, 168, 175, 161	165	6,58	3,98
	**MB 1.3**	**183, 188, 195, 204, 209**	**196**	**10,80**	**5,52**
	Welded seam	212, 215, 201, 196, 195	204	9,20	4,52
	HAZ	175, 172, 170, 180, 181	176	4,83	2,75
	MB 1.4	**181, 183, 180, 179, 188**	**182**	**3,56**	**1,96**
	Welded seam	185, 189, 193, 201, 194	192	5,98	3,11
Samples 1.3 and 1.4	HAZ	192, 193, 191, 178, 193	189	6,43	3,39
	**MB 2.1**	**185, 195, 193, 195, 189**	**191**	**4,34**	**2.27**
	Welded seam	236, 237, 237, 237, 229	235	3,49	1,49
	HAZ	183, 181, 179, 170, 180	179	5,03	2,82
	MB 2.2	**186, 190, 184, 190, 190**	**188**	**2,83**	**1,50**
	Welded seam	213, 221, 239, 216, 224	223	10,11	4,54
Samples 2.1 and 2.2	HAZ	177, 184, 176, 186, 180	181	4,34	2,40
	**MB 2.3**	**186, 184, 187, 190, 197**	**189**	**2,83**	**1,56**
	Welded seam	237, 225, 237, 229, 231	232	5,22	2,25
	HAZ	167, 173, 176, 178, 178	174	4,62	2,65
	MB 2.4	**186, 184, 187, 190, 197**	**189**	**2,83**	**1,56**
	Welded seam	239, 230, 240, 257, 241	241	9,76	4,04
Samples 2.3 and 2.4	HAZ	179, 173, 176, 186, 184	180	5,41	3,01

The five successive measurements were performed, in line, on the characteristic areas (parent metal, welding seam and heat affected zone - HAZ) with distances of about 500μm between fingerprints, using a testing force of 4,903N applied for 10 seconds. In figures 6 ÷ 13 are presented the hardness graphs.

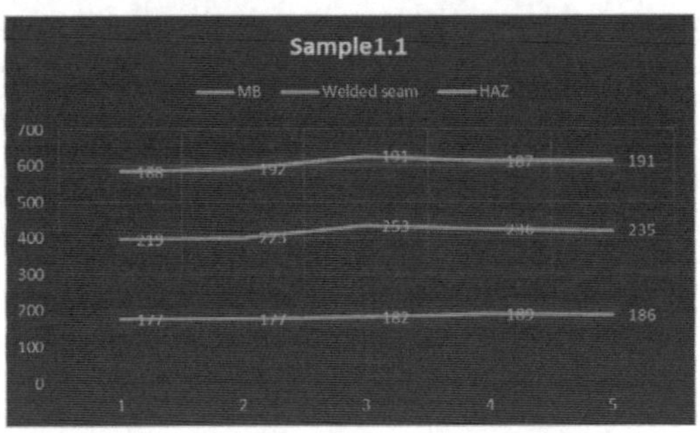

Figure 6. Hardness graph for sample 1.1

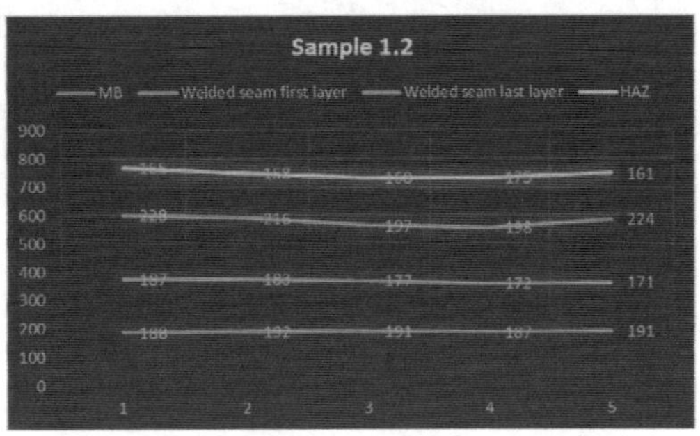

Figure 7. Hardness graph for sample 1.2

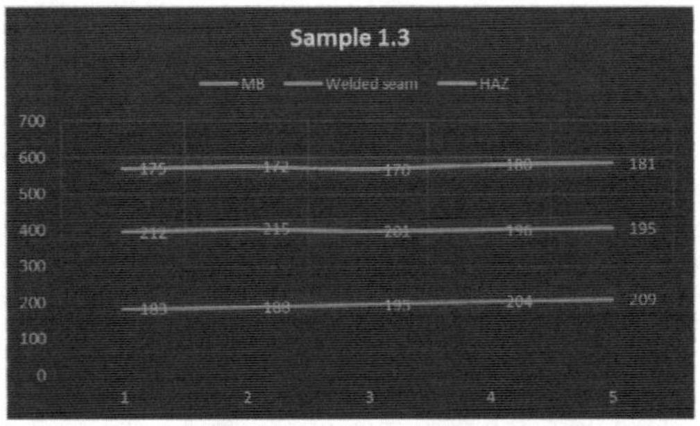

Figure 8. Hardness graph for sample 1.3

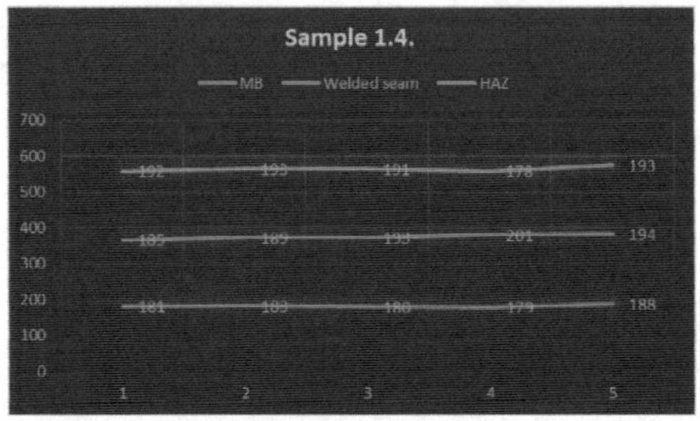

Figure 9. Hardness graph for sample 1.4

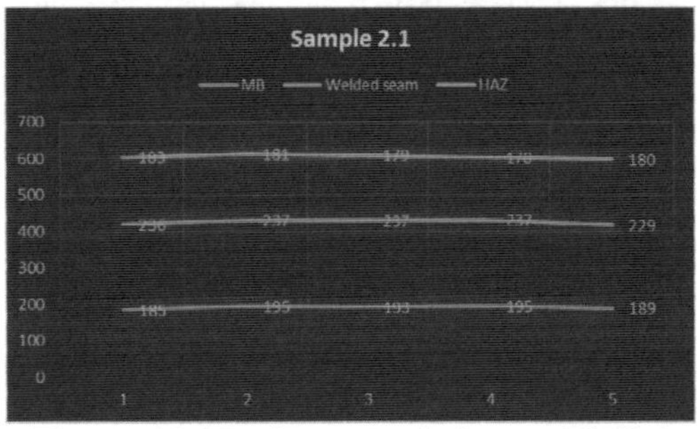

Figure 10. Hardness graph for sample 2.1

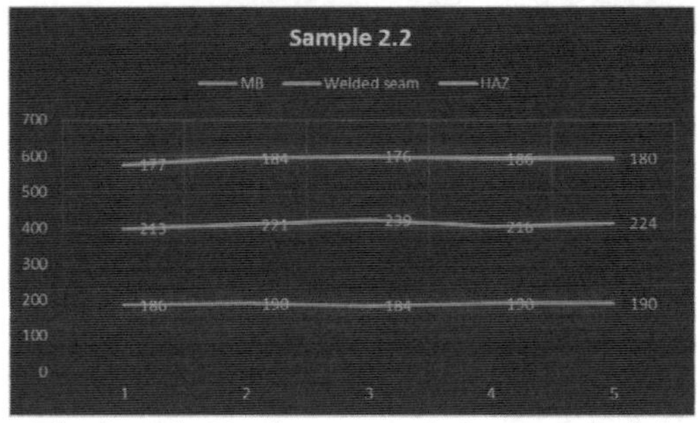

Figure 11. Hardness graph for sample 2.2

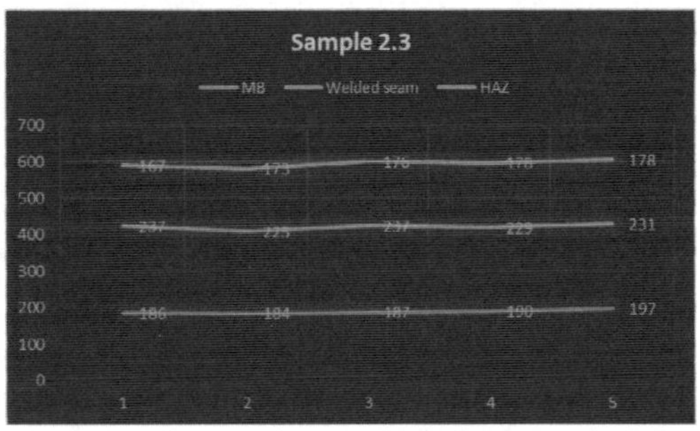

Figure 12. Hardness graph for sample 2.3

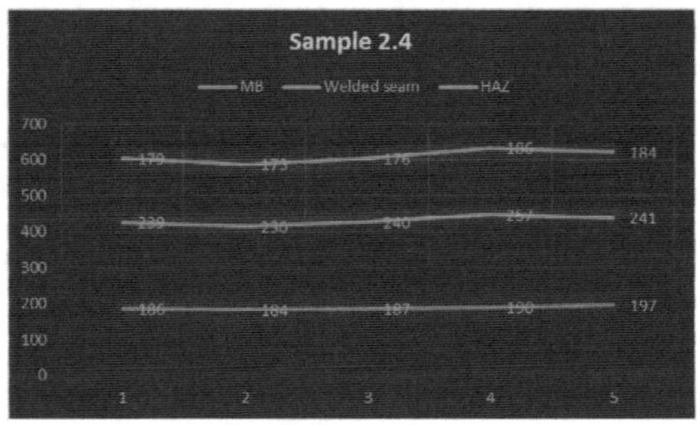

Figure 13. Hardness graph for sample 2.4

Analysing the presented hardness graphs, it is observed that in many of the analysed samples the hardness values in the heat affected zone are lower compared to those in the welded seam or parent metal. Although the probes are made of the same material, there are small differences between the hardness values of the parent metal used. The hardness values in the welded seams performed are higher compared to those of the parent metal, probably due to the formation of hard compounds due to the high cooling rate.

4. Conclusions

Tests were performed on small diameter pipe probes, on which artificial defects were created, which were subsequently repaired by welding using the WIG process. An experimental equipment was made for welding repair of the low-access areas and artificial defects on the probes were repaired using it. The obtained probes were visually and penetrant tested to highlight any imperfections positioned on the surface of the welded joint or communicating with the outside.

Based on this paper content, the following conclusions can be drawn:

✓ Following the welding repair procedure, it can be seen that the higher is the distance from the access end, the lower is the quality of the weld obtained.

✓ The quality of the weld decreases with the increase of the cavity size.

- ✓ Following the direct visual testing, a series of imperfections were identified, but the most serious of these being the lack of penetration.
- ✓ During the inside visual examination, it could be observed that up to a distance of about 100 - 150 mm from the access end, the weld has a good appearance with no serious imperfections.
- ✓ At a distance of 200 mm, metal drops can be observed, but also burns.
- ✓ Following the examination with penetrating liquids, it was observed that the outer surface of the weld has imperfections, the lack of melting being the most common.
- ✓ Following the examination with penetrating radiation it can be seen that the samples showed solid inclusions.
- ✓ In most of the analysed samples the hardness values in the heat affected area are lower compared to those in the welded seam or the parent metal.
- ✓ There are small differences between the hardness values of the parent metals, although the samples are made of the same material (austenitic 316L stainless steel).
- ✓ The hardness values of the welds are considerably higher compared to those of the parent metal, which accredits the idea that it is possible that hard phases may form in the welds during welding or after cooling.

References

[1] Kou S 2003 *Welding metallurgy*, second ed., John Wiley&Sons, Inc., Hoboken, New Jersey.
[2] Mvola B B, Kah P AND Martikainen J 2014 *Dissimilar ferrous metal welding using advanced gas metal arc welding processes*, Rev. Adv. Mater. Sci. **38**, 125-137.
[3] Ranjbarnodeh E, Pouranvari M and Fischer A 2013 *Influence of welding parameters on residual stresses in dissimilar HSLA steels welds*, Metall. Mater. Eng. **19**, 33-43.
[4] Mihai A 2011 Quality Inspection, Non-destructive testing methods, Laboratory guide, Printech.
[5] A 240/A 240M:2004 - *Standard Specification for Chromium and Chromium-Nickel Stainless Steel Plate, Sheet, and Strip for Pressure Vessels and for General Applications*, ASTM International.
[6] EN 10088-2: 2005 *Stainless steels. Technical delivery conditions for sheet/plate and strip of corrosion resisting steels for general purposes.*
[7] SR EN 1011/1:2004 - *Welding. Recommendations for welding of metallic materials. Part 1: General guidance for arc welding.*
[8] Kuppan T 2013 *Heat exchangers Desing Handbook*, CRC Press, Taylor&Francis Group, USA.

Ways for determining the intermediate dimensions when designing the machining technology

L Slătineanu[1], M Coteaţă[1], O Dodun[1], G Nagîţ[1], A Hriţuc[1], I Beşliu-Băncescu[2]

[1] "Gheorghe Asachi" Technical University of Iaşi, Romania
[2] "Ştefan cel Mare" University of Suceava, Romania

hrituc.adelina3295@yahoo.com

Abstract. Among other problems to be solved, the design of a technological process of machining by cutting a part involves establishing the dimensions that can be achieved in the pre-final machining of the same surface of the part. There are distinct ways to calculate the so-called intermediate dimensions and, in particular, to define the dimensions for adjusting the position of the cutting tool tip before starting the actual machining process. The paper analyzes some such ways of establishing the intermediate dimensions, highlighting the conditions under which they can be applied, their advantages and limitations of use. Subsequently, the double-entry matrix method was applied to identify one or more methods with wider possibilities of use. It was concluded that the method of tolerance chart and respectively a method that takes into account a statistical processing of the results accumulated over time are methods likely to lead to an efficient solving the problem of determining the intermediate dimensions.

1. Introduction

Over the last decades, it was considered that one of the main aspects specific to the general concept of *machine manufacturing technology* is the concept of *technological process*. Generally, *a process* defines the sequence of states through which a phenomenon or product passes.

There is the convention to use the concept of *production process* to define the assembly of all the technical-productive, interdependent and interconditioned activities, carried out using the means of work, as well as the totality of the natural processes related to the transformation, direct or indirect, of the objects of the work. The transformation is conceived, designed, organized, conducted and executed of people.

The concept of *industrial process* is the process able to generate changes of the materials and substances and such processes could be the cutting processes, the welding processes, the heat treatment, etc.

In an industrial company, there are *base processes* that supposes a direct transformation of the workpieces in finite products (obtaining the workpieces, machining, assembling, dispatching, etc.) and *auxiliary processes* that includes activities of preparing and helping the development of base processes (manufacturing of tools, jigs and fixture, and tools for quality control, repairing and maintenance of equipment, etc.)

The technological process is that component of the production process related only to the activities of direct, quantitative and/or qualitative transformation, of the object of work [1, 2]. This means, for example, the change of the shape, of the dimensions, of the state of the surfaces, of the chemical composition, of the structure, of the physical-mechanical properties, the location in space, etc.

When the problem of designing the technological process of mechanical machining of a workpiece, the main stages to be passed are generally the following: analysis of the initial information (technical drawing, available machine tools and technological equipment, the available staff and the level of qualification of the operators), selecting the type of workpiece, establishing the technological route, calculating the machining allowances and intermediate dimensions, establishing the machining parameters, calculation of the time norm, economical assessment of the technological process and selecting of the optimal version, etc.

The intermediate dimensions are those dimensions obtained by the successive application of different processing in the case of the same surface. The adequate establishing of the intermediate dimensions is important due to its influence on the machining accuracy and the processes productivity.

Over the years, different opinions were expressed about the way of establishing the intermediate dimensions.

Thus, Whybrew et al. they appreciated that the usual elaboration of tolerance charts takes a long time [3]. They proposed a graph-theoretic approach to developing a tolerance chart that could be used for microcomputer-aided design of those charts.

Pairel et al found that frequently the dimensions obtained by processing are related to the surface that locates the part in the workpiece holder [4]. They have developed a methodology that takes into account the identification of the machining groups that must be performed before extracting workpiece from the workpiece holder and the surfaces to be measured. They appreciated that in this way it becomes possible to use higher tolerances of intermediate dimensions.

Thilak considered that to reduce the machining time, an optimal design of the process plan is necessary [5]. He proposed and developed a model for the simultaneous selection of optimum machining datums and machining tolerance using an evolutionary algorithm.

Within the semestrial design activity at the matters of machine manufacturing technology, establishing the intermediate dimensions is usually a problem that can be solved in various ways. The objective of the research whose results were presented in this paper was to analyze and evaluate some of the known solutions of establishing the intermediate dimensions when approaching the design of the technology of a part mechanical machining. Initially, several methods for calculating intermediate dimensions were briefly analyzed. Subsequently, a way of ordering these methods was used (namely the double-entry matrix method), by taking into account a global analysis, but starting from the usual criteria for evaluating those methods.

2. The connections between the intermediate dimensions and the dimension that characterizes the positioning of the cutting tool tip before processing

The intermediate dimensions are dimensions obtained successively during the operations or machining phases of each surface, by removing the machining allowances (fig. 1). From a theoretical point of view, there is a minimal machining allowance that ultimately ensures the placement of the dimensions of the part inside the prescribed tolerance fields. In some cases, the machining allowance is higher than the minimum and it can be removed by several passes (tool strokes), but within the same machining phase.

The intermediate dimensions may be necessary both for the design of tools, devices and measuring tools required by the technological process of obtaining the part, as well as for the design of molds used to obtain the workpieces.

The calculation of the machining allowances and of the intermediate dimensions is made starting from the dimension b_{max} of the part inscribed in the mechanical drawing, in reverse order to the one applied for obtaining a certain surface. In essence, after calculating the nominal machining allowance A_{cnom} (usually equal to the maximum machining allowance), the maximum size is calculated before applying the last machining operation, for example by means of a valid relation in the case of a flat outer surface of the shape:

$$a_{max} = b_{max} + A_{cnom} \quad (1)$$

where b_{max} is the size obtained at the previous step. Next, the so-called nominal dimension must be determined, this being usually the dimension at which the position of the cutting tool tip is set.

It is obvious that the minimum dimension a_{min} will be determined starting from the maximum size a_{max} and from the tolerance T_p corresponding to the machining by which the analyzed surface is obtained:

$$a_{min}=a_{max}-T_p. \quad (2)$$

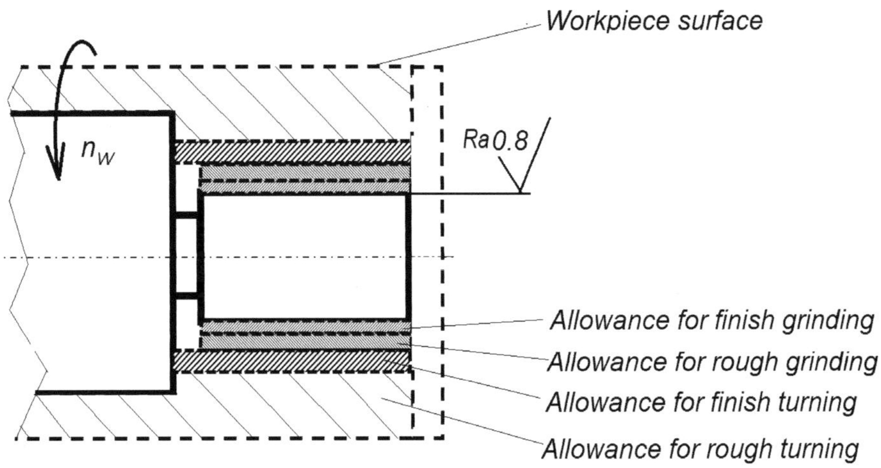

Figure 1. Machining allowances in the case of an external cylindrical surface.

3. Methods for determining intermediate dimensions

At this point, it is possible to consider different situations, mentioned below. It is necessary to mention that the version *A* is valid in the case of individual production, with the individual obtaining of each dimension of the analysed surface, while the other variants can be adopted in the case of automatic obtaining the dimensions, by using a certain adjustment method. of the tool tip to the work dimension [1, 6].

Method A. In the case of individual production, the nominal size can be taken equal to the maximum dimension; once the operator, using the test chip method, has obtained a dimension equal to the maximum allowed dimension, the accuracy requirement valid for the respective surface and for the analysed phase is considered fulfilled.

Method B. In the case of automatic obtaining of the dimensions, the nominal dimension (considered here equal to the positioning dimension of the cutting tool tip) can correspond to the dimension in the middle of the tolerance field (fig. 2):

$$a_{nom}=a_{max}-(1/2)T_p. \quad (3)$$

The argument of such an option considers a possible normal dispersion of the dimensions of the machined surface around this average dimension, which would mean the most convenient situation from the point of view of framing the dimensions of the machined parts in the tolerance field prescribed for the dimension in question.

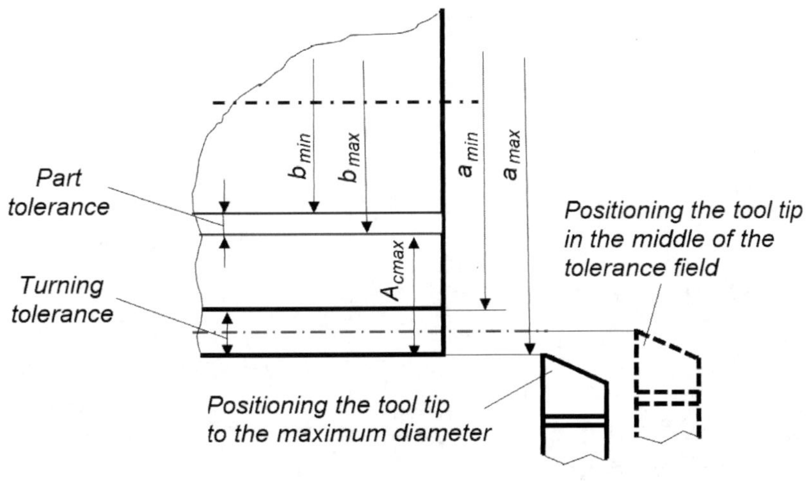

Figure 2. Positioning the turning tool tip when obtaining an intermediate dimension.

Method C. If in the case of automatic obtaining the dimension it is estimated that the wear of the cutting tool tip could have an important role, it is possible to recommend that the nominal dimension be located at 2/3 of the tolerance field T_p in relation to the maximum dimension a_{max}:

$$a_{nom}=a_{max}-(2/3)T_p. \quad (4)$$

It is appreciated that, in this way, even with the evolution of tool wear (cutting along the direction of the dimension to be obtained), it is possible to obtain a maximum number of dimensions of the machined parts within the prescribed tolerance field.

Method D. A fourth variant of calculation of the nominal size a_{nom} takes into account a possible initial expansion of the cutting tool, which would lead to a decrease in the dimension of the machined surface and only then to intervene decisively the wear of the cutting tool tip, which would determine a gradual increase and possibly in accordance with a certain law of evolution (in accordance with the law of wear of the cutting tool) of the dimension of the machined surface. It is now recommended that the nominal size be about one-third of the tolerance range relative to the maximum dimension a_{max}:

$$a_{nom}=a_{max}-(1/3)T_p. \quad (5)$$

Method E. A fifth variant is even more complex, taking into account aspects of statistical nature valid in the case of adjusting the position of the tool tip to the machining dimension by using the test method:

$$a_{pos}=a_{min}+ks \quad (6)$$

in which a_{pos} is the dimension of positioning the cutting tool tip, k is a coefficient determined statistically, as a result of research specific to different machining processes and s is an approximate value of the standard deviation that characterizes the instantaneous dispersion of the dimensions when developing the machining process on a certain machine tool [1, 7, 8].

Method F. Another option for establishing the intermediate dimensions and their related tolerances is applied in some of the economically advanced countries [9, 10]. This version corresponds to a design stage of the technological process called "tolerance charting". After *Part Design Analysis*, the following steps must be completed: *1. Processing*, in which the operations and their contents are described; *2. Strip Layout*, materialized by the elaboration of a set of machining schemes, with the tools tip at the end of the work stroke and with the highlighting of the quotas characteristic of the machined surfaces; *3. Tolerance Chart Form*, when a somewhat standardized form is prepared for further use; *4. Show Strip Layout Data on Tolerance Chart Form*, in which the form is partially completed using certain agreed

IOP Conf. Series: Materials Science and Engineering **1018** (2021) 012019 doi:10.1088/1757-899X/1018/1/012019

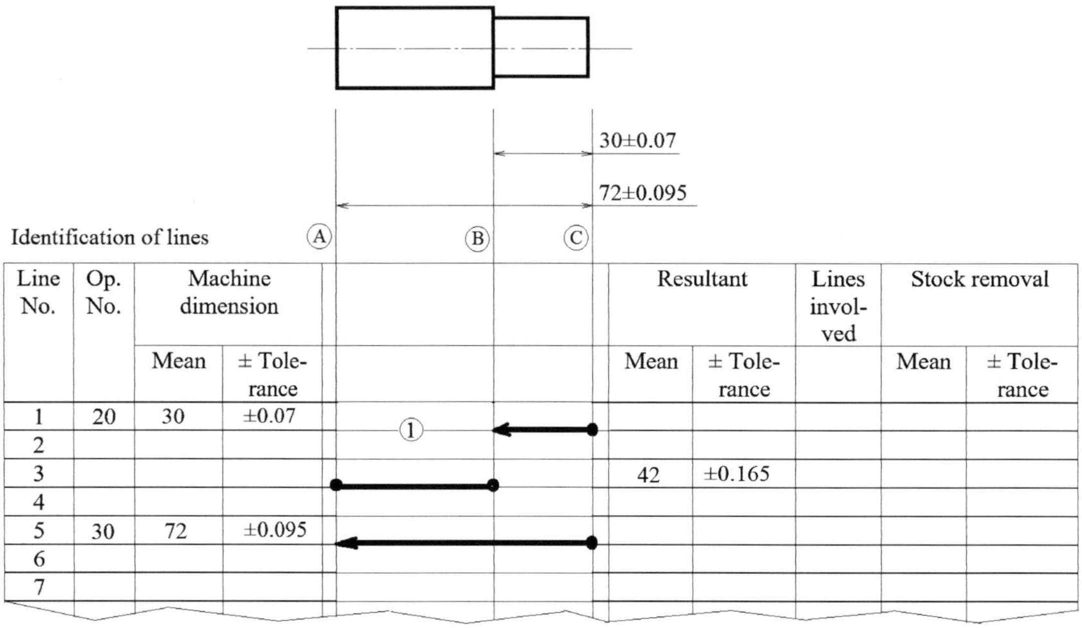

Figure 3. Fragment of the tolerance chart form valid for turning of a cylindrical shaft type part in two-stage (adapted from [9]).

symbols; *5. Calculation of the Resulting Dimensions*, when what we call intermediate dimensions are calculated; 6. *Calculation of the Stock Removal; 7. Indication of the conclusions of Tolerance Charting*, materialized by entering additional information in the standard form; *8. Adding Part Design Dimensions and Tolerances*; *9. Determination of Actual Dimensions and Tolerances and Comparison with the Part Design Specifications; 10. Making Any Required Changes and Corrections*. A fragment of a final form resulting from the application of the steps mentioned above is shown in Figure 3.

4. Comparison of different methods for determining intermediate dimensions

Most of the methods mentioned above are applicable in specific cases (individual or series production, taking into account thermal expansion and tool wear, etc.). An overall evaluation of these methods should take into account several distinct criteria (duration of necessary calculations, need to perform preliminary tests, need to use statistically determined tables, application limits, etc.).

Accepting the risk of a subjective evaluation of these methods for determining the intermediate dimensions, the double-entry matrix method was used to achieve an ordering of the respective methods (Table 1). In this sense, a comparison of the methods was used according to the principle of comparing each with each, granting and entering in Table 1 evaluation marks of type 1-0, when the first method is considered more advantageous, 0-1 then when the second method is appreciated as more convenient and 0.5-0.5, respectively, when the two compared methods are considered to be of equal importance. The entry of the results of the comparisons in Table 1 was made taking into account first of all the information along a column, so that in an additional line of the table it is possible to enter the sum of the marks given to each method. A relative value (a so-called importance coefficient) was used to order the methods, determined as a ratio between the sum of the marks given to each method and the total number of comparisons made. This number of comparisons N_c is determined using the relationship:

$$N_c = n\,(n\text{-}1)/2 \qquad (7)$$

where n corresponds in this case to the number of methods compared. For the 6 methods that were considered, the number of comparisons is equal to $6{\cdot}5/2 = 15$.

In the penultimate line of Table 1, a relative value having the significance of a coefficient of importance was included. Based on this importance coefficient, the order of the methods considered was introduced in the last line of the table. It can be seen that the assessment leads to the finding that a greater applicability seems to correspond to methods *E* and *F*, of course with the consideration of a certain degree of subjectivism in the action of comparing the methods two by two.

Table 1. Use of the double entry matrix method to compare the ways of determining the intermediate dimensions.

Method	*A*	*B*	*C*	*D*	*E*	*F*
A	X	0.5	0.5	0.5	1	1
B	0.5	X	0	0	0.5	1
C	0.5	1	X	0.5	1	1
D	0.5	1	0.5	X	1	1
E	0	0.5	0	0	X	0.5
F	0	0	0	0	0.5	X
Sum of marks	1.5	2.5	1.0	1.0	4.0	4.5
Coeff. of importance	0.1	0.16	0.06	0.06	0.26	0.3
New order of methods	*F – E -B – A – C or D*					

5. Conclusions

In the context of the elaboration of the technology of mechanical machining by cutting of a part, one of the problems that must be solved is that of determining the so-called intermediate dimensions. The correct establishment of the intermediate dimensions is important given their use when adjusting the positions of the cutting tool tip, before starting the actual machining. Over the years, various methods of calculating the intermediate dimensions have been proposed and used, by taking into account certain proper conditions specific to the analyzed machining process. Several such methods have been briefly approached and characterized. Formulating the requirement for an ordering of these methods by applying a global evaluation method, the double-entry matrix method was used. Considering the possible application in the context of educational activities in the field of manufacturing engineering, it was appreciated as more convenient the method of tolerance chart and a method based on statistical processing of practical information accumulated up to a certain moment. In the future, there is the intention to identify and use optimal methods for determining the intermediate dimensions, so as to obtain a minimization of the process time and processing cost.

References

[1] Bohosievici C 1981 *Manufacturing Technologies of Thermal Machines First part* (in Romanian) (Iaşi, Romania: Polytechnic Institute)
[2] Drăghici G 2005 *Design of Processes of Mechanical Machining* (in Romanian) (Timişoara: Publishing House Politehnica)
[3] Whybrew K, Britton G A, Robinson D F and Seermsuti-anuwat 1990 *Int. J. Adv. Manuf. Technol.* **5** 175–83
[4] Pairel É, Goldschmidt E, Vayre B, A Boukar and Pillet M 2013 *Procedia CIRP* **10** 323 – 31
[5] Thilak M, Sivakumar K, Jayaprakash G. 2012 *Int. J. Manufacturing Technology and Management* **25** 135-60
[6] Komissarov V I, Leontiev V I, Starostin V G 1968 *Dimensional Setting of The Universal Cutting Machine Tools* (in Russian) (Moskow: Publishing House Maschinostroenie)
[7] Ishutkin V I 1962 *Setting the machine tools* (in Romanian) (Bucharest, Editura Tehnică)
[8] Mazuru S, Maxim C 2014 *Adv. Mater. Res.* **1036** 292-97
[9] Nelson H D, Schneider Jr G 2001 *Applied Manufacturing Process Planning With Emphasis on Metal Forming and Machining* (Columbus, Ohio, U.S.A.: Prentice Hall)
[10] Dietrich R, Garsaud D, Gentillon S, Nicolas M 1981 *Specific. Machining Methods. Methodology, Production and Standardization* (in French) (Paris: AFNOR)

IOP Conf. Series: Materials Science and Engineering **1018**(2021) 012020 doi:10.1088/1757-899X/1018/1/012020

Research of fuzzy logic application on surfaces roughness prediction under finishing milling process

D.P. Ghencea[1], F.D. Anania[2] and M. Zapciu[3]

[1]University Politehnica of Bucharest, Doctoral School of Industrial Engineering and Robotics, Splaiul Independenței 313, Bucharest, Romania
[2,3]University Politehnica of Bucharest, Department of Robots and Manufacturing Systems, Splaiul Independenței 313, Bucharest, Romania

[1]daniel.ghencea@blackseasuppliers.ro, [2]dorel.anania@upb.ro, [3]miron.zapciu@upb.ro

Abstract. Artificial intelligence systems are usefully tools for estimation of different parameters for industrial and non-industrial applications. In this paper is presented a fuzzy logic application for surface roughness prediction for two material type for a CAM milling process. A minimum number of measurements have been made on a specific material (Al 6061) and were used for prediction of a roughness of another material (Al 7075). The link parameters between these materials, used in Fuzzy logic system, is the Rockwell hardness. In this way many correction parameters of cutting regime can be calculated in a short time with minimum resource (less experiments and tests needed).

1. Introduction

Today the requirements of product quality are getting bigger. High quality products depend of many aspects related with functionality, aspects, strength etc. Some of these aspects are directly related to manufacturing processes and many of these processes are from metal cutting field [1].

In metal cutting process the milling is one of the most used in industry for many fields like aerospace, automotive etc. The milling NC machines, tools and software are developing continuously bringing new flexibility in the cutting process such as possibility of tool orientation (4x, 5x continuous milling), different toolpath over surfaces (like spiral, parallel cut, ISO cutting etc.). Due to these possibilities in machining process the quality of the part obtained can be various through the surface quality (which is influenced by many factors like constant cutting speed, cutting direction, depth of cut, radial step etc.) [2-3].

This research is focus on achieving a constant roughness on different type of material by using similar cutting condition with minimum modification. In order to establish an algorithm and some correction necessary to be made, by the classic way, many experimental tests must be made. This means a lot of time and money for materials, cutting tools, and work hour [4].

By using Artificial Intelligence algorithm, we are trying to make roughness prediction for a variety of material starting from minimum experimental tests for milling processes parameter variation.

The use of artificial intelligence in research on milling free-form surfaces using ball mills is a sensitive topic but one that is on the rise due to IT, both hardware and software [5].

Fuzzy logic prediction of surface quality provides a powerful tool for reducing the number of tests and thus the number of measurements for each test. The use of prediction is a consequence of the growing demand for productivity and product quality. The prediction unifies by the construction model of the membership functions and the inference rules the geometric properties of the surface that depend on the diameter of the cutter, the number of cutting teeth, the cutting speed, the passage and the feed rate [6].

Establishing the fuzzy inference rules (FIS) for the roughness parameter is the most important aspect of the construction of the model under analysis with fuzzy logic because it depends most on the accuracy of the prediction given that the prediction is made for the same material but with different hardness (measured roughness surface Al 7075 and predicted surface roughness Al 6061).

2. Methodology for simulation, prediction and interpretation of results

2.1 Simulation methodology and objectives

The aim of the study is to identify a method to predict as accurately as possible with fuzzy logic the roughness of an aluminum surface finish by milling for different tool trajectories type on a aluminum type (6061) starting from experimental results measured on 7075 aluminum.

The characteristics for Al 6061 and Al 7075 are shown in Table 1 [7], [8].

Table 1 Characteristics for Al 6061 and Al 7075

	Al 6061	Al 7075
Physical Properties		
Density	2.7 [g/cm^3]	2.81 [g/cm^3]
Mechanical Properties		
Hardness, Brinell	95	150
Hardness, Rockwell A	40	53.5
Hardness, Rockwell B	60	87
Tensile Yield Strength	276 [MPa]	503 [MPa]
Modulus of Elasticity	68.9 [GPa]	71.7 [GPa]
Poisson's Ratio	0.33	

For prediction, the average Rockwell A and B harnesses are used, respectively 1.39375.

Milling tests were performed on a First MCV300 3-axis machining center, which has a maximum shaft speed of 8000 [rpm], a shaft power of 7.5 [kW] and a maximum milling feedrate of 10000 [mm/min]. The mass of the First MCV 300 machining center has movements on the X and Y axes, and the axis on the Z axis. The part under test is fixed directly on the table of the First MCV 300 machine using two supports to ensure maximum rigidity. The acquisition of the data set is performed with the MitutoyoSurftest SJ210 instrument which has a measuring range (on the X axis) of 17.5 [mm] and a measuring speed of 0.25 [mm/s] for measuring roughness [9].

2.2 Prediction method

The research focuses on identifying how to construct membership functions for inputs for Al 7075 and outputs (P_6061 and P_7075) to make an accurate prediction.

The application **settings** in, valid for the four ways to build membership functions, are: and method - *min*, or method - *max*, implication - *min*, aggregation - *max*, defuzzification - *centroid (mamdani)*.

The fuzzy logic architecture used (2 inputs and 2 outputs) is shown in Figure 1.

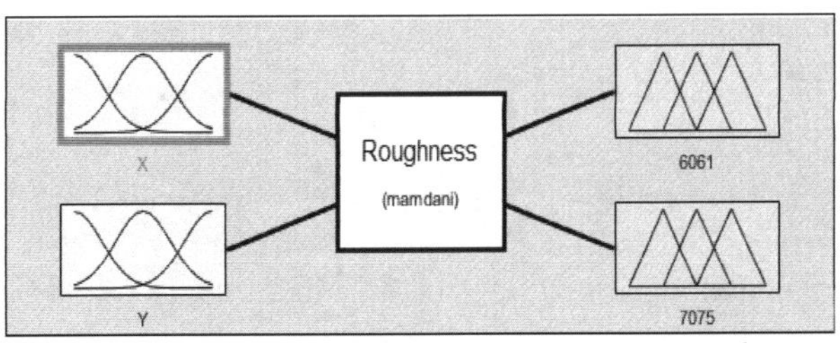

Figure 1 Fuzzy logic architecture used (2 inputs and 2 outputs)

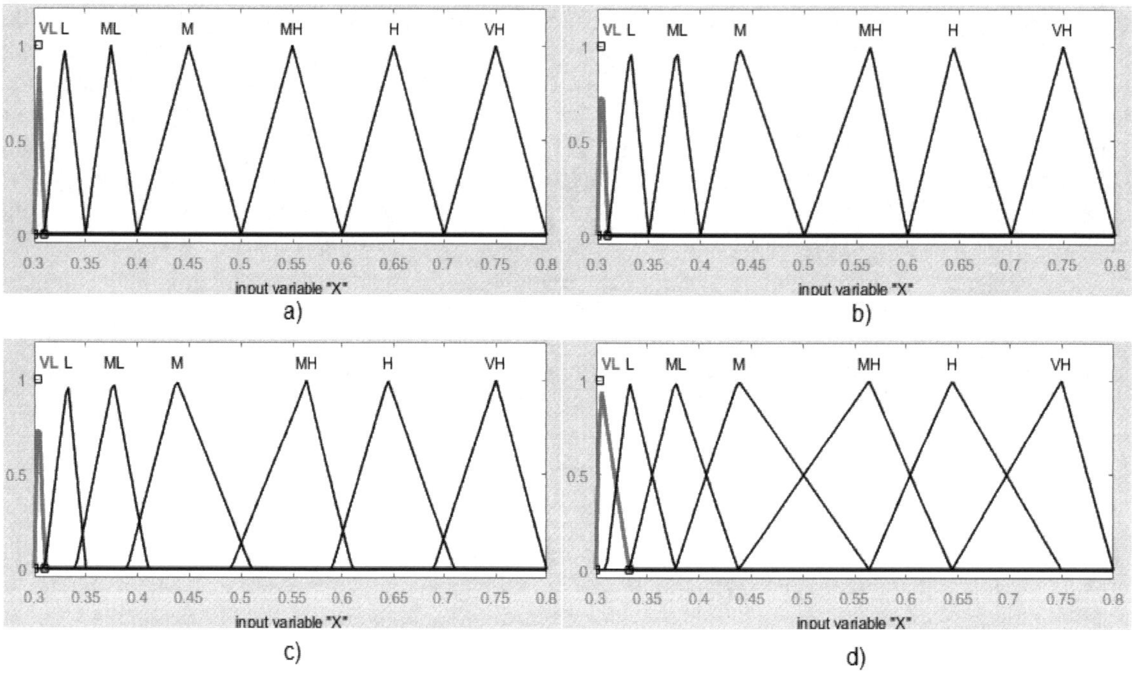

Figure 2 Models of membership functions tested: a) the peak of the triangle is in the middle of the field, b) the peak of the triangle is off-center, c) the peak of the triangle is off-center and the partial overlap, d) the peak top of the triangle is off-center and total overlap

Four ways of constructing triangular membership functions were researched, namely:

- the peak of the triangle is in the middle of the field - C (Figure 2 a));
- the peak of the triangle is off-center - D (depending on the concentration of the data set values in that field) (Figure 2 b));
- the peak of the triangle is off-center and the partial overlap (SP) of the membership functions (20%, respectively: 10% left, 10% right) over the neighboring ones (Figure 1 c));
- the peak top of the triangle is off-center and total overlap (ST) of the membership functions over the neighboring ones (Figure 2 d));

Step 1. Determine the input variables (x and y) and the output variables (average roughness Al 7075 and correction coefficient for average roughness Al 6061) (Figure 1).

Step 2. From the initial data set (35 measurements) for Al 7075 the gross errors represented by the interval ends (min / max for x / y - 4 values) are eliminated. From the initial data set (14 measurements) for Al 6061, chosen to validate the prediction, the gross errors represented by the max interval head for x / y - 3 values are eliminated.

Step 3. The set of 31 values was divided into 7 unequal intervals (as domain and number of values) to which linguistic values were assigned, respectively: VL, L, ML, M, MH, H, VH.

Step 4. Establishing the rules of roughness inference Al 7075 (27 rules - valid for the four ways of building the membership functions).

Step 5. Run the application.

2.3 Interpretation of results and discussions

Following the establishment of the rules of inference of the roughness and running of the application, the 3D surfaces (the four analyzed cases) are obtained, shown in Figure 3. It is observed that the shape of the 3D surface, construction of triangular membership functions, a) and b) for both types of aluminum.

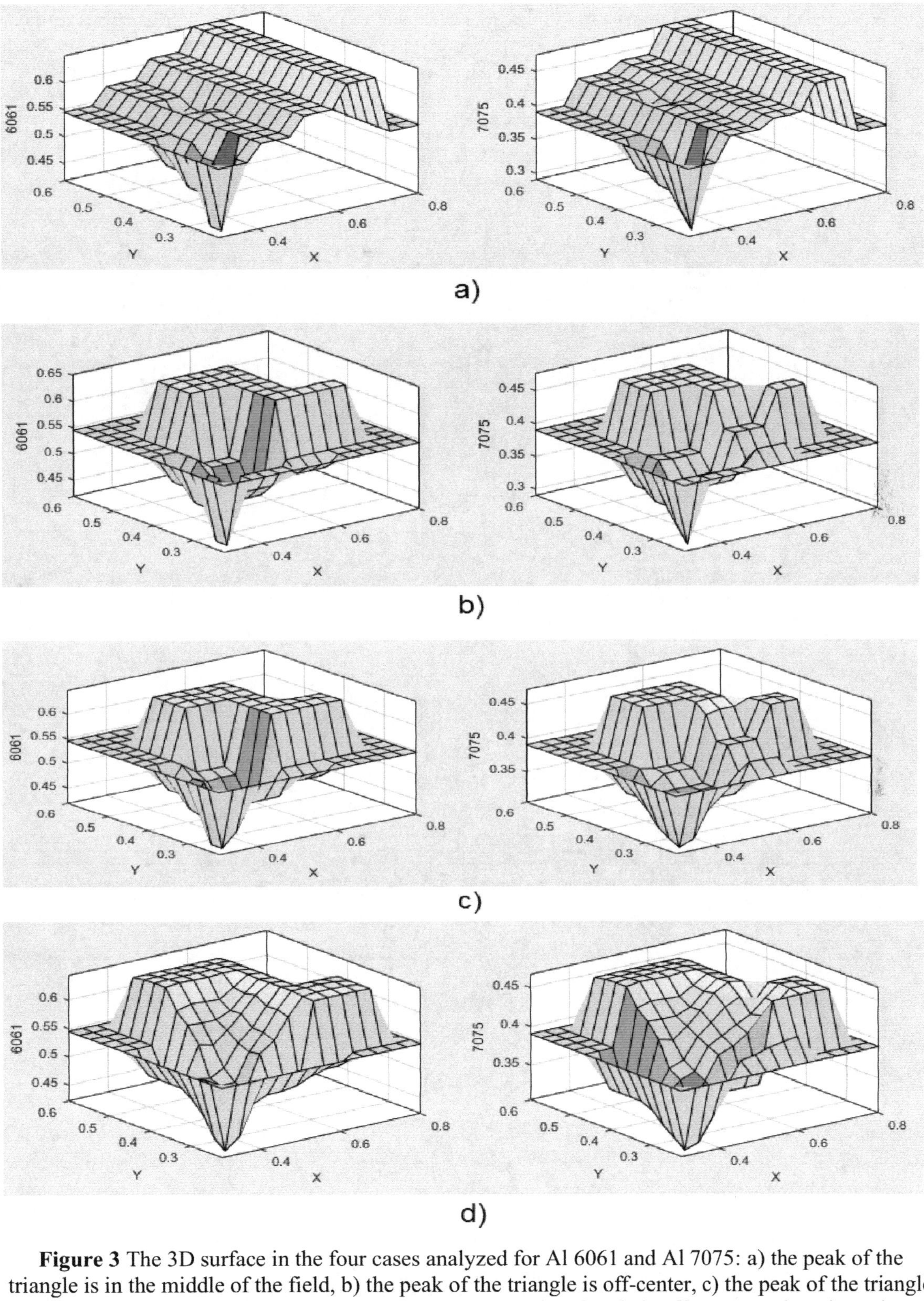

Figure 3 The 3D surface in the four cases analyzed for Al 6061 and Al 7075: a) the peak of the triangle is in the middle of the field, b) the peak of the triangle is off-center, c) the peak of the triangle is off-center and the partial overlap, d) the peak top of the triangle is off-center and total overlap

Supplementing the conditions by partially overlapping the triangular membership functions leads to small surface changes between case b) and c) (Figure 3 b) and c)). Total overlap of membership functions leads to a major change in surface geometry between case c) and d) (Figure 3 c) and d)).

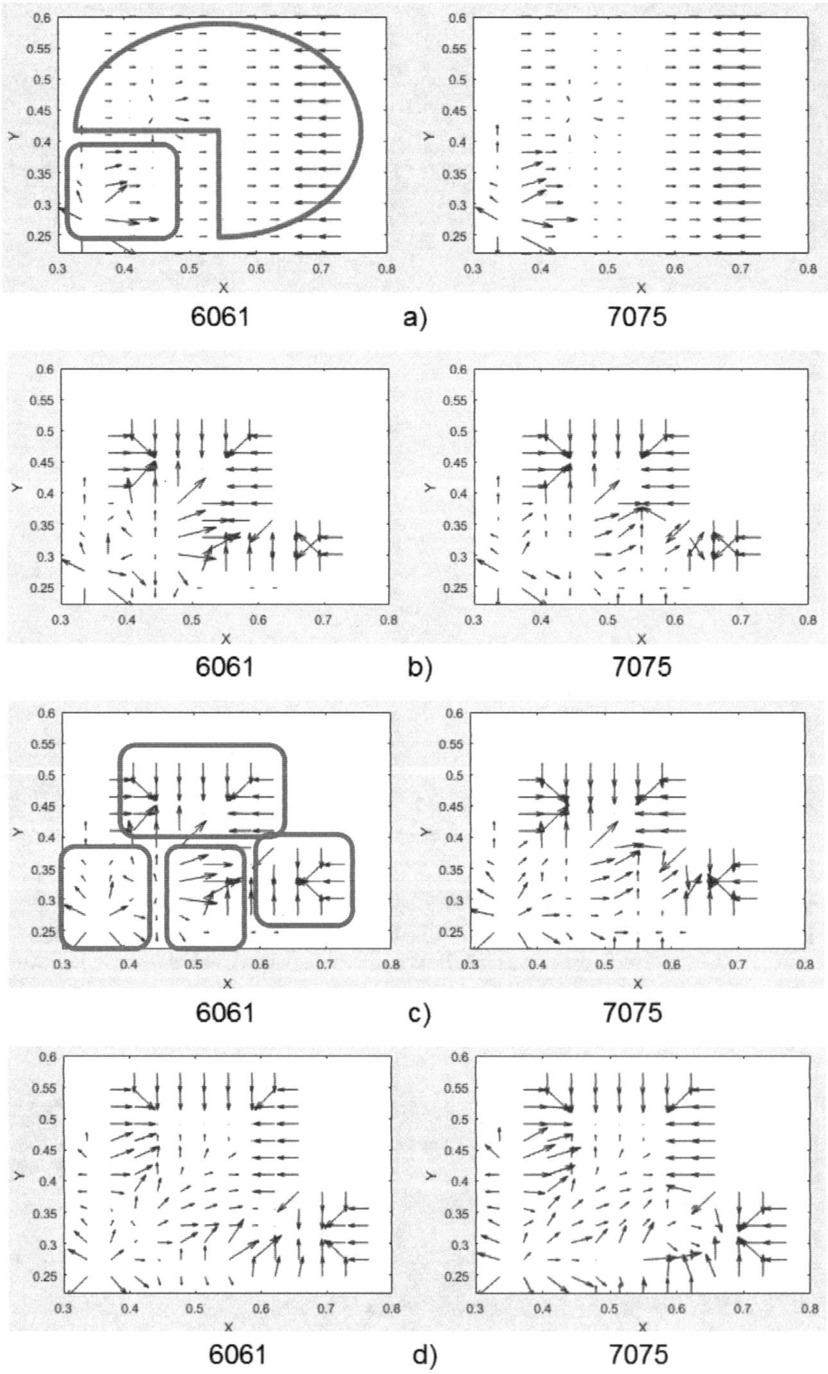

Figure 4 The mode of action of the quiver in the four cases analyzed for Al 6061 and Al 7075: a) the peak of the triangle is in the middle of the field, b) the peak of the triangle is off-center, c) the peak of the triangle is off-center and the partial overlap, d) the peak top of the triangle is off-center and total overlap

From the quiver point of view, the four ways of constructing the membership functions, it is observed that the dynamics of variation has an evolution from the uniform distribution of majority action-reaction type with 2 areas of local solutions, (Figure 4 a) 6061 and 7075) to a concentrated action-reaction distribution with 4 zones of local solutions (Figure 4 c) 6061 and 7075) which is followed by a relaxation distribution of the action-reaction concentration of the 4 zones of local solutions (Figure 4 d) 6061 and 7075).

Results for prediction at the theoretical level

Table 2 Minimum / maximum percentage error for the analyzed cases

6061	C	D	SP	ST	**7075**	C	D	SP	ST
e_{min}	-14.77%	-5.81%	-6.39%	-7.11%	e_{min}	-12.44%	-4.42%	-5.02%	-6.43%
e_{max}	29.42%	29.42%	8.16%	4.80%	e_{max}	28.73%	28.73%	8.24%	9.40%

For the case of the prediction based on the hardness coefficient at theoretical level in the analysis of Table 2 it is observed that the best accuracy is obtained for the triangle with off-center peak and partial overlap (SP - variant c)) of the membership functions (20%, respectively: 10 % left, 10% right) over the neighboring ones.

Table 3 Prediction results for Al 6061 based on measurements for Al 7075

6061	**C**		**D**		**SP**		**ST**	
	Average [%] / **Number of values**							
	(+)	(-)	(+)	(-)	(+)	(-)	(+)	(-)
> 15%	18.58 / **5**		29.42 / **1**					
< 15%	11.71 / **1**	-13.47 / **2**						
< 10%	7.77 / **4**	-8.80 / **1**	6.78 / **3**	-5.81 / **2**	6.62 / **4**	-6.10 / **2**		-6.73 / **2**
< 5%	1.66 / **11**	-1.41 / **7**	1.53 / **12**	-2.14 / **13**	1.54 / **12**	-2.08 / **13**	2.53 / **18**	-2.40 / **11**

As the conditions become stricter in the construction of triangular membership functions, errors of medium and large values disappear, with the majority of small errors centered around ± 6.5% (Table 3). The SP model has the best distribution of error accuracy from the total of the 4 analyzed models.

Table 4 Results obtained by prediction compared to those measured for Al 7075

6061	**C**		**D**		**SP**		**ST**	
	Average [%] / **Number of values**							
	(+)	(-)	(+)	(-)	(+)	(-)	(+)	(-)
> 15%	20.28 / **5**		28.73 / **1**					
< 15%	12.84 / **2**	-12.34 / **2**						
< 10%	7.36 / **2**	-8.89 / **1**	6.53 / **2**		6.60 / **4**	-5.02 / **1**	9.40 / **1**	-6.10 / **2**
< 5%	2.29 / **9**	-2.01 / **10**	2.00 / **14**	-2.35 / **14**	1.97 / **14**	-1.82 / **12**	1.70 / **14**	-2.68 / **14**

Table 4 shows that the accuracy of the SP model prediction improves (in total by 1%) compared to those presented in Table 3, in 3 of the 4 cases.

At a theoretical level the most accurate prediction is obtained for the case where the top of the triangle is off-center and we have partial overlap (SP) of the membership functions over the neighboring ones.

Results for validation of the prediction at a experimental level

For the case of the prediction based on the 11 measurements performed for 6061 (validation by experimental measurements of the prediction) the results presented in Table 5 were obtained.

Table 5 Validation of the results obtained by prediction compared to those measured for Al 6061

6061	C		D		SP		ST	
	Average [%] / **Number of values**							
	(+)	(-)	(+)	(-)	(+)	(-)	(+)	(-)
e > 15%	19.99 / **3**		31.52 / **1**		30.91 / **1**		31.82 / **2**	
e: 7 - 15%	9.56 / **2**	-12.80 / **3**	12.38 / **2**	-12.80 / **3**	12.38 / **2**	-12.80 / **3**	11.22 / **1**	-12.80 / **3**
e: 0 - 7%	3.69 / **2**	-1.71 / **1**	0.93 / **2**	-3.48 / **3**	0.93 / **2**	-3.48 / **3**	0.93 / **2**	-3.48 / **3**
$e_{general\ average}$	4.21 %		0.85 %		**0.79 %**		2.54 %	

It is observed that for the case of the triangle with off-center peak and partial overlap (SP - variant c)) of the membership functions over the neighboring ones, the best results are obtained. Variant c) is extremely close to variant b), the difference being made by:

- error size greater than 15% (31.52% vs 30.91%);
- the difference of the general average of 0.06%.

At the level of validation of the prediction by measurement, the most accurate prediction is obtained also in case the peak of the triangle is off-center and we have partial overlap (SP) of the membership functions over the neighboring ones.

3. Conclusions

The objective of the study to make a prediction of the roughness of a surface of the same material but with a different hardness than the one on which the measurements were performed was achieved.

We have shown that the restrictions that can be imposed in the performance of membership functions and inference rules do not have to be extremely precise because we go over the areas of local solutions tending to a general solution that no longer has the same accuracy. Applying the same set of fuzzy inference rules to the four cases analyzed showed that they are constant regardless of how the membership functions are constructed.

The prediction method based on fuzzy logic AI algorithm has been developed by using a set of experimental test for a milling process with a ball cutting tool for machining a 3D surface on an Al 7075 material [3], [9]. For validation of fuzzy logic algorithm developed, a new set of machining test on a different material AL6061 was made in similar cutting condition. The machining parameter were adjusted according with the material characteristics.

The experimental results were compared with the fuzzy AI logical prediction and an overall accuracy of 99.21% was found for the experimental prediction and 99.12% for the theoretical prediction.

In order to use the prediction fuzzy logic AI algorithm like alternative to experimental cutting measurement more tests are needed to be done on different type of material.

4. References

[1] Pena A E, Anania F D and Mohora C 2017 Method for optimum calculus of machining parameters according to tool trajectories type based on milling process simulation *Tehnički vjesnik* **24**(2) 385-389.

[2] Cosma M 2011 Experimental studies on influence of tool path in 3-axes B.N.E.M. on inclined surfaces at 45 degrees *Acad. J. Manuf. Eng.* **9**(4) 30-35.

[3] Pena A E, Anania F D and Zapciu M 2014 Research regarding software developing for machining parameters optimization calculus *Appl. Mech. and Mater.* **656** 192-199.

[4] Wojciechowski S, Twardowski P and Wieczorowski M 2014 Surface texture analysis after ball end milling with various surface inclination of hardened steel *Metrol. Meas. Syst.* **XXI**(1) 145-156.

[5] Ghencea D P 2018 *Modelare-simulare şi predicţia datelor experimentale specifice sistemelor de fabricaţie utilizând tehnici hibride bazate pe inteligenţă artificială*, Teză de doctorat.
[6] Liu N, Loftus M and Whitten A 2005 Surface finish visualisation in high speed, ball nose milling applications *Int. J. Mach. Tools Manuf.* **45** (10) 1152-1161.
[7] http://asm.matweb.com/search/SpecificMaterial.asp?bassnum=MA6061T6 (accessed 15.08.2020)
[8] http://asm.matweb.com/search/SpecificMaterial.asp?bassnum=MA7075T6 (accessed 15.08.2020)
[9] Anania F D, Pena A E and Zapciu M 2017 Surface quality and machining time optimization based on feedrate correction function of tool trajectories types *Tehnički vjesnik* **24**(4) 987-992.

Acknowledgement
This work has been funded by the European Social Fund from the Sectoral Operational Programme Human Capital 2014-2020, through the Financial Agreement with the title "Scholarships for entrepreneurial education among doctoral students and postdoctoral researchers (Be Antreprenor!)", Contract no.51680/09.07.2019 - SMIS code: 124539.

Correlation between market segmentation, industrial product features and context in design assessment

A E Craciun

Faculty of Industrial Engineering and Robotics, University Politehnica of Bucharest, 060042, Bucharest, Romania

craciun_alexandra@outlook.com

Abstract. When referring to product design assessment, even with a standard evaluation format, the perspective from which the criteria are approached differs strongly from one market category to another, this having a direct impact on the accuracy of the method. In order to achieve an objective assessment, the product must relate to a frame of reference in accordance with the product's destination, taking into account the user specifications and the actual experience of use. This paper represents the study of correlations between the market segment, product features and context and their degree of interdependence. Thus, analysing the relations between sections, the representative influence formulas for each considered feature were deduced, highlighting the extent to which the user's typology and context relate to the product features.

1. Introduction

When addressing design assessment one of the major concerns is the extent to which the assessment can be considered objective. There is the question if the assessment can be approached from an unbiased perspective or is it possible that the method could be quantified and related to a proposed standard assessment system so that the results are as relevant as possible.

Based on a study of the subjectivity of design history Lees-Maffei and Fallan [1] identified the personal influence as a significant factor in design. The debates are even greater when the strong association between design and aesthetics is made. Atalay [2], studying Kant's approach to the theory of aesthetics, addressed the concept of subjectivity and demonstrated at the same time the existence of a component of universal validity. Considering the interpersonal validity, Travanini [3] also talks about a shared criteria of correctness, component of analysis similar to Cheng [4] which recommends criteria-marking scheme in order to make the transition from subjectivity to objectivity in design assessment.

The doubt of subjectivity cannot be completely removed but it can be controlled, as Ulrich and Eppinger [5] state that design is subjective but at the same time, you can determine qualitatively if the design has achieved its objectives. Thus, it remains the obligation of the assessor to fully research the entire context of the product and analyse based on criteria that relate to generally valid standards. A systematic approach that provides quantitative results can be achieved, but the relevance of the results depends on the insight on the correlation between the product and the purpose for which it was designed, the objectivity of the assessment being influenced by the in-depth knowledge of the consumer's profile and the extent to which it is taken into account, since the degree of objectivity can be increased by establishing a more precise structure as a frame of reference.

The consumer behaviour has often played a central role in marketing, the consumer being seen as decision maker [6] for which strategies have been developed to ensure that the product is designed to meet its requirements as its influence determines the marketing strategies [7][8], although there is also the possibility in which marketing ends up influencing the buyer behaviour [9].

In regards to the consumer and its correlation with the purchase decision, Lancaster [10] finds that the features of the product that influence utility are taken into account and not the product in fact. In this case, what can determine a certain feature and what it relates to, as the buyer may not be the actual user of the product and the role of the customer being different from that of the consumer [11]. For example, in the case of a toy intended for infants, the sizing of the product has to take into account the sizing restriction of the user, but in terms of colour scheme, although it must be appropriate for the user, it also influences the buyer (in most cases the parent, or an adult in question) and in the end, affects the purchase decision. Market segmentation can be somehow conditioned by the customer, however, it is the role of the designer to take into account the distinction between the two and to include as much as possible features that satisfy both categories, starting from the user especially if it is the case when a certain feature of the product may affect safety.

2. Sections of analysis – market segment, product features and context

The present paper is meant to study the relations between the user and the product, how product features can be influenced and also the way in which one feature can influence the other and how all this correlates, as the assessment must relate to product typology in accordance with its purpose, being considered all relevant parameters involved in the final design of the product.

2.1 Market segmentation

In order to best meet consumer requirements, market segmentation is used to facilitate the development of customized products for a particular category of consumers. The most common market segmentation categories are demographic such as age or gender, geographic with direct reference to location, psychographic which involves personality traits and what derives from it and behavioural segmentation based on purchase behaviour such as usage rate [12][13][14]. As pointed out by Dolnicar et al. [14], there is no ideal market segment but the most suitable for the given situation, although as proven before certain demographic variable are more significant than others [15]. Thus in the present paper, it was considered the market segmentation with the highest correlation with the product features. Were selected the segment criteria that can be applicable on an extended range of product types and having major impact, demographic such as age, gender, educational level, income and also culture, as a form of geographical and psychographic segmentation related to location with implications in product symbolism.

The product assessment determines whether the product itself is suitable, however, reporting to a proposed market segment is not just a simple reference of the product to one criterion since the segment criteria are also correlated and can influence each other. For instance, the buying habit differs depending on the income level [16], income being a complex variable in market segmentation and it relates also with education [17][18] and with the consumer's location, as developed countries also imply a better standard of living and wealth. Income is also linked to a number of inequalities, such as that in relation to age [19] or the controversial gender pay gap [20]. The relation between gender and education [21] or gender and income highlights the difference between consumer and buyer, although women's incomes are overall lower, they are most often the decision makers when it comes to purchases [22] and it gets even more intricate when it comes to gender in relation to product features. As products with masculine traits are supported to the detriment of feminine ones, men are rejecting the idea of feminine products while women tend to accept products with masculine details [23]. An alternative to the ambiguity of gender traits could be set by the designer itself who would have the power to reduce stereotypes by creating genderless design [24], all the more so as the perception of gender related details tend to become an ever-changing factor.

IOP Conf. Series: Materials Science and Engineering **1018** (2021) 012021 doi:10.1088/1757-899X/1018/1/012021

2.2 Product features

Product features are elements of design and represent the characteristics that define the product, also referred to as products' physical properties [25], aesthetic attributes [26] or even principles of product semantics [27] and generally include colour, line, shape, form, weight, texture etc. as descriptive elements that, altogether, define the product. As a frame of reference in this study, the products' features have been narrowed to four major categories that integrate all relevant characteristics into a compact structure as colour (hue, brightness, saturation etc.), shape (line, contour, form etc.), material (texture, surface etc.) and dimension (size, weight, volume etc.).

The correlations between features are complex, although sometimes obvious when they influence each other directly such as the material in its natural form is already defined by a chromatic scheme or if the weight of the material changes implicitly the weight of the product will change, or the case of shape limitation for material because of geometry restrictions. As it can be seen in Table 1, some correlations are more difficult to detect, especially the perceptual ones in which the way a feature is perceived changes depending on the influence of another one, for example, changing a parameter such as shape can influence the perception on product dimension [28][29].

In the table below are analysed the four proposed features and the type of influence relations between them, that can be both physical and direct or perceptual and indirect and each of these may have a different degree of impact on their relation of interdependence.

Table 1. Influence relations between features (a)

Prior research	Statements and findings	Type of influence	Direction of influence	Considered features	
Bloj M G, Kersten D and Hurlbert A C [30]	The degree of illumination of three-dimensional shapes influence the perception of colour.	perceptual	←	Colour	Shape
Kingdom F A A, Wong K, Yoonessi A et al. [31]	Colour contrast has an impact on the perceived shape and texture.	perceptual	→		
Frayman B J and Dawson W E [28]	The shape of an object and the way it is presented influence the perception of apparent volume.	perceptual	→	Shape	Dimension
Sevilla G and Kahn B E [29]	The complete or incomplete shape of a product has an influence on the perception of sizing.				
Hodgson S N B and Harper J F [32]	The materials may have constraints related to size, shape or geometry.	physical	←	Dimension	Material
Miltenović A, Banić M, Miltenović V [33]	The material used for lightweight design must take into account the constrains related to stiffness and strength.				
Fleming R W [34]	The surface glossiness of the material affects the perception on size.	perceptual			
Kesteren I E H, Stappers, P J and Bruijn J C M [35]	Colour is considered one of the sensory properties in relation to the emotional experience with the material.	physical and perceptual	→	Material	Colour
Piselli A, Colombo S, Faucheu J et al. [36]	The sensory properties of smart materials can affect the sensory language of the product represented in this case by colour, shape and texture.				
Granzier J J M, Vergne R and Gegenfurtner K R [37]	The roughness and glossiness of the material have effect on colour perception.	perceptual			

Table 1. Influence relations between features (b)

Oyama T and Nanri R [38]	The hue and brightness of the colour affects the perception on size.	perceptual	→	Colour	Dimension
Hagtvedt H and Brasel S A [39]	Colour saturation influences the product size perception.				
Weaver P M and Ashby M [40]	The material and the shape are connected as the shape in which a material can be formed is limited by the material itself.	physical	←	Shape	Material
Hodgson S N B and Harper J F [32]	Materials may have constraints related to size, shape or geometry.				
Vangorp P, Laurijssen J and Dutre P [41]	The material can be affected by the perception of shape or illumination.	perceptual	→		

As it can be seen in Table 1, it is relevant not only the type but the direction of influence. Therefore, as presented in Figure 1, all the features are related more or less with each other and even the direction may differ. For example, the relation between material and dimension, as seen before, they are correlated but changing the product's dimension it won't change its material, it is a physical influence in one direction, material to dimension [32][33], however, there is as well a perceptual influence because the material can be defined by a certain colour and implicitly hue, or shine and glossiness etc. and its perception has an influence on product dimension [34].

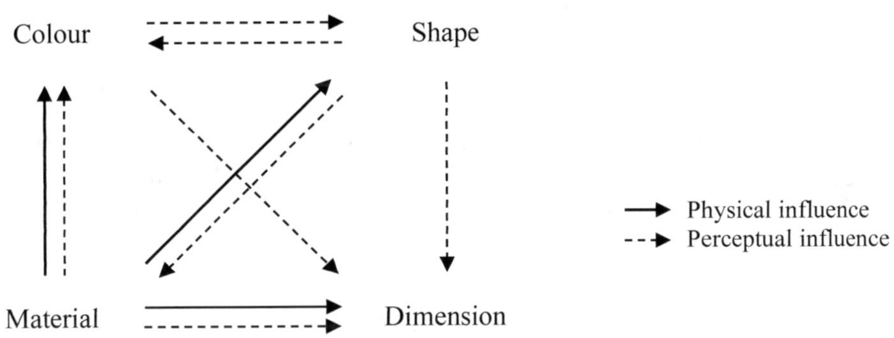

Figure 1. Direction and type of influence between product features.

Considering the relations between product features, there are connections that although do not have a component of influence, there is to be considered the possibility of perceptual interdependence, for example, certain shapes are correlated with different meanings for the materials [42], affective association with a colour differs when it is perceived on different materials [43] and also, there are several recent studies [44] [45] [46] that while contradict Kandinsky [47] in shape-colour associations, they have some common points. In this case it is not highlighted on how changing the shape parameters can influence the perception of colour or vice versa, however, there are preference associations with a higher degree of occurrence, for example the association between circular shapes and the colours red or yellow.

Whether it is a simple link of preference or they even have a direct impact on the product itself, it is important to understand the correlations between features, as they play a major role in product design and the assessment phase and improvement of the product. The assessment highlights which feature needs improvement but if either one in question is modified, the change itself may affect another feature and eventually the entire product, being necessary to acknowledge the relations between them in order to obtain the desired product.

2.3 Product and context of use

Design, furthermore than relating a product to a specific consumer profile considering technical or ergonomic restrictions, it exceeds them and incorporates aesthetic and symbolism elements, therefore the culture through its implications becomes an integral part of the product details and their significance. The cultural meaning of goods [48] is reflected both by symbolism in the relation to the consumer tradition as a part of the market segment and also in the cultural context of the product that involves a historical analysis and past references [49] that provide insights in the way it and its features evolved over time, understanding the specific trends and as well as the evolution of attributes through brand identity and differentiation [50].

The context and the extent to which it is attractive also has the power to change the perception of product quality [51] and ensure the effect on the purchase decision. Above that, in the development stage, the product usability [52] must be fully understood and the product, through its attributes, must take into account the circumstances of use. In addition, there is the situation in which it is evaluated the way the product integrates in the environment and by technotope analysis it is studied the relation of coexistence between the product and the environment [53]. Thus, in this paper, as presented in Figure 2, the context was proposed as a third category of relational analysis and includes a historical branch with the product's cultural implications and its evolution in time and a secondary one with reference to the environment and how the product is correlated to its context of use.

3. Study of correlations between sections

The elements of each proposed section were presented in Figure 2, alongside with the specific correspondent, section A for the user (market segment), B for the product features (these having a central role in the relation between the other two sections) and section C for context.

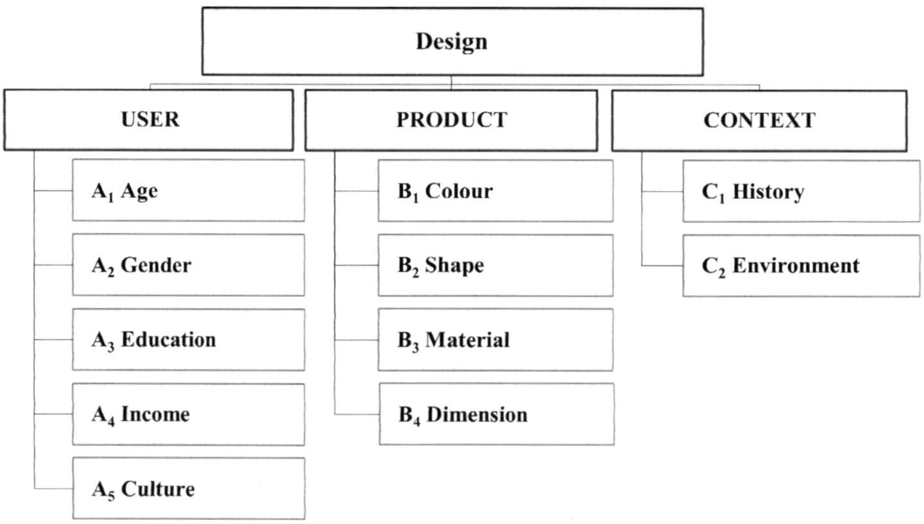

Figure 2. Proposed analysis sections.

In the previous chapter it was established how the elements impact each other within the same section. In order to determine how the product features are related to the market segment or context and to establish the existing connections, in Table 2 are presented previous studies that analyse the relations between sections. The correlations are intricate and the absence of a mention does not represent the absence of the corresponding connection, they just are not part of the prior research in question, as the studies do not contradict but approach the subject from different perspectives.

Table 2. Correlations between sections (a)

Prior research	Statements and findings	A. Market segment	B. Product feature	C. Context
Therrell J A [54]	The product features must be age-appropriate from the perspective of size, shape, material and colour.	A_1	B_1, B_2, B_3, B_4	
Holbrook M B and Schindler R M [55]	The customer preferences are defined by nostalgic influences related to age and time frame.	A_1		C_1
Lesot M J, Bouchard C, Detyniecki M, et. al. [56]	The shape has meaning associations in relation to femininity and masculinity.	A_2	B_2	
Jadva V, Hines M and Golombok S [57]	There are associations between the child's gender and product colour or shape preferences.	A_2	B_1, B_2	
Tilburg M and Lieven T [58]	Product gender can be considered from the perspective of colour, shape, material and size.	A_2	B_1, B_2, B_3, B_4	
Kessous A and Roux E [59]	As a link between age and tradition, nostalgia in approaching products is described as a connection to the past.	A_1, A_5		C_1
Sliburyte L and Skeryte I [60]	Colour perception differs depending on demographic factors such as age, gender and education.	A_1, A_2, A_3	B_1	
Arabi A [61]	Consumer behaviour through the influences of age, gender and culture shows variations in colour selection.	A_1, A_2, A_5	B_1	
Westland S and Shin M J [62]	Colour preferences varies depending on gender and age, ethnicity being also relevant.	A_1, A_2, A_5	B_1	
Tilley A R [63]	Age, gender and ethnicity are relevant in reporting to product size. Their relation to the environmental context must take into account also the scale, movement, contrast and colour.	A_1, A_2, A_5	B_1, B_2, B_4	C_2
Karana E, Hekkert P and Kandachar P [64]	In the selection of the material (through its shape, meaning and other properties) both the typology of the user and the context of use must be considered.	A_1, A_2, A_3, A_5	B_2, B_3	C_2
Panero J and Zelnik M [65]	Anthropometry is influenced by age, gender, culture and socioeconomic status. Also, the dimensions are related to the type of activity of the context of use.	A_1, A_2, A_3, A_4, A_5	B_4	C_2
Crilly N, Moultrie J and Clarkson P J [66]	The personal characteristics and the culture of the consumer are related to the visual attributes and the context of use of the product.	A_1, A_2, A_5	B_1, B_2, B_3, B_4	C_2
Choate J M [67]	There are correlations between age, education, culture and socio-economic status in relation to colour preferences.	A_1, A_3, A_4, A_5	B_1	
Bloch P H [68]	The consumer characteristics and the cultural traits have influence on the products' form (described as shape, scale, material, colour etc.).	A_1, A_2, A_3, A_4, A_5	B_1, B_2, B_3, B_4	
Zuo H, Hope T, Castle P et. al. [69]	In approaching the material, both the category of subjects and the environmental context are taken into account.	A_1, A_2, A_3, A_4, A_5	B_3	C_2
Viková M, Vik M and Kania E [70]	Colour preferences vary and are influenced by cultural references.	A_5	B_1	
Bluntzera J B, Ostrosib E, Sagot J C [71]	There are correlations between the shape of cars and the cultural identity of a country.	A_5	B_2	

Table 2. Correlations between sections (b)

Zhang Y, Feick L and Price L J [72]	Preferences for angular versus rounded shapes differ in relation to cultural influences.	A_5	B_2	
Karana E [73]	The meaning of the material, even by the form in which it is presented, is influenced both by culture and history.	A_5	B_2, B_3	C_1
Xue H and Woolley M [74]	Through cultural nostalgia, the designer has the opportunity to use product attributes traced in the past in order to evoke familiarity.	A_5		C_1
Wastiels L and Wouters I [75]	In the selection of the material in architecture, are relevant both the material's physical properties and the processes involved as well as the meaning of the cultural and use context.	A_5	B_3	C_1, C_2
Dumitrescu A, Ulmeanu M E and Crăciun A E [53]	Colour is a relevant factor in the harmonious relation between the product and its context of use.		B_1	C_2
Brooker G and Stone S [76]	The elements of form and structure of the interior space are associated in relation with the history and context of the environment.		B_2	C_1, C_2
Tseng Y S and Ho M C [77]	Integrating product elements that invoke the past is a design strategy meant to generate emotional response from the consumer.		B_1, B_2, B_3	C_1

The relations between the sections were structured and displayed in the proposed correlation matrix (Figure 3), in which was presented the graphic and numerical notation of the sections' relations. The connections were noted according to the degree of interdependence, where a relation of rank 3 is considered of high impact, rank 2 of medium impact and rank 1 for minimum impact, being also the possibility of no correlation. The missing link is between age and gender, although there is a well-known gender difference in life expectancy, it cannot be determined a direct connection between gender and age relevant to the proposed subject.

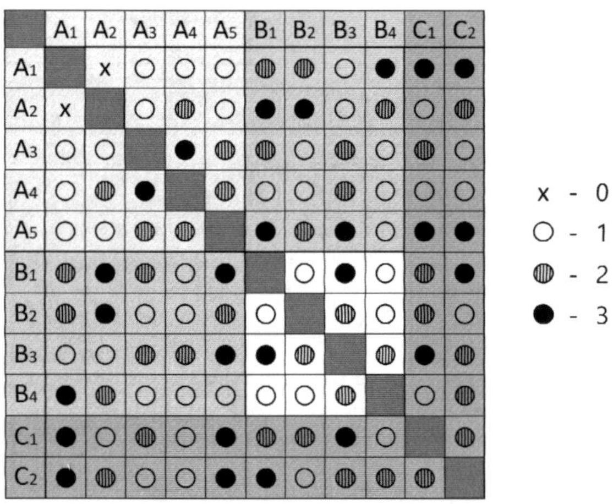

	A_1	A_2	A_3	A_4	A_5	B_1	B_2	B_3	B_4	C_1	C_2
A_1		0	1	1	1	2	2	1	3	3	3
A_2	0		1	2	1	3	3	1	2	1	2
A_3	1	1		3	2	2	1	2	1	2	1
A_4	1	2	3		2	1	1	2	1	1	1
A_5	1	1	2	2		3	2	3	1	3	3
B_1	2	3	2	1	3		1	3	1	2	3
B_2	2	3	1	1	2	1		2	1	2	1
B_3	1	1	2	2	3	3	2		2	3	2
B_4	3	2	1	1	1	1	1	2		1	2
C_1	3	1	2	1	3	2	2	3	1		2
C_2	3	2	1	1	3	3	1	2	2	2	

Figure 3. Correlation matrix between the three sections (graphic and numerical version).

When analysing the matrix, it can be seen that although two elements are defined by the maximum correlation, each in turn can be strongly influenced by another element, for instance, the connection between the product dimension and user age, as the age also has a rank 3 link to the context of use, both through the environment that is specific to it and through the temporal affinity influences that can be found in the product details, since it is a strong connection between the user age and historical frame.

4. Product features in relation to the market segment and context

In order to deduce the implications and the way the product features are influenced, following the interpretation of the matrix (Figure 3) and the influence relation between features (Figure 1), the representative formulas for each feature were concluded as follow, equation (1) for colour, equation (2) for shape, equation (3) for material and equation (4) for dimension, as the corresponding indices are presented in Figure 2.

For each feature, the first relation represents its relation to the market segment and context, meaning the elements on which it is conditioned in its analysis, and the second one represents the correlation of influence within the same section and how it can affect another feature or can be affected in turn. In which case, it was noted with minus if it has influence on another feature (implicitly the other feature is influenced by it), with plus if it is influenced by another feature or plus and minus if it is bidirectional in the relation of influence.

4.1 Colour

The colour is primarily correlated with gender-specific associations [57][58][60][61][62][66][68], cultural meaning [61][62][63][66][67][68][70] and also environment [53][63][66] as the colour of the product is related to the context. Secondary, colour schemes should be age appropriate [54][60][61][62][63][66][67][68], not necessarily as a restriction as there are chromatic preferences depending on the age segment. Additionally, colour it is related with the educational level [60][67][68] and history [77] as some shades are specific for certain historical periods, for instance, as art-deco is associated with golden metallic details. Although with less importance, the income [67][68] is also related to colour with associations regarding socio-economic status.

In its relation with the other features, the colour can be generated directly by the natural shade of the material [35][36][37], association that can be excluded if artificial dyes are used to cover the surfaces. In terms of connection with shape, the colour may influence [31] or be influenced by it [30] as it may affect also the perception on size [38][39], for instance, the light-coloured details seem larger and the dark ones seem smaller by comparison.

$$B_1 = \begin{cases} 3(A_2 + A_5 + C_2) + 2(A_1 + A_3 + C_1) + 1A_4 \\ 3B_3 \pm 1B_2 - 1B_4 \end{cases} \tag{1}$$

4.2 Shape

The shape of the product is firstly related to gender [56][57][58][63][64][66][68], there are gender preferences and among others, rounded shapes are associated with femininity and those defined by straight lines with masculinity. Secondary, the shape is related to ergonomic limitations bound to age [54][63][64][66][68] or ease of use and also cultural symbolism [63][64][66][68][71][72][73] and historical associations [73][76][77] most often given by design trends. Lastly, there are shape preferences according to the consumer's socio-economic status [64][68] and also associations between shape and environment [63][64][66][76] all the more so from an architectural point of view.

Physically, shape has technical limitations in relation to the material [32][40] and it can as well affect the perception of it and its reflectance [41]. Furthermore, the perception on shape can be influenced by colour [31] and it can influence it [30] as well as the dimension [28][29].

$$B_2 = \begin{cases} 3A_2 + 2(A_1 + A_5 + C_1) + 1(A_3 + A_4 + C_2) \\ \pm 2B_3 \pm 1B_1 - B_4 \end{cases} \tag{2}$$

4.3 Material

Material is strongly related to cultural symbolism [64][66][68][69][73][75] and the product's historical [73][75][77] evolutionary analysis. Of medium rank, is the relation with education [64][68][69] and associated preferences and also income [68][69] (for example, the price for the same product can differ with different materials, that could be the case for jewellery, textile products etc.) and finally context, since the product material should be adapted to the environment [64][66][69][75] both by colour and texture. Furthermore, although it is less important in this situation, similar to the shape case, the material too must be age-appropriate [54][64][66][68][69] especially if it involves safety in use. As far as gender [58][64][66][68][69] is linked to material, there are specific associations, as is the case with metallic products that are considered masculine.

In relation to the other features, as mentioned, material can define the colour of the product [35][36][37], may physically limit the form [32][40] and be perceptually influenced by it [41], however in its relation to dimension [32][33][34], the material determines the entire weight of the product, which has a direct impact, but does not directly affect its size.

$$B_3 = \begin{cases} 3(A_5 + C_1) + \ 2(A_3 + A_4 + C_2) + 1(A_1 + A_2) \\ \qquad -3B_1 \pm 2B_2 - 2B_4 \end{cases} \tag{3}$$

4.4 Dimension

The product dimension has an important relation with age [54][66][68] and implicitly with anthropometric limitation [63][65]. In the case of other ergonomic recommendations, gender [58][63][65][66][68] is relevant too in association to dimension. Regarding the size of the product and context, it's recommended that the product is scaled to the dimension of the room or any other environment of use [63][65][66]. In its relation to education and income, there are associations with socio-economic status [65][68] and it's not necessarily a good or a bad thing depending on the situation. Additionally, there is a link between the size of the product and socio-cultural meaning [63][65][66][68] as well as historical research that is relevant in fully understanding the way the product changed over time and evolved both technologically and through its aesthetic attributes, in this case the relation in time with the product size.

Within the same section, dimension is influenced through its weight [32][33] and size perception [34] by the material. In addition, as previously presented, the perception on product size differs also due to colour [38][39] or shape variations [28][29].

$$B_4 = \begin{cases} 3A_1 + 2(A_2 + C_2) + 1(A_3 + A_4 + A_5 + C_1) \\ \qquad 2B_3 + 1(B_1 + B_2) \end{cases} \tag{4}$$

Thus, when approaching an industrial product from the perspective of design assessment, for each attribute, it is recommended to consider it depending on the product typology, taking into account, gradually, the degree of interdependence of its relation to the market segment and context.

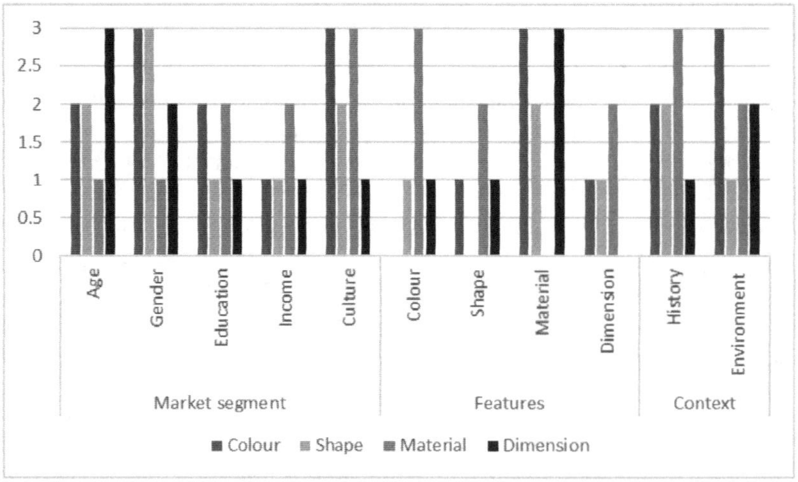

Figure 4. Product features in relation to market segment and context.

5 Conclusions

Within the paper, the proposed sections and subordinated elements were studied in turn, the market segment (age, gender, education, income and culture), the product features (colour, shape, material and dimension) and the context (history and environment), being analysed both the relations within the same section and the general correlations between them. The results were centralized and presented in the matrix of correlations along with the degree of impact between elements, from which the corresponding formulas for attributes were deduced, being presented the way that a product feature is influenced. Thus, the purpose of this paper is to help designers, either in product development or in assessment both after and before the final production (in the concept phase) by approaching the features based on the hierarchy of influence elements within the market segment and the product's context of use, being established the extent and the direction in which a feature is affected. Improving the product by modifying a certain feature, the proposed study highlights also what other feature is affected or what can influence it in turn, in order to achieve the desired results.

The presented study is not an actual assessment method and does not approach the industrial products from the perspective of the importance of aesthetics, ergonomics, symbolism etc. [78] However, it takes those principles into account and represents a foundation in design, since the proposed matrix of correlations is set as a frame of reference of the product in relation to its purpose and the perspective of use. As Dumitrescu states, except for functionality, product attributes cannot be considered absolutely positive or negative [79], the design assessment must account for the purpose of the product and the values of the market segment [80].

In regard to the study limitations, were considered the sections elements with major impact in the interdependence relation, thus in order to increase complexity, the market segment can be approached from an elaborate perspective that includes more from the other market segmentation categories with implications in detail in lifestyle or purchase behaviour. Furthermore, although they were implicitly included in the product features, the subordinate features could be studied individually such as considering the texture separately and not as an integral part of the material, as may be the case for brightness or gloss and colour, or other relevant subordinate features.

References

[1] Lees-Maffei G and Fallan K 2014 It's personal: subjectivity in design history *Design and Culture* **7** pp 5-27
[2] Atalay M 2007 Kant's aesthetic theory: subjectivity vs. universal validity *Percipi* **1** pp 44-52
[3] Travanini C 2008 *Aesthetic Value Judgements and the Challenge to Objectivity* vol **8** ed Dorsch F and Ratiu D E (Fribourg: Proc. of the European Society for Aesthetics) pp 515-526
[4] Cheng L C 2014 *From Subjectivity to Objective Evaluation: A Techno-Rationalist Approach of Assessment Design for Art & Design Education* (Penang: Int. Colloquium Of Art And Design Ed. Res.) pp 195-200
[5] Ulrich K T and Eppinger S D 2014 *Diseño y Desarrollo de Productos* (New York: McGraw-Hill, 5th edition) p 223
[6] Solomon M, Bamossy G, Askegaard S and Hogg M 2006 *Consummer Behavior, A European Perspective* (London: Prentice Hall, 3rd ed.)
[7] Peter J P and Olson J C 2010 *Consumer Behavior and Marketing Strategy* (Boston: McGraw-Hill/Irwin, 9th edition)
[8] Barmola K and Srivastava 2010 The role of consumer behaviour in present marketing management *Productivity* **51**(3) pp 268-275
[9] Correard S and Amrani H E 2011 *The Impact of Marketing on Customer's Behaviour, Influence or Manipulation?* (Halmstad: Dissertation paper Halmstad University)
[10] Lancaster K J 1966 A new approach to consumer theory *The J. of P. E.* **74**(2) pp 132-157
[11] Datta S 2016 Customer vs consumer – a different perspective *Int. J. in Mgt. and Soc. Sci.* **4**(9) pp 109-114

[12] Tynan C A 1987 Marketing segmentation *J. of Mktg. Mgt.* **2**(3) pp 301-335
[13] Martin G 2011 The importance of marketing segmentation *Am. J. of Bus. Ed.* **4**(6) pp 15-18
[14] Dolnicar S, Grün B and Leisch F 2018 *Market Segmentation Analysis* (Electronic book retrieved from https://www.econstor.eu/bitstream/10419/182365/1/978-981-10-8818-6.pdf) pp 42-45
[15] Patel J and Bansal A 2018 Effect of demographic variables on e-marketing strategies: A review *Int. J. of Ac. Res. and Dev.* **3**(1) pp 311-321
[16] Ramya N and Mohamed S A 2016 Factors affecting consumer buying behavior *Int. J. of App. Res.* **2**(10) pp 76-80
[17] Wolla S A and Sullivan J 2017 Education, income, and wealth *Federal Reserve Bank of St. Louis - research.stlouisfed.org*
[18] Blanden J and Gregg P 2004 Family income and educational attainment: a review of approaches and evidence for Britain *Oxford Rev. of Econ. Pol.* **20**(2) pp 245-263
[19] Wang C, Wan G, Luo Z and Zhang X 2017 Aging and inequality: the perspective of labor income share *ADBI Wrk. Pap. Ser.* (764)
[20] Grybaitė V 2006 Analysis of theoretical approaches to gender pay gap *J. of Bus. Econ. and Mgmt.* **7**(2) pp 85-91
[21] Smyth E 2007 Gender and education *Int. St. in Educ. Inequality, Theory and Policy* **1** pp 135-153
[22] Moss G 2016 *Gender, Design and Marketing* (New York: Routledge)
[23] Neale L, Robbie R and Martin B A S 2016 Gender identity and brand incongruence: when in doubt, pursue masculinity *J. of Strategic Marketing* **24**(5) pp 347-359
[24] Cakiroglu I 2017 Genders of products: creating genderless design *Conf. Senses & Sensibility'17: Design Beyond Borders and Rhizomes* (Funchal: Research Centre in Design and Communication)
[25] Blijlevens J, Creusen M E H and Schoormans J P L 2009 How consumers perceive product appearance: the identification of three product appearance attributes *Int. J. of Design* **3**(3) pp 27-35
[26] Johnson K W, Lenau T and Ashby M F 2003 The aesthetic and perceived attributes of products *Int. Conf. on Engg. Design ICED (Stockholm)* ed. Norell M (Glasgow: The Design Society)
[27] Sunde H 2017 Product semantics, how design can affect user response and behavior *Norwegian Univ. of Sci. and Tech. NTNU* pp 1-13
[28] Frayman B J and Dawson W E 1981 The effect of object shape and mode of presentation on judgments of apparent volume *Perception & Psychophysics* **29**(1) pp 56-62
[29] Sevilla G and Kahn B E 2014 The completeness heuristic: product shape completeness influences size perceptions, preference, and consumption *J. of Mktg. Res.* **51**(1) pp 57-68
[30] Bloj M G, Kersten D and Hurlbert A C 1999 Perception of three-dimensional shape influences colour perception through mutual illumination *Nature* **402** pp 877-879
[31] Kingdom F A A, Wong K, Yoonessi A and Malkoc G 2006 Colour contrast influences perceived shape in combined shading and texture patterns *Spatial Vision* **19**(2-4) pp 147-159
[32] Hodgson S N B and Harper J F 2004 Effective use of materials in the design process – more than a selection problem *The 7th Int. Conf. on Engg. and Product Design Ed.* (Delft: Delft University of Technology)
[33] Miltenović A, Banić M and Miltenović V 2014 Role and importance of lightweight design in the product development *COMETa 2nd Int. Sci. Conf.* (Jahorina: University Of East Sarajevo) pp 529-538
[34] Fleming R W 2014 Visual perception of materials and their properties *Elsevier - Vision Res.* **94** pp 62-75
[35] Kesteren I E H, Stappers, P J and Bruijn J C M 2007 Materials in products selection: tools for including user-interaction in materials selection *Int. J. of Design* **1**(3) pp 41-55
[36] Piselli A, Colombo S, Faucheu J, Delafosse D and Curto B 2018 Educational tools to teach design students the dynamic behaviours of smart materials *5th Int. Conf. for Design Ed. Res.* (Ankara: Middle East Technical University)

[37] Granzier J J M, Vergne R and Gegenfurtner K R 2014 The effects of surface gloss and roughness on color constancy for real 3-D objects *J. of Vision* **14**(2) pp 1-20
[38] Oyama T and Nanri R 1960 The effects of hue and brightness on the size perception *Japanese Psy. Res.* **2**(1) pp 13-20
[39] Hagtvedt H and Brasel S A 2017 Color saturation increases perceived product size *J. of Cons. Res.* **44**(2) pp 396–413
[40] Weaver P M and Ashby M 1996 The optimal selection of material and section-shape *J. of Engg. Design* **7**(2) pp 129-150
[41] Vangorp P, Laurijssen J and Dutre P 2007 The influence of shape on the perception of material reflectance *ACM Transactions on Graphics* **26**(3) pp 1-9
[42] Karana E, Weelderen W and Woerden E J 2007 The effect of form on attributing meanings to materials *Proc. in ASME Int. Design Engg. Tech. Conf.* vol 37 (Las Vegas: Computers and Information in Engineering Conf.)
[43] Ulusoy B and Olguntürk N 2017 Understanding responses to materials and colors in interiors *Wiley Periodicals* **42**(2) pp 261-272
[44] Dreksler N and Spence C 2019 A critical analysis of colour–shape correspondences: examining the replicability of colour–shape associations *i-Perception* **10**(2) pp 1-34
[45] Chen N, Tanaka K, Matsuyoshi D and Watanabe K 2015 Associations between color and shape in Japanese observers *Psy. of Aesthetics, Creativity, and the Arts* **9**(1) pp 101-110
[46] Wuerger S and Makin A 2013 The IAT shows no evidence for Kandinsky's color-shape associations *Frontiers in Psychology* **4**(616) pp 1-7
[47] Kandinsky W 1914 *The Art of Spiritual Harmony* (Boston and New York: Houghton Mifflin Company)
[48] Mccracken G 1986 Culture and consumption: a theoretical account of the structure and movement of the cultural meaning of consumer goods *J. of Cons. Res.* **13**(1) pp 71-84
[49] Derbyshire J and Giovannetti E 2017 Understanding the failure to understand New Product Development failures: Mitigating the uncertainty associated with innovating new products by combining scenario planning and forecasting *Tech. Fcst. & Soc. Chg.* **325** pp 334-344
[50] Farhana M 2012 Brand elements lead to brand equity: differentiate or die *Inf. Mgmt. and Bus. Rev.* **4**(4) pp 223-233
[51] Schnurr B, Brunner-Sperdin A and Stokburger-Sauer N E 2017 The effect of context attractiveness on product attractiveness and product quality: the moderating role of product familiarity *Mktg. Letters* **28**(2) pp 241-253
[52] Chamorro-Koc M and Popovic V 2008 Context-of-use and the design of user-product interactions: exploring causal relationships *Proc. of the Design Res. Soc. Conf.* (Sheffield: Sheffield Hallam University)
[53] Dumitrescu A, Ulmeanu M E and Crăciun A E 2020 Testing the technotope concept *UPB Sci. Bulletin, Series D* **82**(3) pp 241-250
[54] Therrell J A 2002 Age determination guidelines: Relating children's ages to toy characteristics and play behavior *Cons. Product Safety Commission* (SUA: Electronic paper retrieved from https://www.cpsc.gov/s3fs-public/pdfs/blk_pdf_adg.pdf) pp 8-10
[55] Holbrook M B and Schindler R M 1996 Market segmentation based on age and attitude toward the past concepts, methods, and findings concerning nostalgic influences on customer tastes *J. of Bus. Res.* **37** pp 27-39
[56] Lesot M J, Bouchard C, Detyniecki M and Omhover J F 2010 Product shape and emotional design - an application to perfume bottles *Int. Conf. on Kansei Engg. and Em. Res.* (Paris: Arts et Métiers ParisTech)
[57] Jadva V, Hines M and Golombok S 2010 Infants' preferences for toys, colors, and shapes: sex differencesand similarities *Arch. of Sexual Behaviour* **39**(6) pp 1261-1273
[58] Tilburg M and Lieven T 2013 You sexy thing: product design as a source of product gender and product value *41st Conf. of the Eur. Mktg. Assoc.* (Istanbul: Istanbul Technical University)

[59] Kessous A and Roux E 2008 Nostalgia as connection to the past: A semiotic analysis *Qual. Mkt. Res.* **11**(2) pp 192-212
[60] Sliburyte L and Skeryte I 2014 What we know about consumers' color perception *Proc. Soc. and Beh. Sci.* **156** pp 468-472
[61] Arabi A 2017 Influence of colors on consumer behavior "Conceptual and theoretical approaches" *Acad. Brâncuşi, Economy Series* (3) pp 163-170
[62] Westland S and Shin M J 2015 The relationship between consumer colour preferences and product-colour choices *J. of the Int. Colour Assoc.* **14** pp 47-56
[63] Tilley A R 1993 *The Measure of Man and Woman: Human Factors in Design* (New York: Watson-Guptill Publications) pp 10-55
[64] Karana E, Hekkert P and Kandachar P 2010 A tool for meaning driven materials *Elsevier – Materials and Design* **31**(6) pp 2932-2941
[65] Panero J and Zelnik M 1979 *Human Dimension and Interior Space* (London: Architectural Press Ltd) pp 73-113
[66] Crilly N, Moultrie J and Clarkson P J 2004 Seeing things: consumer response to the visual domain in product design *Elsevier - Design Stud.* **25**(6) pp 547-577
[67] Choate J M 1977 *Color Preferences Relative to Demographic Factors, Personality, and Self-Concept With Implications for Homemaking Education* (Lubbock: Dissertation paper Faculty of Texas Tech University)
[68] Bloch P H 1995 Seeking the ideal form: product design and consumer response *J. of Mktg.* **59**(3) pp 16-29
[69] Zuo H, Hope T, Castle P and Jones M 2001 An investigation into the sensory properties of materials *In Proc. of The 2nd Int. Conf. on Affective Human Factors Design* (London: Asean Academic Press) pp 500-507
[70] Viková M, Vik M and Kania E 2015 *Cross-Cultural Variation of Color Preferences* (Liberec: Technical University of Liberec) pp 285-288
[71] Bluntzera J B, Ostrosib E and Sagot J C 2014 Car styling: a CAD approach to identify, extract and interpret characteristic lines *24th CIRP Design Conf.* vol 12 (Belfort: Elsevier) pp 258-263
[72] Zhang Y, Feick L and Price L J 2006 The impact of self-construal on aesthetic preference for angular versus rounded shapes *Soc. for Pers. and Social Psy.* **32**(6) pp 794-805
[73] Karana E 2010 How do materials obtain their meanings? *METU JFA* **27**(2) pp 271-285
[74] Xue H and Woolley M 2013 Creatively designing with/for cultural nostalgia: designers' reflections on technological change and loss of physicality *Proc. Conf. IASDR* (Tokyo: Int. Assoc. of Soc. of Design Res.)
[75] Wastiels L and Wouters I 2008 Material considerations in architectural design: a study of the aspects identified by architects for selecting materials *Proc. Conf. Design Res. Soc.* (Sheffield: Hallam University)
[76] Brooker G and Stone S 2007 *Form and Structure in Interior Architecture* (Lausanne: VA Publishing SA)
[77] Tseng Y S and Ho M C 2011 Designing the personalized nostalgic emotion value of a product *Design, User Experience, and Usability. Theory, Methods, Tools and Practice* pp 664-672
[78] Crăciun A E 2019 Strategic outlook in industrial design assessment based on product category *IOP Conf. Ser.: Mater. Sci. Eng.* **682** 012006
[79] Dumitrescu A 2013 *Tratat de Design* (București: Editura Academiei Române) p 610
[80] Dumitrescu A and Crăciun A E 2019 Testing criteria for a complex assessment method for industrial design *34th IBIMA Conf. WOS:000556337401026* pp 653-664

IOP Conf. Series: Materials Science and Engineering 1018(2021) 012022 doi:10.1088/1757-899X/1018/1/012022

Brief review of methods and techniques used in Learning Factories in the context of Industry 4.0

G C Neacsu*, **I G Pascu, E L Nitu and A C Gavriluta**

University of Pitesti, Faculty of Mechanics and Technology, Manufacturing and Industrial Management Department, Targul din Vale Street No.1, Romania

*georgiana.neacsu@upit.ro

Abstract. The rapid evolution of science and technology has forced the processes and systems of production, training and learning to adapt to innovation and the digital era of industrial technology. In the case of production processes and systems, innovation and digitalization refer to the transformation of production units into smart factories. In the case of training and learning processes and systems, innovation and digitalization lead to the creation of learning factories. Thus, this process involves the use of modern methods and techniques to develop innovative training, learning and production processes and systems. This paper is an analysis of recent studies and seeks to briefly present the evolution of industrial production processes and systems and Industry 4.0 but also the concept of learning factory. The paper also presents a review of the main methods and techniques used in innovative industrial training, learning and production processes and systems. Finally, some conclusions related to the use of these methods and techniques are summarized.

1. Introduction

1.1. The evolution of industrial production processes and systems

The evolution of production processes began in the 600s with the spread of water mills. Over the years, the most known production processes have remained: craft production, which began in 1850, followed by American production in 1900, which is no longer used, then in 1913 mass production, 1955 Lean production, then mass customization in 1980, global manufacturing in 2000 and sustainable production in 2010 [1], presented in figure 1 [1,2].

The evolution of production systems, represented in figure 2 [3], began around the 1900s with the appearance of the assembly line, followed by the Toyota production system (TPS) in 1960, then flexible manufacturing systems (FMS) in 1980, reconfigurable manufacturing systems (RMS) in 1990 and Cloud-based manufacturing systems (CBMS).

1.2. The evolution of Learning Factory

The history of learning factories (LF) began in the late 1980s or early 1990s in Germany, with the CIM learning factory, largely unknown today [4]. The first form of learning factories appeared since the middle ages, in the form of crafts in workshops, taught by craftsmen to disciples [5]. After the end Second World War, Japan established and developed systematic quality systems and quality circles in 1962. In the beginning of the 1970s Germany established learning in small groups with the concept of the so-called Lernstatt (a composition of the German words "lernen" = to learn and "Werkstatt" =

factory/workshop). In 1980, there was another change in the way work processes were improved in which application-oriented learning methods related to specific complex situations were addressed. This development of workplace learning has been facilitated by the emergence and spread of new information and communication technologies, but also by new organizational concepts. At the beginning of the 1990s, new theoretical approaches to application-oriented learning were developed [6]. Figure 3 [6] shows the historical development of learning approaches, from the first form to the present.

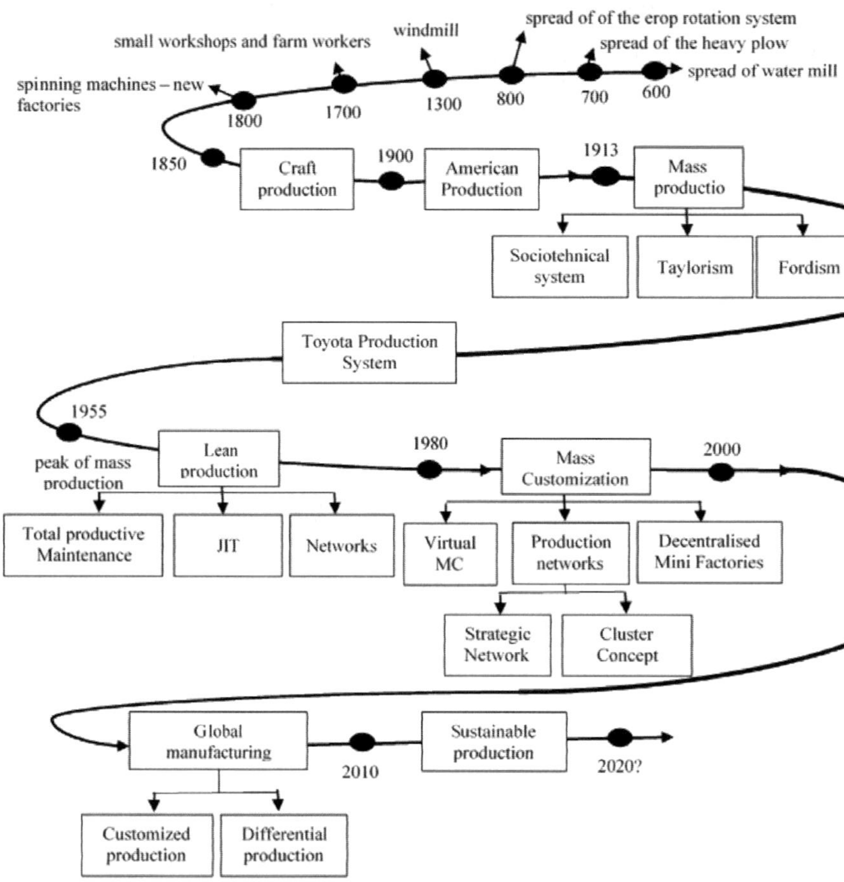

Figure 1. The evolution of production processes, adapted after [1] and [2]

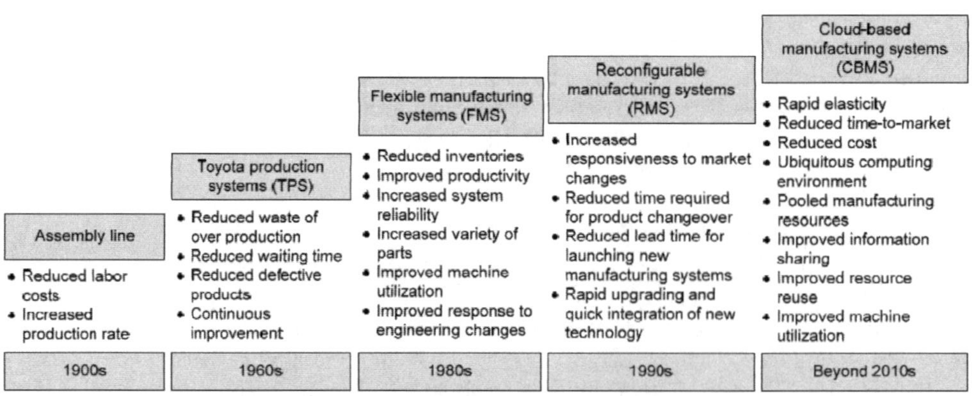

Figure 2. The evolution of production systems [3]

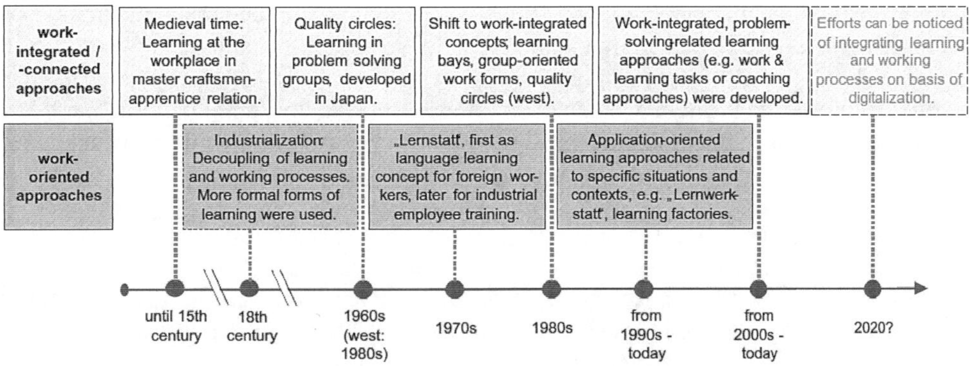

Figure 3. Historical development of work-related learning approaches [6]

1.3. The industrial revolution

The term Industry 4.0 officially appeared in 2011 at industrial trade fair "Hannover Messe" [7]. The prefix 4.0 shows us that Industry 4.0 is a direct successor to the other three industrial revolutions developed in previous years.

Thus, the first industrial revolution appeared at the end of the 18th century and refers to the introduction of mechanical production equipment driven by the force of water or steam. The second industrial revolution refers to the use of electric drive of production equipment and the realization of a mass production based on the division of labor and developed in the late nineteenth century. Later in the twentieth century came the third industrial revolution that focused on the use of electronic systems and information technology in production automation [8]. In figure 4 [9] the evolution of the technologies of the four industrial revolutions is presented.

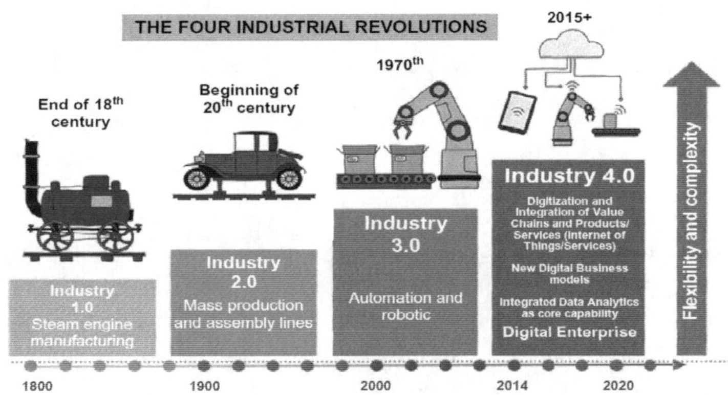

Figure 4. The four industrial revolutions [9]

This paper aims to present the concept and a classification of learning factory and a short review of the main methods and techniques used in innovative industrial training, learning and production processes and systems.

2. Learning Factory

2.1. The concept of Learning Factory

The learning factory can be seen as an idealized replica of an industry section/workshop in which informal, non-formal and formal learning can take place. Thus, learning factories contribute greatly to a healthy connection between industry, research and education [6].

The literature describes various positive effects of the learning factory concept such as: enhancement of the quality of education and training in general [10], creation of possibilities for research, innovation, and technology transfer [11], catalyst for business creation [12], development of soft skills and interdisciplinary competences [13], strengthening of attitude and work philosophy [14], motivation to learn and motivation and confidence to apply [15], overcoming problems of traditional teaching methods, strengthening of the link between industry and academia [16], strengthening the link between education and research [17], improving the prestige of industry and manufacturing and design education [18].

2.2. Learning Factory classification

Analyzing the existing learning factories, it can be stated that they are most frequently found in the following areas: learning factories in education, in training and learning factories in research. In learning factories in the field of education, in general, several ways of learning are approached: open, steered, and closed student projects; active learning in learning factories; action-oriented learning in learning factories; experiential learning and learning factories; game-based learning in learning factories and gamification; problem-based learning in learning factories; project-based learning in learning factories; research-based learning in learning factories [6]. The learning factory, for educational scope, can be used for stand-alone student projects and industry partnered student projects [19], see figure 5 [6].

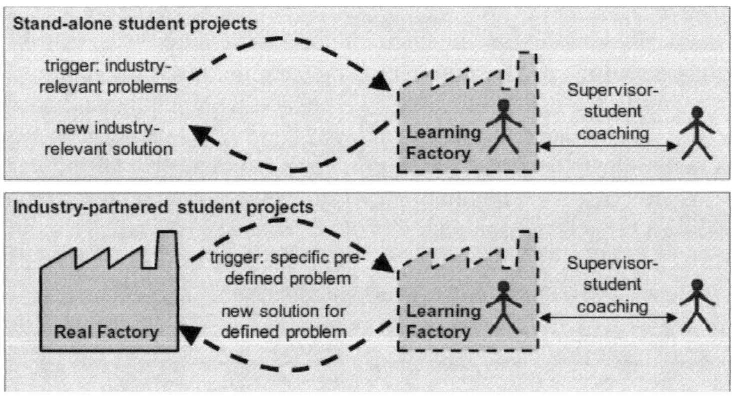

Figure 5. The use of learning factories in education in connection with stand-alone and industry - partnered projects [6]

In learning factories used for learning, in most cases, the main goal is to competency development [16], can be used as a part of change management approaches [20], and in recent years, in context with digitalization and Industry 4.0, in the field of demonstration of new technologies and innovations [21].

In the field of research, the link with learning factories is made by learning factories as research objects and learning factories as research enablers [6]. The concept is presented in figure 6 [22].

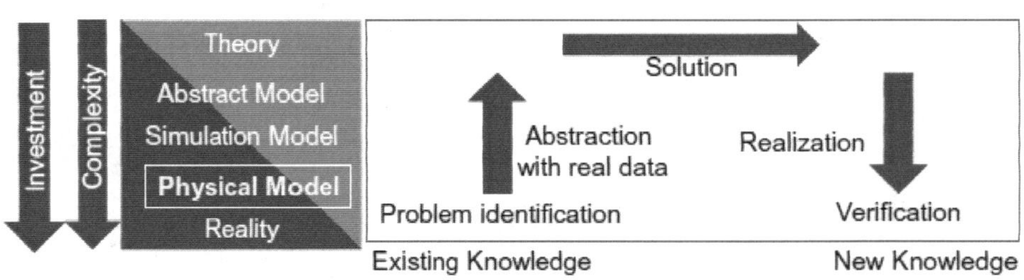

Figure 6. Learning factories as research enablers [22]

This concept describes how research problems are identified and solutions are tested and verified with the help of the physical model learning factory, at low cost and complexity compared to research in real production factories.

Table 1 gives an overview of established learning factories existence and presented in various scientific articles.

Table 1. Overview on existing learning factories

Name of Learning Factory	Location	Activity type			Article
		Education	Training	Research	
Alberta Learning Factory	University of Alberta, Edmonton, Canada	X		X	[23]
AM model factory	FAPS/ KTmfK, Nürnberg, Germany		X	X	[24]
Anglo American Training Center	Johannesburg, South Africa		X		[25]
BERTHA	Bremen Institute for Mechanical Engineering, University Bremen, Germany	X	X	X	[26]
CubeFactory	TU Berlin, Germany		X		[27]
FMS Training Center	Laboratory of Mechanical Engineering and Industrial Systems, Tampere University of Technology, Finland	X			[28]
Industrial Engineering Laboratory	IPS, TU Dortmund, Germany	X			[29]
Learning Factory at the Campus Velbert/ Heiligenhaus	University of Applied Sciences Bochum Heiligenhaus, Germany	X	X		[30]
Learning Factory Split	FESB, University Split, Croatia	X			[31]
Live Training Center	Live Training Center, Bruchsal, Germany		X		[32]
Model Factories	McKinsey, U.S.A		X		[33]
Product Development Process Learning Factory	Laboratory for Computer Integrated Design and Manufacturing, Methodist University of Piracicaba, Brazil	X			[34]
Railway Operation Research Center	Chair of Railway Engineering, TU Darmstadt Germany	X	X	X	[35]
SEPT Learning Factory	Booth School of Engineering Practice and Technology, McMaster University, Canada	X	X	X	[36]
Textile Learning Factory 4.0	ITA Aachen, Germany		X	X	[37]

Name of Learning Factory	Location	Activity type			Article
		Education	Training	Research	
The PuLL Learning Factory	Competence Center for Production and Logistics, University of Applied Sciences Landshut Germany	X		X	[38]
XPRES Laboratory	KTH Stockholm, Sweden	X		X	[39]
Learning Factory EAFIT University	University EAFIT, Medellín, Columbia	X			[40]
Lean Learning Factory	University of Pitesti, Romania	X	X	X	[41]

3. Methods and techniques used in innovative industrial training, learning and production processes and systems

Once with development of the concepts of Learning Factory and Industry 4.0, methods, techniques and tools in the area of training, learning and production processes and systems have also been developed. The most common, in the studied literature, are summarized in table 2.

Table 2. Methods, techniques and tools

Methods	Techniques and tools	Application domain	Article
Virtual / augmented / mixed reality	Unity game engine, questionnaires	education	[42]
	Holographic devices, real-time production tracking		[43]
	HTC-Vive, Jack 8.4., Process Simulate, module: SCARE and Jack	production	[44]
	Unity 3D, Microsoft HoloLens, Software CAD	product design	[45]
Digital assistant	RTLX	industrial	[46]
	C++ and Qt	industrial	[47]
Experiment-based learning	Kanban, Industry 4.0 technologies: Iot and internet of things (iiot), cyber-physical systems	research	[36]
	Systems CAD/CAM/CAE, software Teamcenter, NX, Microsoft Office	education	[34]
	Fused deposition modeling (FDM), selective laser melting (SLM) for metals, and selective laser sintering (SLS)	industrial	[24]
Project-based learning	Flow analysis, MATLAB, RFID, cyber-physical systems	education	[48]
	ERP, cyber-physical systems, printer Lego-3D		[23]
Problem-based learning	Simulation	education	[29]
	Just-in-Time, Push-Pull production, Tact time, Kanban, Kaizen, VSM		[31]
Intuitive learning	Print 3D, recycle	industrial	[27]

IOP Conf. Series: Materials Science and Engineering **1018** (2021) 012022 doi:10.1088/1757-899X/1018/1/012022

A brief description of the specific methods and techniques used in Learning Factories in the context of Industry 4.0 and their application situations are given below.

3.1. *Virtual/ augmented/ mixed reality*

The concepts of virtual reality, augmented reality and mixed reality are among the most common innovative methods used in the development of training, learning and production processes and systems.

Virtual reality (VR) is considered a paradigm that through computers and human-computer interfaces creates the effect of a three-dimensional world [49] in which the user can obtain sensory experiences in a similar way to the real world [50]. Phrase virtual reality has been proposed, in 1989, by Jaron Lanier [51].

In the paper [42], Darque Pinto studying the impact of using virtual reality in learning a foreign language compared to the conventional audio method. Thus, the experiment involved 12 students, aged between 12 and 15 years old, from a private school in Portugal. The materials used in the study were delivered in audio and VR format, a knowledge retention objective test (in English), and a sense of presence and a satisfaction questionnaire (in Portuguese). The simulation contains two scenarios created by using the Unity game engine. The first scenario is a formal dialogue, which takes place in an office, while the second scenario refers to an informal dialogue, which takes place in a pub. Specific props and characters were introduced in both scenarios. The tools used in the research were various questionnaires related to presence, satisfaction and verification of knowledge. The questionnaire used for measure the sense of presence was Igroup Presence Questionnaire (IPQ) [52], the one for the satisfaction of the participant was After-Scenario Questionnaire (ASQ) [53], and for to evaluate students' knowledge retention an eight questions multiple-choice test was created by the English teachers of the school. Each study participant tested each scenario, alternatively. Following the analysis of the results for the presence and satisfaction, it was found that the VR learning mode is preferred by the participants (figure 7, [42]).

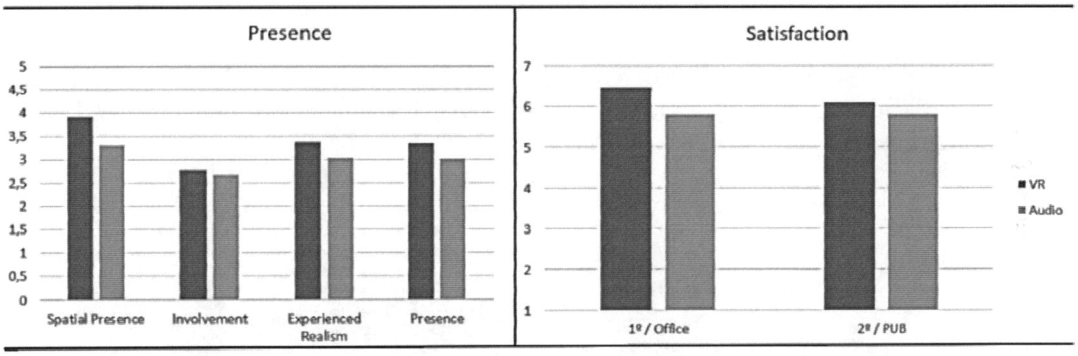

Figure 7. Graph that shows the results of the questionnaires [42]

In the case of knowledge retention in none of the conditions it was reached the minimum value of 50%, but it could be observed in the informal scene, virtual reality had more than 40% of correct answers. The results are represented in figure 8 [42].

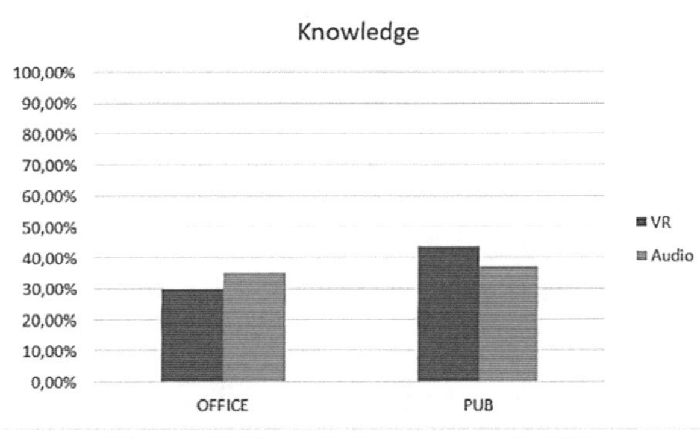

Figure 8. Knowledge test graph [42]

In the paper [44], Martin Dahl presents another approach to the use of virtual reality, namely by expanding the virtual commissioning (VC) with VR concepts. In the experiment was used an existing robot cell within the Production System Laboratory (PSL) at Chalmers University of

Technology, Sweden. The robot cell is made up of four ABB-robots, one KUKA-robot and a pallets conveyor. Virtual model of the cell has been created in Process Simulate, a software provided by Siemens and includes a human operator who must test the sensors, buttons and safety equipment. In order to achieve this and to add to the virtual function VR concepts, different tools were used: HTC-Vive, Jack 8.4., Process Simulate and two modules: SCARE and Jack. The HTC Vive device consists of one HMD, which allows the display of two different images are shown for the user, one for the left eye and one for the right eye, and two hand held controllers can be used for 360° motion capture. At the beginning of the simulation the user is outside the robot cell from where he can operate the virtual door opening button. After opening the doors, the safety stop will be activated and the operator will be able to check if the system is working properly. Figure 9 [44] is an image of the robot cell in the virtual model (figure 9b) and the real cell (figure 9a). The expansion of VC with VR elements has helped to move from observation to interaction with the virtual production system.

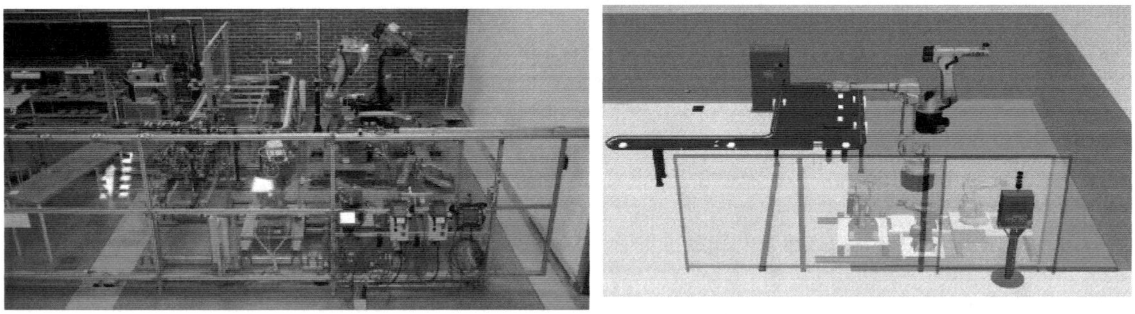

a) the real cell b) the virtual model of the cell

Figure 9. Comparison between the real- and virtual production cell [44]

Augmented Reality (AR) can be defined as a system that includes a combination of the real world and the virtual world, interactive in real time, registered in 3D and their overlap in the real world [54]. The term augmented reality first appeared in 1992 in the work of Tom Caudell [55].

An approach to augmented reality is presented by Dimitris Mourtzis in paper [45], in which he developed an application that uses augmented reality to evaluate and improve the product design process (figure 10, [45]).

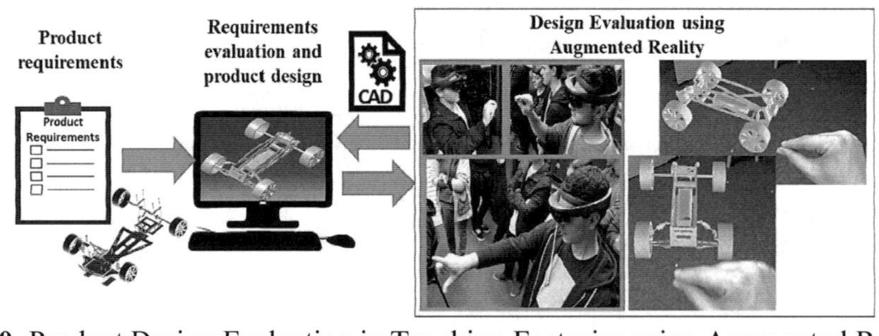

Figure 10. Product Design Evaluation in Teaching Factories using Augmented Reality [45]

The augmented reality tool used in this case is the Microsoft HoloLens headset, which offers the capability to see the designed geometry in a realistic scale, in the field of view of the designer, while also interacting with the virtual object like it is real, using grab and rotate moves. Thus, the designer gains a realistic experience from experience. The development of applications on this headset was possible thanks to the use of the Unity 3D platform. The geometry of the object is extracted directly from the CAD. The user can grab, move and rotate the virtual object to be able to check the design of the product from different perspectives and in this way the defects that could not be seen in the CAD software will be easily detected. Thus, the integration of the application that uses augmented reality has brought visible improvements in the results of the participants in the delivery factory, the errors being reduced by 12% and the assembly process of the products being 10% faster.

Mixed reality (MR) is a technology that, from a historical and conceptual point of view, has evolved from virtual reality [56] and that offers the user an experience that incorporates both the real and the virtual world [57]. Max Juraschek presents in paper [43] the benefits of using mixed reality in a learning factory. Thus, he states that the existence of mixed reality in learning factories allows the creation of an experience that combines the real world with the virtual one using both real and virtual objects. Juraschek argues that combining the real world with the virtual one in learning factories improves the learning process. The paper presents two potential applications of mixed reality that can be used in learning factories and are represented in figure 11, [43].

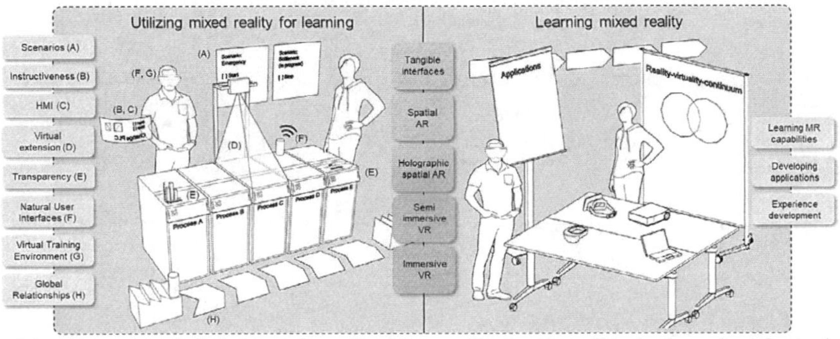

Figure 11. The potential fields of applications of mixed reality in learning factories [43]

Thus, Juraschek considers that one of the applications can be about learning the notions of mixed reality in the learning factory by intertwining didactics with technology. The second application refers to the use of mixed reality in the learning process, he believes that combining digital content with the real world can help make it easier to understand processes and methods.

3.2. Digital assistant

A digital consent system supports the interaction of the operator with the work equipment for the proper performance of work tasks [58]. The concept of digital assistance system is addressed by Hendrik Oestreich in the paper [46] in which a study of the effectiveness of learning with digital assistance is performed compared to traditional learning by explanation in a real assembly scenario, represented in figure 12, [46].

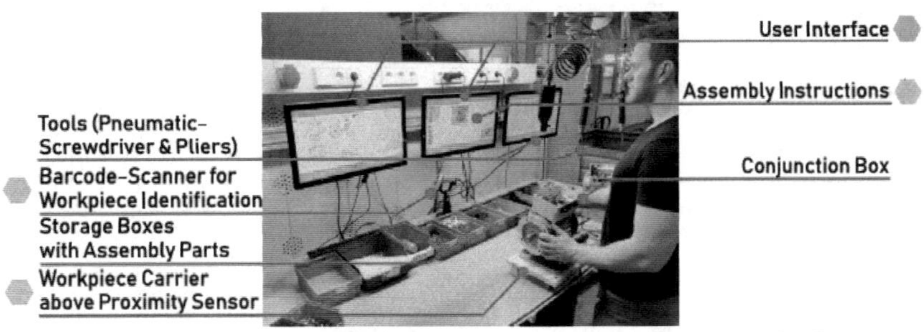

Figure 12. Workbench with the assistance system [46]

This assembly station equipped with a digital assistance system consists of three work areas. Proximity sensors are integrated in the support on which the assembly is placed and thus on the screen specific to each work area are displayed the instructions and tasks, specific to each stage, which must be performed in order to move to the next work area. Switching to the next work area is done by pressing a confirmation button on the screen. The study was conducted on a number of 13 people for learning with digital assistance and on a number of 3 people for traditional learning with a teacher. The study lasted 5 days and each participant had to complete a questionnaire at the first and last experience. The study found that learning with the help of the digital system is as effective as in the case of traditional learning.

Another approach about digital assistant is presented by Volkan Aksu in [47]. This shows how a digital assistant can support the quality of work of people with disabilities and independence in carrying out tasks. Based on some thinking tests of the needs, requirements and problems of users with disabilities created a hardwarebased prototype which supports users in stages of the cutting, glueing and folding, necessary for the production of a jewelry box. Thus, to guide the user in the cutting process, a 3D mould was made, equipped with remote controlled LEDs controlled by Arduino and Bluetooth technology and video instructions played on mobile devices (figure 13a [47]). For playing the instructions were used software with a user-friendly graphical interface using C++ and Qt. The passing the next step is done by pressing, by the user, a button on the mould. For the glueing task, an automatic glue dispenser with timer control and foot pedal it was used (figure 13b). To perform the last step, two identical mould were used (figure 13c [47]). In this study took part five male participants, with ages from 20 to 21 and different levels of cognitive disability.

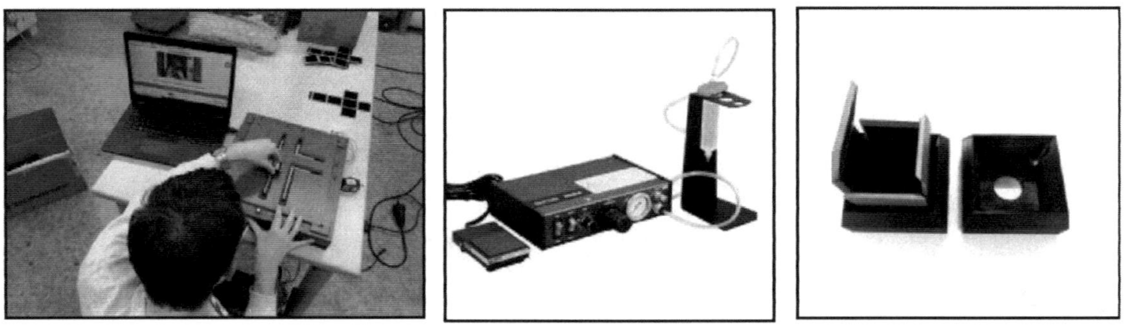

a) mould 3D for cutting with video instructions b) automatic glue dispenser c) mould for folding

Figure 13. Assistant system [47]

Finally, the results on task efficiency and subjective evaluation were analyzed. Regarding the efficiency of the tasks, the efficiency of the participants in performing the cutting task regarding the time, accuracy and completeness, the results being presented in figure 14 [47].

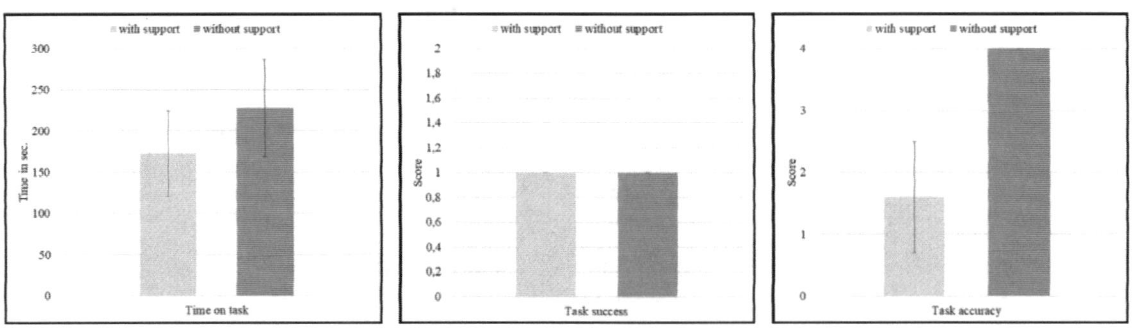

Figure 14. Comparison of mean values and standard deviation for the variables time on task, task success and task accuracy [47]

The subjective evaluation of the participants' interaction was measured according to the mental effort and the use of the system (figure 15).

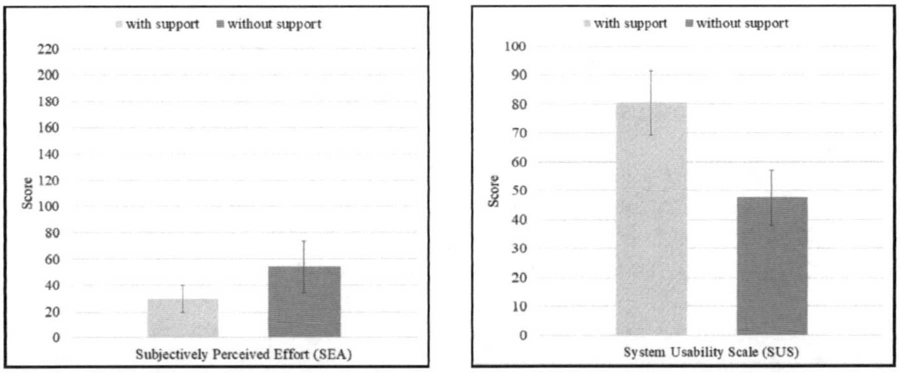

Figure 15. Comparison of Subjective Perceived Effort (SEA) and of System Usability Scale (SUS) in "with-support" and "without-support"-condition [47]

This paper shows that a care system can improve the quality of work and increase employment opportunities for people with disabilities.

4. Conclusions

The rapid evolution of processes and systems of production, training and learning, determined the development of methods, techniques and tools to keep up with new technology. This review presents the implementation of methods and techniques in different scenarios of production simulation, learning and training, applied in industry or educational institutions. The most common methods in the in the studied literature are: virtual reality, augmented reality, mixed reality, experiment-based learning, project-based learning, problem-based learning, intuitive learning and digital assistance.

In the case of using virtual reality, augmented and mixed, it was found that both in the learning scenario and in the production simulation scenario, these methods are much more agreed by users, compared to conventional methods, due to the experiences they offer. In most cases of using learning methods, users had the opportunity to experience real scenarios in production, in order to improve different skills and abilities. The use and implementation of these methods, techniques and tools could be a major support in the development of innovative production, training and learning processes and systems.

The study will continue by developing and/or integrating of some methods and techniques used in innovative industrial training, learning and production processes and systems in the existing Lean Learning Factory [41] from the University of Pitesti.

References

[1] Mourtzis D and Doukas M 2014 The evolution of manufacturing systems: From craftsmanship to the era of customisation *IGI Global*

[2] Koren Y 2010 *The global manufacturing revolution: Product-Process-Business Integration an Reconfigurable Systems*

[3] Wu D, Rosen D W, Wang L and Schaefer D 2014 Cloud-Based Manufacturing: Old Wine in New Bottles? *Procedia CIRP* **17** 94 – 99

[4] Reith S 1988 Außerbetriebliche CIM-Schulung in der „Lernfabrik" *Springer-Verlag Berlin Heidelberg*

[5] Schaper N 2000 *Gestaltung und Evaluation arbeitsbezogener Lernumgebungen*

[6] Abele E Joachim M and Michael T 2019 *Learning Factories - Concepts, Guidelines, Best-Practice Examples*

[7] Sabine P 2017 The Vision of Industrie 4.0 in the Making – a Case of Future Told, Tamed, and Traded *Nanoethics* **11** 107–121

[8] Dorel B 2016 A patra revoluţie industrială a început. Este pregătită România pentru a face faţă sfidărilor acestei noi revoluţii? *Revista de politica ştiintei şi scientometrie* **5** 194 – 201

[9] Munirathinam S 2020 Industry 4.0: Industrial Internet of Things (IIOT) *Advances in Computers*

[10] Abele E Eichhorn N and Kuhn S 2007 Increase of productivity based on capability building in a learning factory *11th International Conference on Production Engineering* Zagreb

[11] Entingh D J Andrews C J Kenkeremath D C Mock J E and Janis F T 1987 *Guide book for technology transfer managers: Moving public R & D to the marketplace* Washington.

[12] Ecodesign circle: Learning factory ecodesign, available at https://www.ecodesigncircle.eu/resources-for-you/learning-factory-ecodesign, accessed: 2020-03-24

[13] Kreimeier D Morlock F Prinz C Krückhans B Bakir D C and Meier H 2014 Holistic learning factories - A concept to train lean management, resource efficiency as well as management and organization improvement skills *47th CIRP Conference on Manufacturing Systems*

[14] Badurdeen F Marksberry P Hall A and Gregory B 2010 Teaching lean manufacturing with simulations and games: A survey and future directions *Simulation & Gaming* **41** (4) 465–486

[15] Tisch M Hertle C Abele E Metternich J and Tenberg R 2015 Learning factory design: A competency-oriented approach integrating three design levels *International Journal of Computer Integrated Manufacturing* 1355–1375

[16] Tisch M Hertle C Cachay J Abele E Metternich J and Tenberg R 2013 A systematic approach on developing action-oriented, competency-based learning factories *46th CIRP Conference on Manufacturing Systems* 580–585

[17] Blume S Madanchi N Böhme S Posselt G Thiede S and Herrmann C 2015 Dien Lernfabrik - Research-based learning for sustainable production engineering *5th CIRP Sponsored Conference on Learning Factories* 126–131

[18] Abele E Metternich J Tisch M Chryssolouris G Sihn W ElMaraghy H A Hummel V and Ranz F Learning factories for research, education, and training *5th CIRP-Sponsored Conference on Learning Factories* 1–6

[19] Jorgensen JE Lamancusa JS Zayas-Castro JL Ratner J 1955 The Learning Factory *Proc. of the Fourth World Conference on Engineering Education*

[20] Wagner C Heinen T Regber H and Nyhuis P 2010 Fit for change—Der Mensch als Wandlungsbefähiger. *Zeitschrift für wirtschaftlichen Fabrikbetrieb* 722–727

[21] Erol S Jäger A Hold P Ott K and Sihn 2016 W Tangible industry 4.0: A scenario-based approach to learning for the future of production *6th CIRP-Sponsored Conference on Leanring Factories* 13–18

[22] Seifermann S Böllhoff J Metternich J and Bellaghnach A 2014 Evaluation of workmeasurement concepts for a cellular manufacturing reference line to enable low cost automation for lean machining *47th CIRP Conference on Manufacturing Systems* 588–593

[23] Ahmad R 2018 Alberta Learning Factory for training reconfigurable assembly process value stream mapping *8th Conference on Learning Factories 2018 – Avanced Engineering Education & Training for Manufacturing Innovation* 237-242

[24] Yoo I S 2016 Model Factory for Additive Manufacturing of Mechatronic Products: Interconnecting World-Class Technology Partnerships with Leading AM Players *6th CIRP Conference on Learning Factories Procedia CIRP* **54** 210 – 214

[25] Makumbe S 2018 Effectiveness of using Learning Factories to import Lean principles in mining employees *8th Conference on Learning Factories 2018 – Advanced Engineering Education & Training for Manufacturing Innovation Procedia Manufacturing* **23** 69-74

[26] Schreiber S 2016 BERTHA - A flexible learning factory for manual assembly *6th CIRP Conference on Learning Factories Procedia CIRP* **54** 119 – 123

[27] Muschard B 2015 Realization of a learning environment to promote sustainable value creation in areas with insufficient infrastructure *5th Conference on Learning Factories 2015 Procedia CIRP* **32** 70 – 75
[28] Toivonen V 2018 The FMS Training Center – a versatile learning environment for engineering education, *8th Conference on Learning Factories 2018 – Advanced Engineering Education & Training for Manufacturing Innovation Procedia Manufacturing* **23** 135 - 140
[29] Steffen M 2012 *Problem Based Learning in Industrial Engineering Education at TU Dortmund University*
[30] Faller C and Feldmüller D 2015 Industry 4.0 learning factory for regional SMEs *5th CIRP Sponsored Conference on Learning Factories Procedia CIRP* **32** 88–91
[31] Veza I Gjeldum N and Mladineo M 2015 Lean learning factory at FESB - University of Split *5th CIRP-Sponsored Conference on Learning Factories Procedia CIRP* **32** 132–137
[32] Reichert and Markus 2011 Qualification of employees in the development department with SEW live training center in E. Abele, J. Cachay, A. Heb, & S. Scheibner (Eds.) *1st Conference on learning factories* 100-117 Darmstadt
[33] Hammer M 2014 Making operational transformations successful with experiential learning *CIRP General Assembly* Nantes
[34] Schützer K Rodrigues L F Bertazzi J A Durão L F C S and Zancul E 2017 Learning environment to support the product development process *Procedia Manufacturing* **9** 347–353
[35] Streitzig C and Oetting A 2016 Railway operation research centre —A learning factory for the railway sector *Procedia CIRP* **54** 25–30
[36] Elbestawi M 2018 SEPT Learning Factory for Industry 4.0 Education and Applied Research *Procedia Manufacturing* **23** 249-254
[37] Küsters D Praß N and Gloy Y S 2017 Textile learning factory 4.0—Preparing Germany's textile industry for the digital future *Procedia Manufacturing* **9** 214–221
[38] Blöchl S J and Schneider M 2016 Simulation game for intelligent production logistics—The PuLL® learning factory *Procedia CIRP* **54** 130–135
[39] Sivard G and Lundholm T 2013 XPRES—A digital learning factory for adaptive and sustainable manufacturing of future products *3rd Conference on Learning Factories* 32–154 Augsburg
[40] Baena F 2017 Learning Factory: The Path to Industry 4.0 *7th Conference on Learning Factories Procedia Manufacturing* **9** 73 – 80
[41] Nitu E L and Gavriluta A C 2019 Lean Learning Factory at the University of Pitesti *IOP Conf. Series: Materials Science and Engineering* **591**
[42] Pinto D 2019 Virtual Reality in Education: Learning a Foreign Language, *WorldCIST'19* 589-597
[43] Juraschek M 2018 Mixed Reality in Learning Factories *Procedia Manufacturing* **23** 153-158
[44] Dahl M 2017 Virtual Reality Commissioning in Production Systems Preparation *22th IEEE International Conference on Emerging Technologies and Factory Automation (ETFA)*
[45] Mourtzis D 2018 Augumented Reality supported Product Design towards Industry 4.0: a Teaching Factory paradigm *Procedia Manufacturing* **23** 207-212
[46] Oestreich H Töniges T Wojtynek M and Wrede S 2019 Interactive Learning of Assembly Processes using Digital Assistance *9th Conference on Learning Factories 2019 Procedia Manufacturing* **31** 14–19
[47] Volkan A Sascha J Björn K and Carsten R A 2019 Digital Assistance System Providing Step-by-Step Support for People with Disabilities in Production Tasks 2019 *Springer International Publishing AG* 775–785
[48] Stephan S 2017 Learning in the AutFab – the fully automated Industrie 4.0 learning factory of the University of Applied Sciences Darmstadt *7th Conference on Learning Factories Procedia Manufacturing* **9** 81–88
[49] Bryson S 1996 *Virtual Reality in Scientific Visualization*
[50] Yoh M 2001 The Reality of Virtual Reality *Proceedings of the Seventh International Conference on Virtual Systems and Multimedia*

[51] Ebersole S 1997 *A brief history of virtual reality and its social applications* University of Southern Colorado
[52] Schubert T Friedmann F and Regenbrecht H 2001 The experience of presence: factor analytic insights *Presence Teleoperators Virtual Environ* **10** (3) 266–281
[53] Lewis J R 1995 IBM computer usability satisfaction questionnaires: psychometric evaluation and instructions for use *International Journal of Human-Computer* **7** (1) 57–78
[54] Azuma R T 1997 A survey of augmented reality *Presence-Teleoperators and Virtual Environments* 355–385
[55] Caudell T P and Mizell D W 1992 Augmented reality: An application of heads-up display technology to manual manufacturing processes, System Sciences, *Proceedings of the Twenty-Fifth Hawaii International Conference* **2** 659–669
[56] Kunz A and Fjeld M 1994 Mixed Reality: A Survey *Computer Science*
[57] Milgram P and Kishino F 1994 A taxonomy of mixed reality visual displays, *IEICE Trans, Information and Systems* **77** 1321-1329
[58] Keller T Bayer C Bausch P and Metternich J 2019 Benefit evaluation of digital assistance systems for assembly workstations *52nd CIRP Conference on Manufacturing Systems Procedia CIRP* **81** 441–446

IOP Conf. Series: Materials Science and Engineering 1018(2021) 012023 doi:10.1088/1757-899X/1018/1/012023

A brief review of the methods and techniques used in the innovative internal logistics processes and systems

I G Pascu*, **G C Neacsu** , **E L Nitu and A C Gavriluta**

University of Pitesti, Faculty of Mechanics and Technology, Manufacturing and Industrial Management Department, Targul din Vale Street No.1, Romania

*iuliana.pascu@upit.ro

Abstract. In the last period, logistics knew a strong technological evolution that has changed the way it operates and determined its transformation into innovative logistics, especially in the case of internal logistics. Innovation in internal logistics processes and systems consists in the use of different Industry 4.0 and digitization technologies. So, it is necessary to use modern methods and techniques to support the transformation of internal logistics processes and systems into innovative ones. This paper presents an analysis of recent studies, briefly presenting the evolution of the concept of logistics, the concepts of process and logistics system and Industry 4.0. Also, the most common methods and techniques used in the realization of innovative internal logistics systems are presented and some conclusions are presented regarding their use.

1. Introduction

1.1. The evolution of the logistics concept

The concept of logistics has two origins: mathematical and military. The term "logistikos" was first used by the Greeks in 428 BC, and in the early 20th century, the British mathematician and logistician Bertrand Russell highlighted a link between logistics and mathematics logistics [1]. On the other hand, in the military field, the term logistics refers to the flow of objects in time and space. Today, the term is used in industry and refers to the methods of transport, storage and handling of objects [2]. Logistics is a process that ensures the flow of materials, the supply of the organization and its activities to the customer. It can be represented schematically as in Figure 1 [3]. According to the explanatory dictionary of the Romanian online language [4], it defines the concept of logistics as a set of methods and means that deal with the organization of the operation of a service carried out in an enterprise.

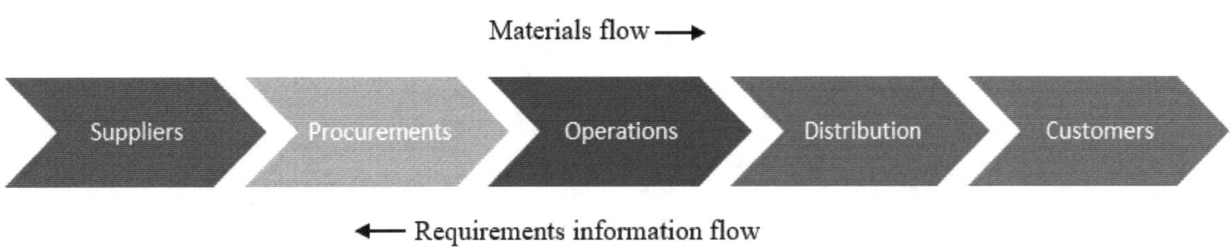

Figure 1. The logistics process [3]

IOP Conf. Series: Materials Science and Engineering 1018(2021) 012023 doi:10.1088/1757-899X/1018/1/012023

1.2. Internal logistics processes and systems

The logistics process includes the flow of goods, services and information related to the movements of goods and services, called logistical information. The internal logistics process in a production organization involves logistics flows, including materials, ongoing activities and final products [5].

The logistics system is the set of means by which it acts to perform, physically, all manufacturing operations, transport, but also storage, allowing the movement of materials from suppliers to beneficiaries, after they have undergone the necessary transformations to become finished product. Logistics systems carry out activities and processes of supply, transport, handling, storage, control and distribution. Through these activities the supply of raw materials and materials, the manufacture of the product and the distribution of the finished product are achieved [6].

According to the degree of involvement of the human operator, the internal logistics systems are classified as follows:

- classical systems: are those systems where the human operator performs all handling activities;
- mechanized systems: are those systems in which the human operator intervenes on the system for a manual control;
- automated systems: are those systems in which the human operator cannot intervene and all functions are performed by systems [7].

Today, internal logistics is playing an increasingly important role in the industry. Handling equipment means the handling of materials from the warehouse to various operations within the enterprise until the final product is ready for delivery to the customer [8].

This paper aims to present the concepts of logistics, Industry 4.0 and Logistics 4.0 and a short review of the main methods and techniques used in innovative internal logistics processes and systems.

2. Industry 4.0 and Logistics 4.0

2.1. Industry 4.0 concept

The term Industry 4.0 was first introduced at the "Hannover Fair" in 2011. The first industrial revolution was the introduction of mechanical manufacturing equipment powered by the force of water or steam. The second industrial revolution concerns the use of electricity, which allowed the use of the conveyor belt and the assembly line. The third industrial revolution brought automation of production through the use of electronic systems and communication technologies [9;10]. In figure 2 [11] the evolution of the technologies of the four industrial revolutions is presented.

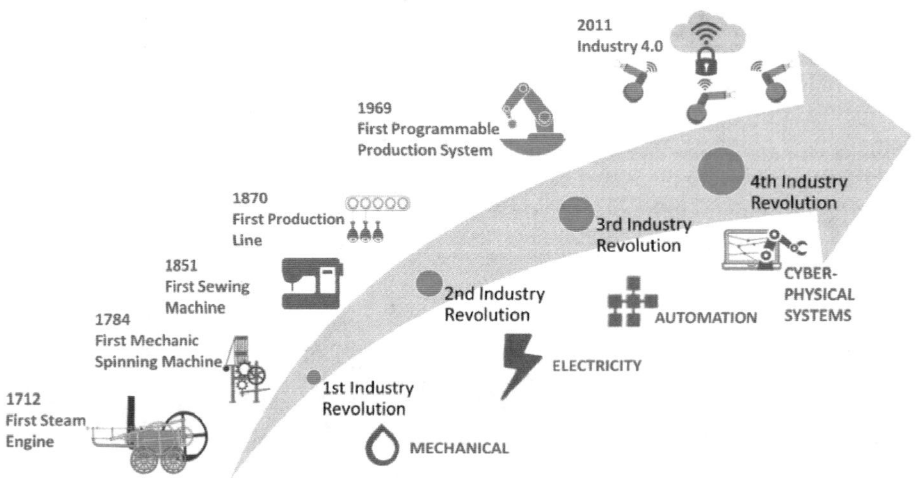

Figure 2. The evolution of the four industrial revolutions [11]

Overall, Industry 4.0 encompasses the development and integration of innovative information and communication technologies in industry. The main goal is to create smart grids and process along the chain, thus enabling them to use organizational processes more efficiently, to improve customer benefits by providing them with new products and services.

The Industry 4.0 concept is based on new technologies, such as: internet of things (IoT), big data, cyber-physical systems (CPS), robots; augmented reality, cloud, machine learning, additive manufacturing, real-time optimization. Figure 3 shows the main technologies of the Industry 4.0 concept [12].

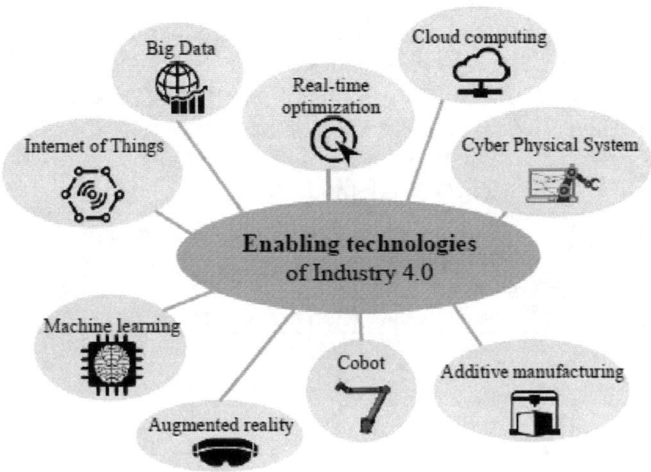

Figure 3. Industry 4.0 Technologies [12]

As Industry 4.0 emerged, some companies began to use advanced control tools activated by real-time monitoring systems that allow the development of precise planning models to enable managerial decision-making. Although we are seeing a growing trend in the automation of human labour in almost every industry, human workers still play an important role in production and logistics systems [13].

2.2. Logistics 4.0 concept

Today, most businesses are undergoing a digitalization process with the fourth industrial revolution. The focus of digital transformation is mainly on production, so terms like "Factory of the Future" or "Smart Factory" are used similarly to this concept. However, there are many reasons to consider the impact of digitization in logistics and the importance of the supply chain for Industry 4.0 [14].

There are several equipment and types of automation commonly used in logistics:

- Automated loading and unloading systems;
- Automatic guided vehicles;
- Automated storage systems;
- Conveyor belts;
- Robots;
- IT systems for planning, managing operations and systems;
- RFID systems;
- Bar code systems;
- Item selection devices [15].

We use the term "Logistics 4.0" to refer to the combination of the use of logistics with the innovations and applications added by CPS. Logistics 4.0 is bound by the same conditions as smart services and products. It can then be considered an approach based on the technology used to define "Smart Products" and "Smart Services" defined as "Smart Logistics". Smart products and services are those that can perform tasks normally performed by humans. In addition, they make it possible to delegate activities so that employees can focus on tasks that need more intelligence than automated processes or the intelligence that a simple smart product or service can have.

"Intelligent logistics" is a logistics system that can improve flexibility, adjust to market changes and will bring the company closer to customer needs. This will make it possible to improve the level of services, optimize production and lower storage and production prices. "Intelligent logistics" will change according to current technology, is time-dependent and therefore defines the state of technology [16].

3. Methods and techniques used in internal logistics processes and systems

With the development of the Industry 4.0 concept, the methods, techniques and tools in the area of internal logistics processes and systems have been developed. The most common technical methods and tools in the studied literature, used in the development of innovative internal logistics processes and systems are presented in Table 1.

Table 1. Methods, techniques and tools

Method	Techniques and tools	Article
Logistics flow optimization	Simulation	[18, 20]
	Mathematical modelling	[20]
Development of autonomous logistics systems	Navigation techniques, proximity sensors	[22]
	Navigation techniques (MY3 wheels)	[23]
	Microprocessor, RFID reader	[24]
Augmented / Virtual reality	3D image	[27, 31]
	Roivis, Virtual Retinal Display	[31]

A brief description of the main features of each innovative method and application situations are given below.

3.1. Logistics flow optimization

Simulation is defined as a process by which a model of the real system is built. One of the most recognized software is Tecnomatix Plant Simulation developed by Siemens PLM to model, simulate, analyse, visualize and optimize production systems and processes, material flow and logistics operations. Plant Simulation can display both 2D and 3D production sequences [17]. Recently, simulation is also used for research and development purposes in universities. The simulation allows the imitation of a suggested solution to achieve the company's objectives, i.e. to increase production efficiency.

In his work, Viharos [18] uses a general model of discrete event simulation developed by Siemens Software shown in the figure 4 [18]. The purpose of developing this simulation model is to plan and evaluate reconfigurable assembly systems. He has developed a heuristic algorithm that can manage a wide variety of systems composed of robotic stations. With this algorithm, it was able to set the number of orders in real time and reduce the time to make products. The purpose of this algorithm is to reduce the manufacturing time of the products.

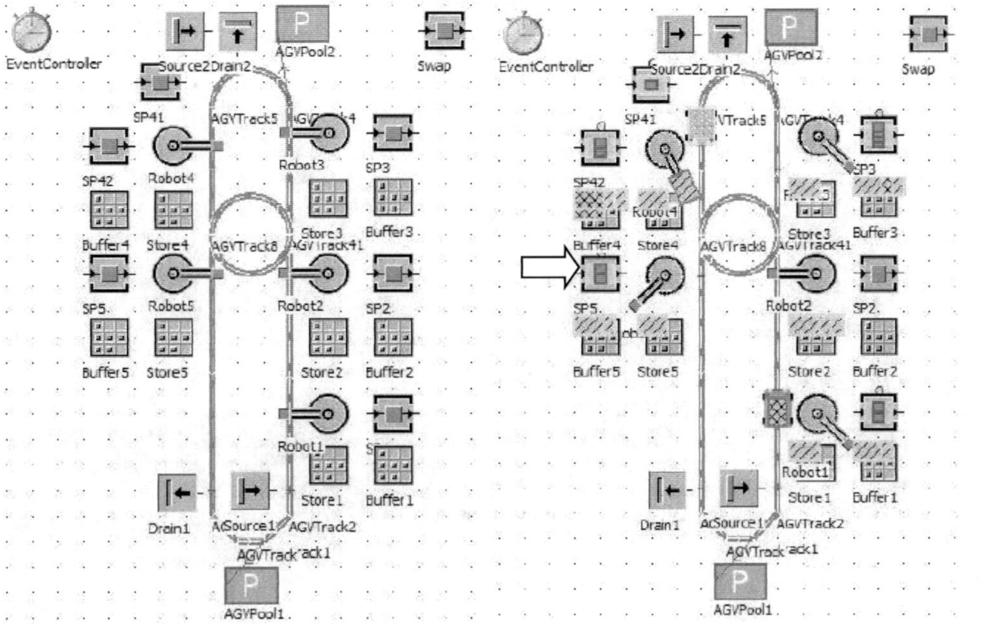

Figure 4. Example of a model of an automatic layout system in Plant Simulation [18]

Mathematical modelling is the process of developing the mathematical model. A mathematical model describing a system that uses mathematical concepts and language. Mathematical models are used in the analysis of logistics systems. A model helps to explain a system, studies the effects of different components and makes predictions about their behaviour. Mathematical modelling has become increasingly important for logistics systems as they have become more complex [19].

For example, Vavrik [20] proposes a logical mathematical model used in performing the simulation to optimize the transport route. This mathematical model, shown in figure 5, is performed before collecting process data and according to this mathematical calculation the number of vehicles required can be determined.

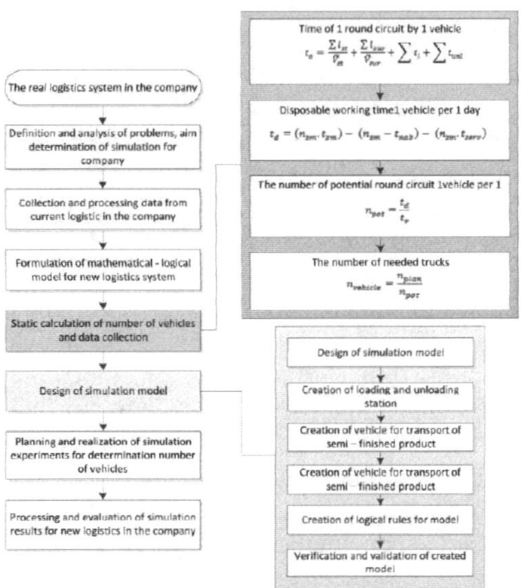

Figure 5. Mathematical model developed by Vavrik [20]

Figure 6 [20] shows a simulation model in Plant Simulation in both 2D and 3D. The main purpose of the experimental simulation in this paper was to estimate an essential number of automatically guided transport vehicles that serve the operations performed in production. To make the simulation model, Vavrik went through the following steps: creating tracks for moving the vehicle; creation of the loading and unloading station; the creation of the vehicle for the transport of the semi-finished product and the creation of logical rules. Thus, by using the simulation and adjusting some parameters, it was possible to observe a decrease in the speed of the old man by 1 m / second on a straight line compared to 1.2 m / second, and on the area with curves 0.6 m / second compared to 0.8 m / sec.

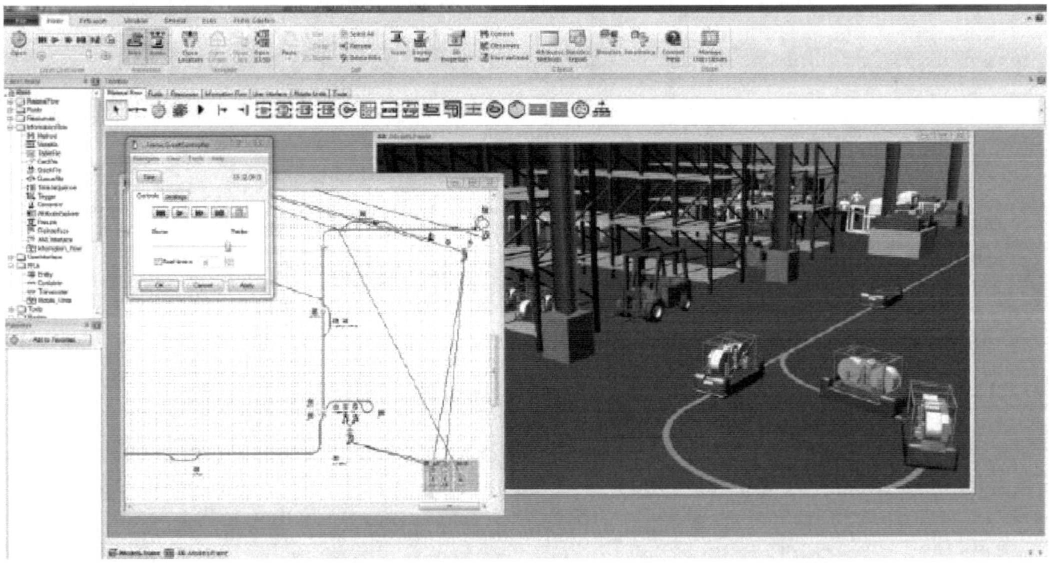

Figure 6. 2D and 3D simulation model in Plant Simulation [20]

3.2. Development of autonomous logistics systems

An autonomous system is a system that can perform certain functions without the intervention of an operator. Each autonomous system uses intelligent techniques such as: artificial intelligence [21], navigation techniques, proximity sensors, RFID reader.

For example, Jaiganesh [22] provides a model for designing an advanced autonomous guided vehicle using a programmable logistics controller (PLC). The differentiating feature of this vehicle is that it includes a robot arm, at the top, to streamline the transfer of materials and reduce costs. This vehicle includes a control device consisting of a DC motor, proximity sensors, programmable logic controller, a navigation system and a robotic arm. The control device is used to monitor the movement and speed of the vehicle. The DC motor was used due to its linear characteristics and ease of speed adjustment. Proximity sensors containing inductive proximity switches that have the ability to detect metal objects without physical contact were used to detect objects in the vicinity of the autonomous guided vehicle and station vehicles. A PLC was used to monitor and control the vehicle. As a navigation technique, it has been proposed to use a magnetic strip on the floor that guides the vehicle and which poses it to accelerate or decelerate or even stop. In this paper, the robotic arm is seen as an automated material handling device. The mode of operation of the proposed vehicle is: to reach the destination, the vehicle receives a signal from the common network, and with the help of proximity sensors it will be able to stop at each station. After stopping at the station, the sensor will read if the material is available or not. Figure 7 [22] shows a proposal for the construction of an automatic guided vehicle with robotic arm.

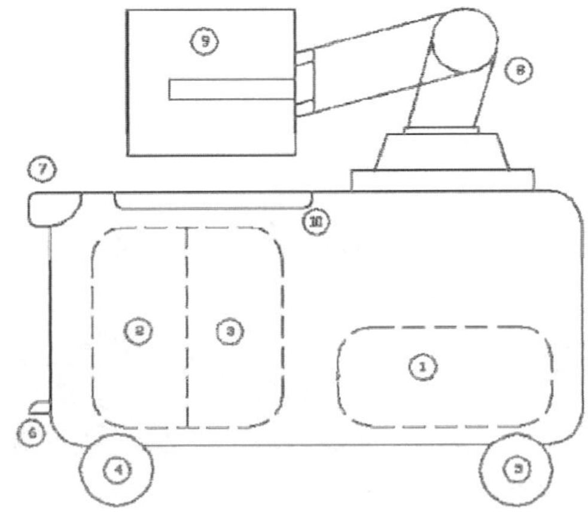

Components:
1. PLC operated brushless DC motor;
2. PLC Components;
3. Battery;
4. Front wheel;
5. Rear wheel;
6. Proximity sensor;
7. Photo sensor;
8. Robotic arm;
9. Material to be shifted;
10. Provision for accommodation of material.

Figure 7 Scheme of the AGV proposal construction [22]

In his work, Yu [23] designed an omnidirectional prototype AGV using MY3 wheels. These MY3 wheels consist of two balls with equal diameters on a common shaft. The balls are sectioned into four spheres and each sphere can rotate freely around its own axis shown in figure 8 [23]. The MY3 omnidirectional wheel prototype consists of a mobile platform and three shelves shown in figure 9 [23]. On this mobile platform there are four wheels that are positioned evenly at a 90 ° angle to be able to perform the movements. The first shelf is used to transport materials, the middle shelf is used to transport the battery, and the lower shelf is used to transport the engine that drives the four MY3 wheels.

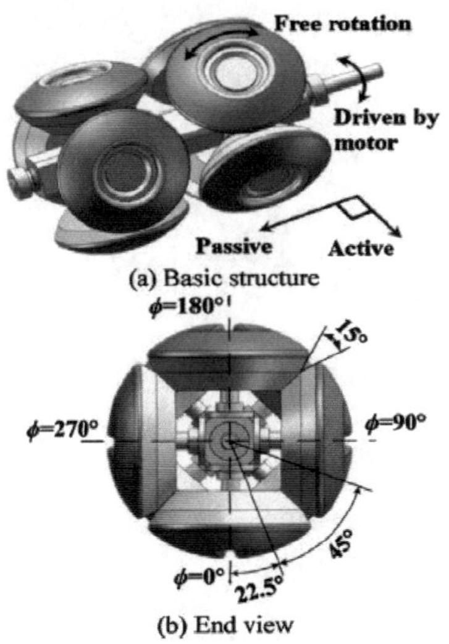

Figure 8. MY3 wheels [23]

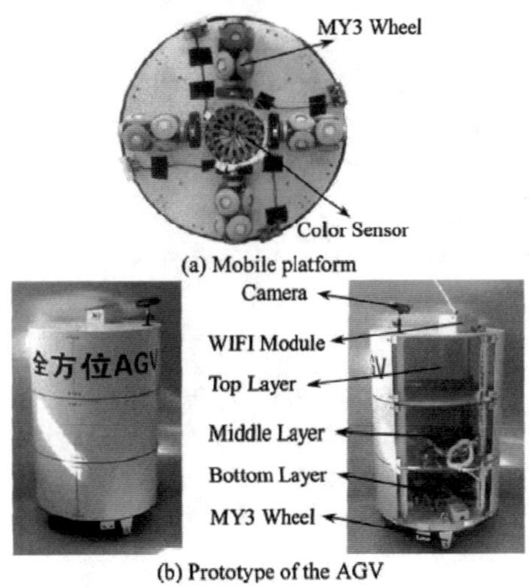

Figure 9. MY3 omnidirectional wheel prototype [23]

In another paper, Mehami [24] made two different automatically guided systems for transporting objects named Karl and Jimmy. The Jimmy system, shown in figure 10 [24], is a robot composed of an mCore microprocessor board. This system consists of an RFID reader used for movement, a frontal sonar to avoid collisions and a flat surface (platform) for transporting objects. Its travel speed is 0.3 m/s. In Jimmy's case, RFID tags were used to change lanes and stop.

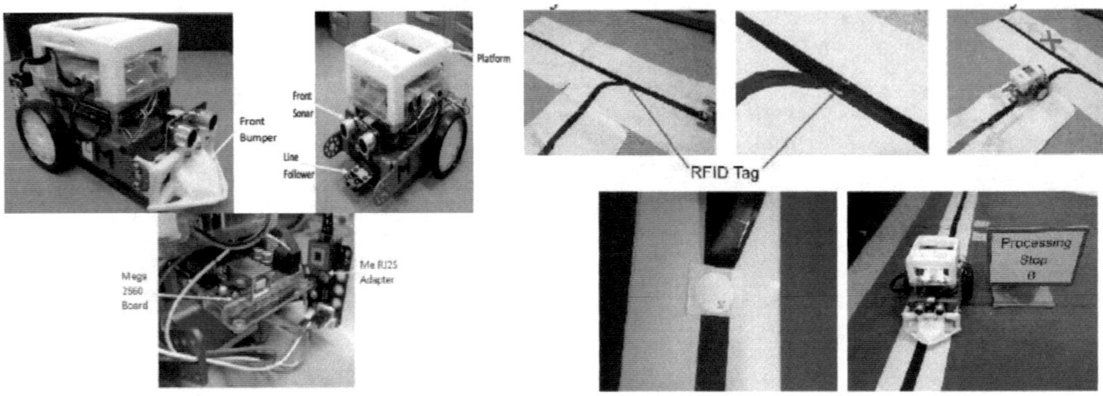

Figure 10. Jimmy system [24]

The Karl system, shown in figure 11 [24], is a robot that has a clamping arm equipped with a MegaPi microprocessor board. This system consists of two RFID readers, one attached to the arm with clamp to identify the products and the other reader is attached to the base to move, switches to limit the rotation of the arm and body. Its travel speed is 0.1 m / s.

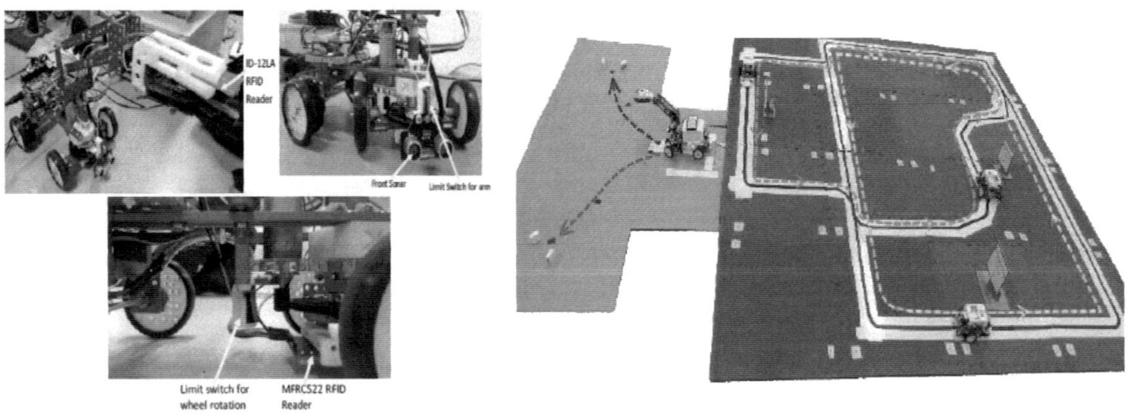

Figure 11. Karl system [24]

3.3. Augmented and virtual reality

The term ***augmented reality*** first appeared in 1992 in the work of Tom Caudell [25]. Augmented Reality (AR) can be defined as a system that includes a combination of the real world with the virtual world, interactive in real time, recorded in 3D and their overlap in the real world [26].

Zheng [27] proposed for development a portable visual device based on binocular vision and deep learning to achieve rapid detection and recognition of cable brackets. The drinkable visual inspection device based on AR technology is used to guide the operator in the cable assembly process and contains a computing unit, a four-wheel mobile loading unit and binocular cameras that can be adjusted up and down, so as shown in Figure 12 [27]. Deep learning and visual recognition are used to read the text on the cables, and the results are sent to wearable devices for visual assembly guidance.

The AR-based cable assembly process consists of two steps:

- The first step is to design the assembly process that is done by the process engineer. In this step, the process engineer uses the AO software tool, generating an XML document for each operation to obtain all the information needed for the AR-based process. The interface of this software is shown in Figure 13.a, and Figure 13.b shows the detailed contents of the XML file automatically generated by the software.
- The second step is to visualize the assembly process.

Figure 12. Interface of AR assisted assembly system [27]

The final interface of the AR-assisted assembly system is shown in Figure 14. Product and process information is displayed at the top left of the figure, and the 3D image of the assembly to be made is displayed at the bottom left.

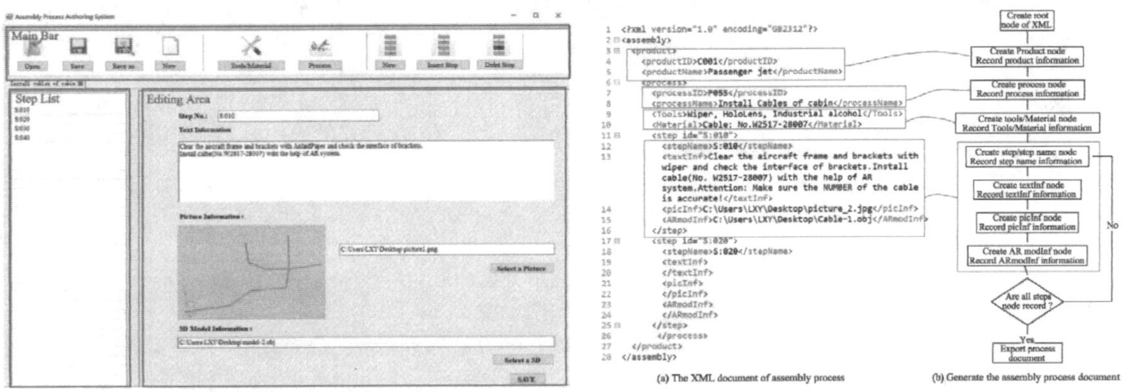

a) System interface b) Detailed XML document

Figure 13. AO software used in AR assisted assembly system [27]

Figure 14. Interface of AR assisted assembly system [27]

Following the study, the author found an improvement in assembly efficiency and better quality of the assembled product. In the future, it aims to develop a Meta-Learning system.

Virtual reality was proposed in 1989 by Jaron Lanier [28] and is considered a paradigm that through computers and human-computer interfaces creates the effect of a three-dimensional world [29] in which the user can get sensory experiences in a way similar to the real world [30].

In his paper [31], Rupert Reif addresses the use of virtual and augmented reality in various logistics applications such as: logistics planning, operator support when choosing orders, simulation and training.

The use of augmented reality in the case of planning is done with the help of the tool "roivis", with which the image of the existing system can be integrated various virtual objects (Figure 15, [31]), without interrupting production to perform tests with real objects.

When choosing commands, augmented reality is used to project orientation data (3D arrows showing the path to the storage location and pickup point) and process, in the operator's field of view using an HMD (Figure 16, [31]). Virtual Retinal Display was used for the display, and the interaction is performed through speech control. Thus, search times are reduced and errors are avoided due to optical media.

Figure 15. Real system on which virtual objects are superimposed [31]

Figure 16. Choosing commands with the help of AR [31]

The virtual reality system is connected to a virtual reality glove and a treadmill, which together form a captivating framework due to the fact that the real processes of taking orders can now be reconstructed in the virtual environment (figure 17, [31]). The virtual reality glove is used to take over the goods from the storage place and the tape aims to generate a virtual walk through the warehouse. Thus, when the warehouse worker takes a step forward, the system identifies the movement and the tape starts. Following the implementation of this system, it was possible to observe a reduction in costs and an efficiency of the employee training process.

Figure 17. Integrated human planning and simulation of virtual ordering systems [31]

An evaluation of the choice of orders was also made using AR and VR by comparing four different types of takeover technologies:

- selection of paper orders with selection lists;
- pick-by-voice system;
- pick-by-voice system based on augmented reality;
- virtual reality simulation environment for taking orders.

According to the results, the highest motivation during order processing was during the use of the virtual environment (VR), followed by the augmented reality (AR) system, figure 18.a [31]. Another result based on the sensation of the subjects regarding the cognitive load of the subjects during the process of choosing the orders is presented in figure 18.b [31]. Thus, this study highlighted that the choice of orders based on virtual and augmented reality are innovative methods in the field of logistics that are accepted and provide high motivation to users.

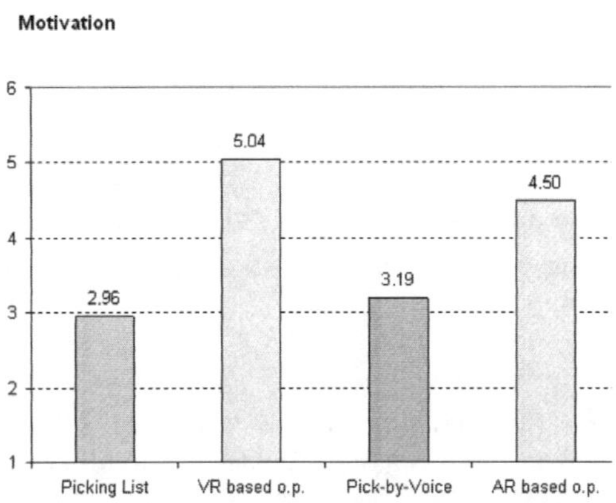

a) motivation of the subjects regarding the different types of taking orders

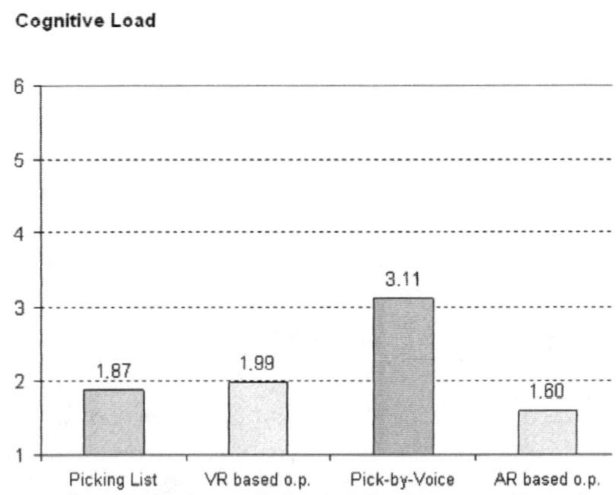

b) the sensation of the subjects regarding the cognitive load during the different selection processes

Figure 18. The results of using different order taking techniques [31]

4. Conclusions

The globalization of production and the increasingly fierce competition in the field of production systems have imposed a continuous evolution of internal logistics processes and systems and, thus, have determined the development of innovative methods and techniques in this field as well.

The methods and techniques most common in the studied articles, in connection with the development of innovative internal logistics processes and systems, are: optimization of logistics flows, development of autonomous logistics equipment and use of techniques based on augmented / virtual reality. In the case of optimizing logistics flows, the most used technique is simulation with discrete events, one of the frequently used software being Plant Simulation. In the case of the development of autonomous logistics equipment, the most used techniques are those based on navigation systems and proximity sensors. In the case of virtual and augmented reality, they are based on the combination of the real world with elements of the virtual world. These methods use visual, auditory and olfactory sensors to develop innovative internal logistics processes and systems.

The study will continue through the development and/ or integration of methods and techniques specific to innovative internal logistics processes and systems in the existing Lean manufacturing laboratory at the University of Pitesti.

5. References

[1] Zgaya H and Hammadi S 2016 Logistics Engineering *Logistics Engineering and Health* 1-53
[2] Jaekel F 2019 Fundamentals of Logistics Systems *Cloud Logistics*
[3] Kain R and Verma A 2018 Logistics Management in Supply Chain – An Overview *ICMPC Materials Today: Proceedings* **5** 3811-3816
[4] Dex Online Available: https://dexonline.ro/ [Accessed:20.09.2020].
[5] Nguyen P and Crase L 2008 Organizational Logistics Processes: A Literature Review and an Exploratory Investigation of International Multimodal Transport in Vietnam *Asia Pacific Management Review* **13** 403-418
[6] Ghiani G ,Laporte G and Musmanno R 2013 *Introduction to Logistics Systems Management*
[7] Inman R ,Blumenfeld D, Huang N and Li J 2010 Designing production system design for quality: Research opportunities from an automotive industry prospective *Int. J. Prod. Res.* **41** 1953-1971
[8] Baptista F and Pinto G 2019 Intralogistics and industry 4.0: designing a novel shuttle with picking system *Procedia Manuf.* **38** 1801–1832

[9] Fouad F 2019 The Fourth Industrial Revolution is the AI Revolution An Academy Prospective, *IJITCS* **8** (5) 155-167
[10] Dorel B 2016 A patra revoluţie industrială a început. Este pregătită România pentru a face faţă sfidărilor acestei noi revoluţii? *Revista de politica ştiintei şi scientometrie* **5** 194 – 201
[11] Gokalp E, Gokalp M and Eren E 2018 Industry 4.0 Revolution in Clothing and Apparel Factories: Apparel 4.0, *Industry 4.0 from the MIS Perspective* 169-184
[12] Bortolini M, Ferrari E and Gamberi M 2017 Assembly system design in the Industry 4.0 era: a general framework, *IFAC-PapersOnLine* **50** 5700-5705
[13] Sgarbozza F and Grosse E 2020 Human factors in production and logistics systems of the future, *Annual Reviews in Control Volume* **49** 295-305
[14] Kayikci Y 2018 Sustainability impact of digitization in logistics, *Procedia Manuf.* **21** 782-789
[15] Granlund A and Wiktorsson M 2014 Automation in Internal Logistics: Strategic and Operational Challenges *Int. J. Logist. Syst. Manag.* **18**(4) 538-558
[16] Barreto L Amaral A 2014 Industry 4.0 implications in logistics: an overview *Procedia Manuf.* **13** 1245-1252
[17] Ada Computers, https://www.adacomputers.ro/solutii-tecnomatix/ [Accessed: 30.06.2020].
[18] Viharos A Nemeth I 2018 Simulation and scheduling of AGV based robotic assembly systems, *IFAC PapersOnLine Volume* **5** 1415-1420
[19] Zvirgzdina B Tolujevs J 2014 The use of mathematical models for logistics systems analysis *Proceedings of the 14th International Conference Reliability and Statistics in Transportation and Communication* 63-64
[20] Vavrik V, Gregor M and Grznar P 2017 Computer simulation as a tool for the optimization of logistics using automated guided vehicles *Procedia Eng.* **192** 923-928
[21] Kim J, Lee D and Yang J 2020 Conceptual design of autonomous emergency operation system fornuclear power plants and its prototype *Nucl. Eng. Technol.* **52** 308-322
[22] Jaiganesh V Kumar D 2014 Automated Guided Vehicle with Robotic Logistics System *GCMM Procedia Eng.* **97** 2011-2021
[23] Yu S and Ye E 2016 Development of an omnidirectional Automated Guided Vehicle with MY3wheels, *Proceesings of the 1st Czech-China Scientific Conference* 364-368
[24] Mehami J Nawi M 2018 Smart automated guided vehicles for manufacturing in the context of Industry 4.0 *NAMRC 46 Procedia Manuf.* **26** 1077-1086
[25] Caudell T P Mizell D W 1992 Augmented reality: An application of heads-up display technology to manual manufacturing processes, System Sciences, *Proceedings of the Twenty-Fifth Hawaii International Conference on, vol.* **2**. *IEEE* 659–669
[26] Azuma R T 1997 A survey of augmented reality, *Presence-Teleoperators and Virtual Environments* 355–385
[27] Zheng L Liu X 2020 A smart assistance system for cable assembly by combining wearable augmented reality with portable visual inspection *VRIH* **2** (1) 12-27
[28] Ebersole S 1997 *A brief history of virtual reality and its social applications* University of Southern Colorado
[29] Bryson S 1996 Virtual Reality in Scientific Visualization *Communication of the ACM,* **39** 62-71
[30] Yoh M 2001 The Reality of Virtual Reality *Proceedings of the Seventh International Conference on Virtual Systems and Multimedia*
[31] Reif R Walch D 2008 Augmented & Virtual Reality applications in the field of logistics *The Visual Computer* **24**(11) 987–994

New design of precise deployable reflector. Technologies for manufacturing of physical model

V I Bujakas[1] **and M D Glotov**

P. N. Lebedev Physical Institute of Russian Academy of Science, 119991, Leninsky pr. 53, Moscow, Russia

E-mail: bujakas@yandex.ru

Abstract. The new design of petal space antenna is considered in the paper. Various versions of solid petal-type mirrors have been considered and investigated in a number of articles and space projects. The classical scheme of petal type deployable space reflector was proposed and developed by Dornier Corporation within FIRST space project and was used in the Radioastron project to create a 10-meter antenna of space radio telescope. However, the classical petal mirror design has two significant drawbacks. Firstly, in the open state, the petals are cantilevered on the central mirror, what leads to low rigidity of the structure in open state. Secondly, in the classical design, the accuracy of the reflecting surface of the open mirror strongly depends on small errors in the operation of the deployment system. To overcome these drawbacks a new design of petal type mirror and a new system for precise opening of the mirror were proposed and studied. To test a new technical solution, a physical model of a deployable reflector was developed. The model contains a central mirror, a set of petals, hinges, locks and actuators for opening the mirror. 3D printing technology, CNC (computer numerical control) milling, plastic injection molding and carbon fiber technologies have been tested for central mirror and petals of the model fabrication. The results of computer and physical simulation are presented in the paper.

1. Introduction

Solid petal-type space antennas were developed and investigated in a number of works and space projects [1-4]. The classical design of a petal deployable reflector was proposed by the Dornier Corporation during the development of the "FIRST" space project ("FIRST" - far infrared space telescope) [5, 6]. Later, the similar design was used for the 10 m antenna of “Radioastron” space telescope, which operates at cm wavelengths [7].

However, the classical design of the petal mirror has two significant shortcomings. First, in the open state, the petals are cantilevered on the central mirror, which with large sizes of the petals, leads to low rigidity of the structure in the operating state. Second, in the classical design, the accuracy of the reflecting surface of the mirror strongly depends on small errors in the operation of the deployment system. As the result is a loss of the mirror quality in the short wavelength range. The paper proposes a new design of the petal-type mirror, free from these shortcomings.

[1] bujakas@yandex.ru

2. Petal-type deployable reflector

2.1. Classical design of petal reflector

Within the classical design, a petal-type reflector is a transformable structure and contains a central mirror and a set of petals. In the folded state, the petals are disposed in the vertical position above the central mirror. Each petal is connected to the frame of the central mirror by a cylindrical hinge. In the open state, the petals and the central mirror form a large parabolic reflector. Stages of the mirror deployment are presented in figure 1.

The main shortcoming of the classical petal reflector is the following. Actuators for opening the mirror are at the base of the petals. The length of the petal, which depends on the aperture, is 3.5-5 meters. As a result of the deployment, it is necessary to align the vertices of adjacent petals with a high accuracy. For an operating wavelength $\lambda = 1.35$ cm, the vertices of adjacent petals should be aligned with accuracy $\Delta = \lambda / 16 = 0.084$ mm. In the millimeter region of the spectrum, this accuracy is an order of magnitude higher. Moreover, the alignment is carried out without any feedback. Technically, this is a difficult task. The result is a loss of the mirror quality in the short wavelength range. For example, the antenna aperture efficiency (area_usage_factor) of the "Radioastron" telescope in orbit at a wavelength of 1.35 cm turned out to be 0.1 [7]. This means that in the short wavelength range, the 10 meter reflector works like a precise 3 meter mirror. The goal of our research is to improve the petal mirror in the short wavelength range.

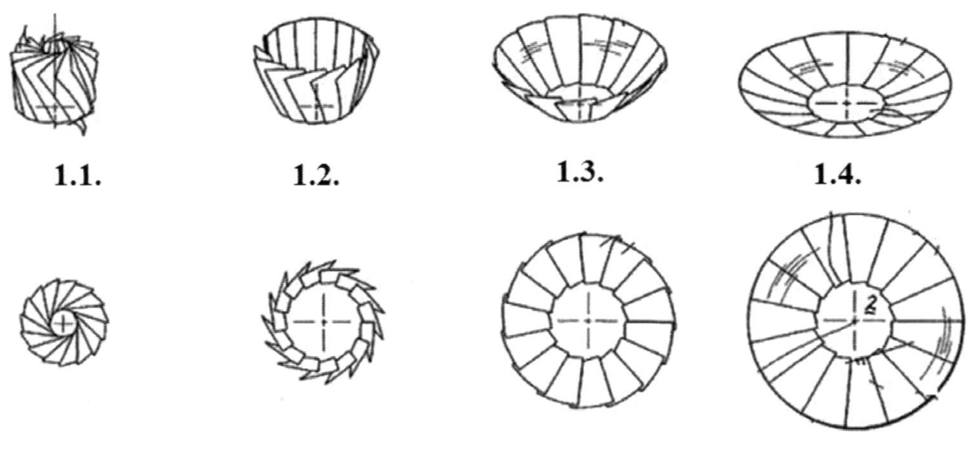

Figure 1. Stages of the classical petal-type mirror deployment

Earlier we proposed [8, 9] new technical solutions aimed at improving the quality of the transformable petal-type reflector. The present work extends these studies.

2.2. New design of petal reflector

In order to avoid the described shortcomings, we propose to change the mirror opening pattern, namely, to place the left vertex of one petal on the outer edge of the neighboring and to open the mirror by synchronously moving the vertices along the edges (see figure 2). This approach will provide high precision alignment of the vertices of adjacent petals at the final stage of the deployment.

At the same time, the adjacent petals of the open mirror turn out to be connected along the external contour, which increases the stiffness of the open reflector. For this purpose, it is proposed to use small actuators and place them in the region where the vertices and edges are combined.

However, it is impossible to use directly the above-mentioned idea - the cylindrical hinges of the classical opening scheme impede the required mutual movement of the petals. In order to make this unstressed movement possible, it is proposed to replace the cylindrical hinges of the classical scheme with the spherical hinges and install additional spherical hinges at the points of connection of adjacent petals – figure 2.

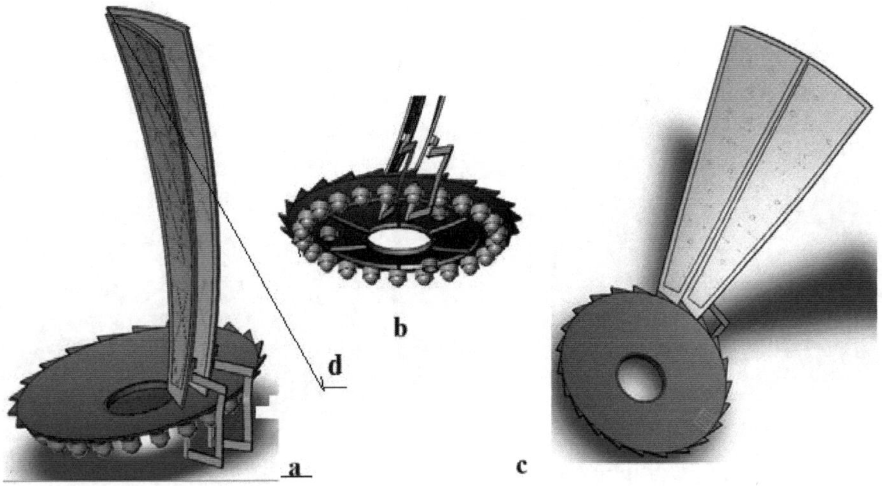

Figure 2. New design of the reflector: **a., c.** petals in the transport and working positions, **b.** spherical hinges on the back side of the central mirror, **d.** spherical hinge on the edge of the petal.

Main requirements for the kinematics of a transformable mirror are as follows: at each moment of the deployment, the structure must remain:

a. geometrically unchangeable and
b. stress-free (statically determinable).

It is easy to check that the *necessary condition* for fulfilling these requirements (the generalized Maxwell's rule) is satisfied in the construction [9]. Indeed, on the one hand, each petal connected to the central mirror by a spherical hinge has three degrees of freedom relative to the central mirror. On the other hand, each spherical hinge connecting the vertex of the petal with the edge of the adjacent petal introduces three kinematic constraints into the structure. Thus, the total number of kinematic constraints in the structure is equal to the number of degrees of freedom, and the *necessary condition* of geometrical unchangeability and static determinability is satisfied.

Further, a mathematical model of deployment kinematics was built, and it was shown that the *sufficient condition* for static determinability at each moment of deployment in the structure is fulfilled. This means that during the deployment, the composite mirror remains unstressed and geometrically unchangeable *indeed.*

3. Computer and physical simulation

To check the proposed technical solution the computer model of the new design in the Solid Works (SW) package was built. In the computer model, the mirror is opened by synchronous movement of the vertex of each petal along the edge of the adjacent petal. Simulation confirmed the possibility of unstressed opening of the composite reflector. Concerning final accuracy see [9]. Fragments of simulation are shown in figure 3.

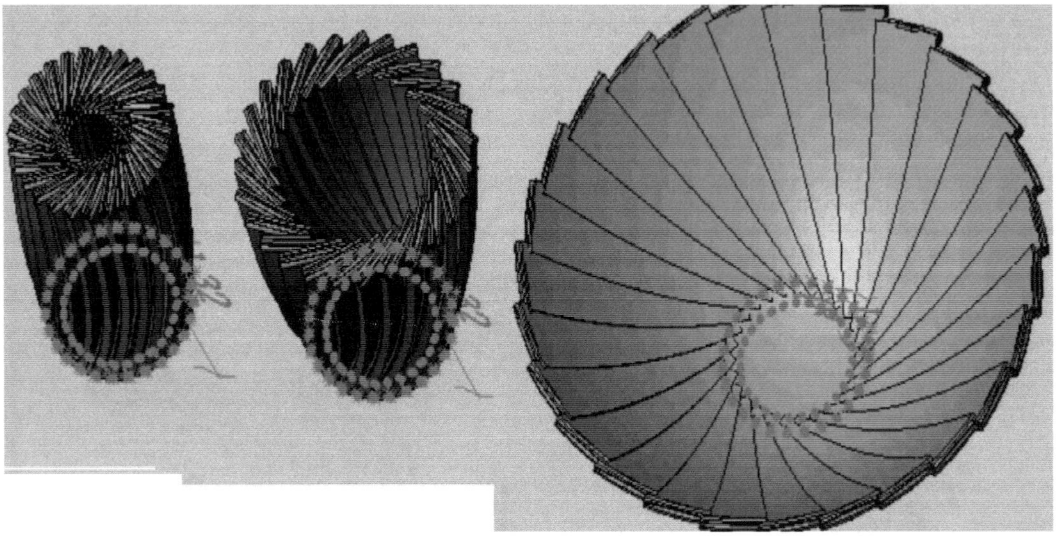

Figure 3. Computer simulation of petal-type mirror deployment.

The preliminary results of physical simulation are shown in figures 4 and 5.
The physical model includes a central mirror, a set of petals, a block of spherical hinges located on the back of the central mirror, and small-sized actuators (see figures 5b, 5c). The spherical hinges of the central mirror are used to attach the petals to the central mirror (see figures 4a, 4c).

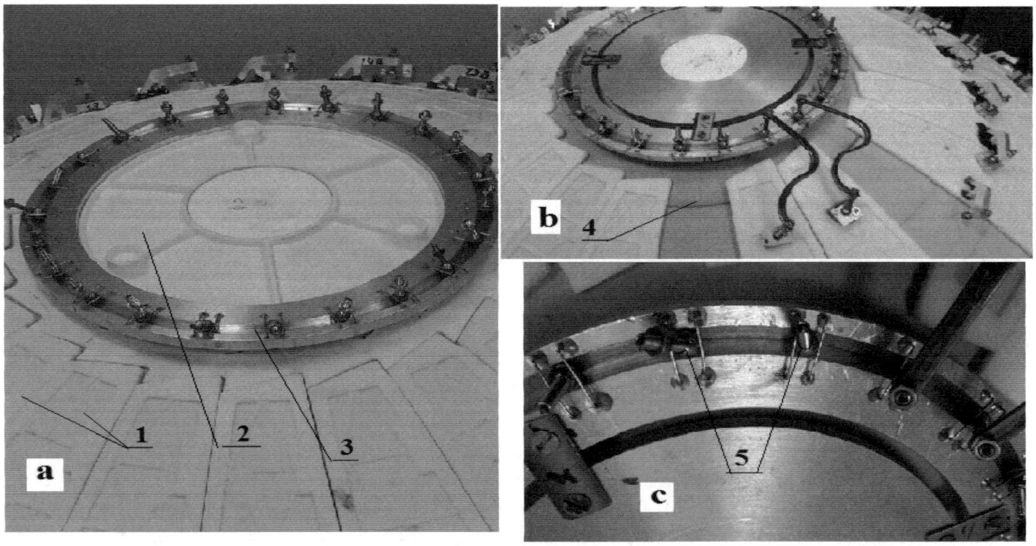

Figure 4. Assemblage of a composed mirror on a template. a. Petals (1), Central Mirror (2), Block of spherical hinges of central mirror (3) on a parabolic template (4) for assemblage. b. The connection of the petals and the central mirror on the template (4). c. Spherical hinges (5) on back side of central mirror.

The actuator moves along the edge of the petal using a worm gear (see figure 5b). A spherical hinge connecting the apex of one petal with the edge of an adjacent petal is located on the sidewall of the actuator (see figure 5a).

IOP Conf. Series: Materials Science and Engineering 1018(2021) 012024 doi:10.1088/1757-899X/1018/1/012024

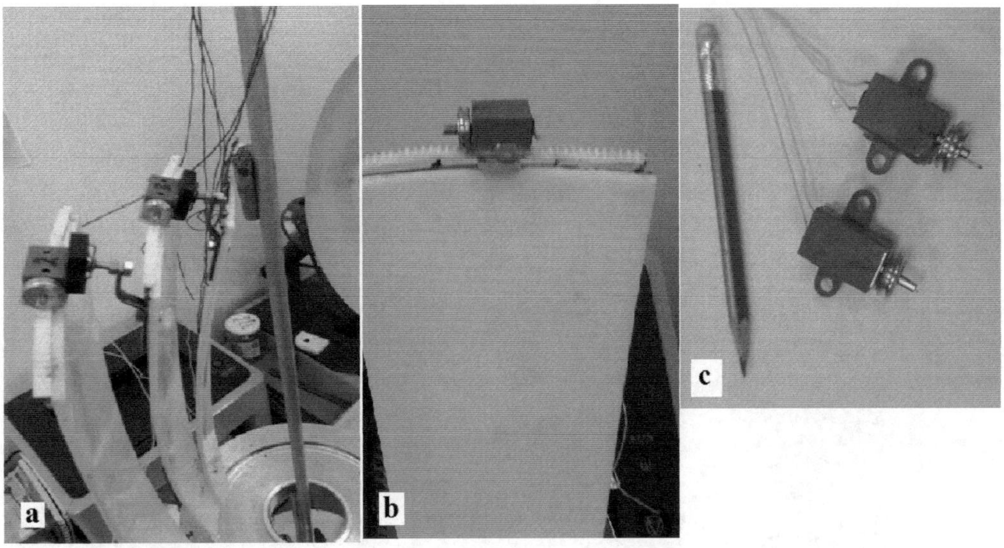

Figure 5. Linear actuators on the outer edge of the petal. **a**. Petals of deployable mirror in transport position. **b**. Actuator on the edge of the petal. **c**. View of the actuators.

Physical modelling is carried out in two stages. First, the petals and the central mirror are laid on the parabolic template (see figures 4 a, 4 b) and the connecting elements of the structure are precisely glued. Then the open mirror is disassembled, mounted in the transport position (see figure 5 a), and the opening system is worked out. Currently work on physical simulation continues.

Figure 6. CNC milling technology for central mirror and petals of the model fabrication. a - petals fabrication, b, c - central mirror and petals made with CNC milling technology

4. Technologies for manufacturing of physical model

3D printing technology, CNC milling (Fig. 6), plastic injection molding and carbon fiber technology (see figure 7) have been tested as possible candidates for the center mirror and petals fabrication.

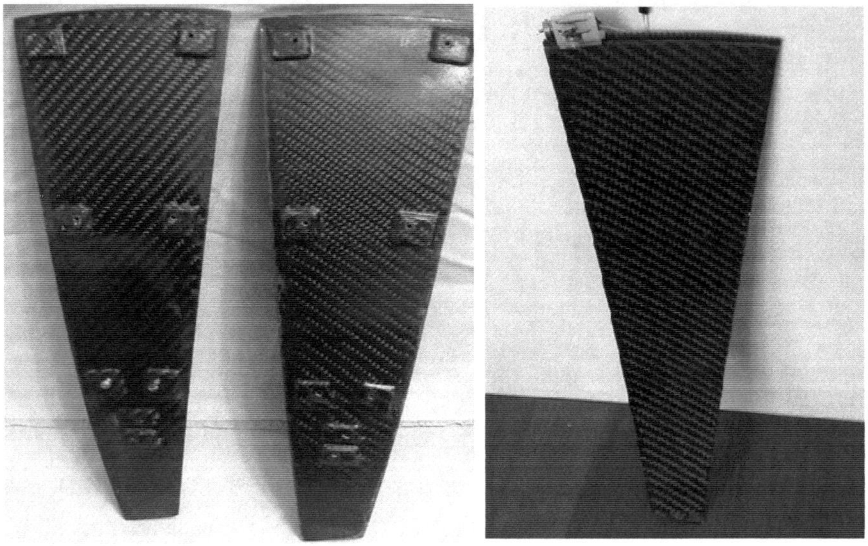

Figure 7. Carbon fiber technology for petals fabrication

At final stage of physical simulation, the carbon fiber technology was used for petals fabrication, milling technology was used to made central mirror and 3D printing technology was used to manufacture the elements of the deployment system.

5. Conclusions

To increase surface accuracy and rigidity of large deployable space mirror a new design of petal-type reflector was proposed and developed. The proposed new technical solution will make it possible to create space radio telescopes operating in a shorter wavelength region of the spectrum than the previous instruments.

References

[1] Kardashev N. S and Slysh V.I 1988 . The Radioastron Project *Proceedings of Symposium of International Astronomical Union 433*
[2] Kardashev, N. S et al 2017 *Astronomy Reports.* **61** 310
[3] He Huang et al 2018 *Acta Astronautica.* **148** 310
[4] Tao An et al 2020 *Adv. Space Res.* **65**(2) 850
[5] Dornier 1987 FIRST Technology study *Final Dornier Report RP-FA-D003*
[6] Westphal M 1990 *United States Patent* **4.899.167**
[7] Kardashev, N. S et al 2013 *Astronomy Reports.* **57(3)** 153
[8] Bujakas V I 2014 *Multibody Mechatronic Systems* Proceedings of the MUSME Conference (Mexico) Springer 234
[9] Bujakas V. I. and Kamensky A 2016 *Microactuators and Micromechanisms* Proceedings of MAMM-2016, Ilmenau, Germany, Springer 135
[10] Tarnai T 1980 *Int. J. solids struct.* **16** 347

Application of software solution for solving engineering design optimization problems

M Madic[1], P Janković[1], M Trifunović[1] and M Kovačević[2]

[1] Faculty of Mechanical Engineering in Niš, University of Niš, Aleksandra Medvedeva 14, 18000 Niš, Serbia
[2] Independent researcher

E-mail: madic@masfak.ni.ac.rs

Abstract. Engineering design problems involving a set of continuous, discrete and integer design variables and complex, non-convex objective functions and linear and non-linear constraints represent optimization problems with considerable complexity which are not trivial to solve. In recent years a number of optimization methods, particularly meta-heuristic optimization algorithms, were highlighted as effective optimization tools to deal with this type of engineering design problems. Despite certain advantages, their stochastic nature may be insufficient in dealing with various kinds of variables, constraints and objective functions. This paper discusses the application of developed software solution for solving engineering design optimization problems which is based on deterministic approach, i.e. the use of exhaustive iterative search algorithm. The use of the developed software solution is validated using four standard engineering design problems reported in the referential literature. In all case studies, the determined optimization solutions are equally good or better than those reported from other researchers using algorithms representative of the state-of-the-art in the area.

1. Introduction

The performance of a complicated engineered product, process, or system could be affected by a high number of parameters (or decision variables) interacting in complicated ways. Modelling and optimization of such design problems is, thus, an active area of research [1]. Whether it is preliminary design, detailed design or prototype system fabrication, optimization concepts and methods are helpful at every stage of the process [2].

Engineering design optimization problems are omnipresent in engineering and industry. In mechanical engineering, while designing a given product or system designers are faced with certain technological requirements, i.e. objectives that final product should be meet (such as strength, deflection, weight, wear, corrosion resistance), while considering a number of design parameters. For example, in tire design it should consider a number of parameters such as belt angle, belt cord spacing, thickness of tread layer, number of cap plies, elasticity of tread compound, etc. and at the same time satisfy as much as possible several objectives such as dry/wet handling and traction, endurance, wear resistance, ride comfort, rolling resistance, aquaplaning, weight, etc. Applications in engineering design range from the design of individual structural members to the design of separate pieces of equipment to the preliminary design of entire production facilities. For purposes of optimization, the shape or structure of the system is assumed to be known, and the optimization problem reduces to that of selecting values of the unit dimensions and operating variables that will yield the best value of the selected performance criterion [3]. Design optimization of a gear train, radial ball bearing, Belleville spring, multi-plate disc clutch brake, robot gripper, pressure vessel, tension/compression spring, screw jack, clamps, etc. represent typical engineering design optimization problems.

Formulated as classical mathematical optimization problems, these problems usually consider continuous, integer and discrete design variables, multi-dimensional, complex, non-convex and nonlinear objective functions as well as linear and nonlinear constraints of inequality or equality type. Constraints are very important in engineering design problems, since they are normally imposed on the statement of the problems and sometimes are very hard to satisfy, which makes the search difficult and inefficient [4]. These are often written either as simple bounds such as the ranges of material properties or as nonlinear relationships including maximum stress, maximum deflection, minimum load capacity and geometrical configuration [5].

For searching of complex feasible optimization domains with multiple local optimums a number of optimization algorithms and methods were proposed in literature. The most popular methods include traditional optimization methods (direct and gradient-based methods) and in recent time meta-heuristic algorithms such as genetic algorithm (GA), harmony search algorithm (HSA), cuckoo search algorithm (CSA), particle swarm optimization (PSO), ant colony optimization, artificial bee colony, firefly algorithm, simulated annealing, etc. As noted by Gandomi et al. [5], the efficiency of meta-heuristic algorithms can be attributed to the fact that they imitate the best features in nature, especially the selection of the fittest in biological systems which have evolved by natural selection over millions of years.

Most traditional optimization methods are designed to work with continuous variables only, in dealing with design problems having discrete variables, these methods do not work well [6]. Since derivative information is not used, the direct search methods are usually slow, requiring many function evaluations for convergence. On the other hand, gradient-based methods quickly converge to an optimal solution, but are not efficient in non-differentiable or discontinuous problems or in cases when the objective function and constraints have multiple or sharp peaks and also when initial solution is not adequately set [7]. In the case of meta-heuristic algorithms, the solution of non-linear problems becomes quite difficult especially when they are heavily constrained [8]. Although these heuristic methods have the advantages of easy implementation for complex problems, they cause convergence difficulties and cannot guarantee the global optimality of the solution. Moreover, the probability of finding the global solution decreases when the problem size increases [9]. Also, as noted by Kovačević et al. [10], specific meta-heuristic algorithm parameters settings have a strong influence on the final solution and there is no universal rule for setting the algorithm parameter values.

Most of the aforementioned methods and algorithms can handle single and multi-objective design optimization problems. Despite numerous optimization methods, every method has certain advantages and disadvantages for implementation in real-life. However, there exists no universal method which is the "best" choice for solving engineering design optimization problems [11]. Above all, some optimization methods can be difficult to use for engineers who are not experts in optimization theory [12].

The idea of this paper was to apply the developed software solution for solving engineering design optimization problems, based on the use of deterministic approach, i.e. application of exhaustive iterative search algorithm. Moreover, the developed software solution incorporates also simple stochastic search algorithm, i.e. Monte Carlo method which can be applied in combination with exhaustive iterative search algorithm. The main purpose of this software prototype is the validation of the engineering design optimization solutions obtained by meta-heuristic algorithms.

2. Software solution

Software solution "BRUTOMIZER" is a software tool that can be used for solving engineering design optimization problems. It enables the mathematical definition of the mathematical model, definition of a number of constraints (linear and nonlinear of inequality or equality type) and definition of the design (search) space with possible values for each variable. Currently, one deterministic algorithm, i.e. exhaustive iterative search algorithm and two stochastic algorithms, i.e. discrete and continual Monte Carlo algorithms are available for the purpose of solving optimization problems. In contrast to existing software solutions for optimization, "BRUTOMIZER" enables:

- simple mathematical model definition, without the need to write any code,
- simple definition of design optimization problems with or without constraints,
- simple definition of design variables space and possible values for each variable,
- ability to determine a specified number of solutions: in the case of maximization, minimization, or in the case of targeting specified mathematical model value,
- analysis of a number of optimization solutions,
- guaranteed optimality of the determined optimization solutions,
- parameter free optimization approach with user friendly interface.

Software solution "BRUTOMIZER" is developed by Marko Kovačević and Miloš Madić and is freely available [13]. Snapshot of user interface of the developed software solution while solving case study 1 is shown in figure 1. The left side of figure 1 shows forms for definition of decision variable's space, number of possible combinations as well as for definition of the objective function, i.e. mathematical model. The right side of the figure 1 shows optimization window with optimization results along with applied optimization method and optimization time.

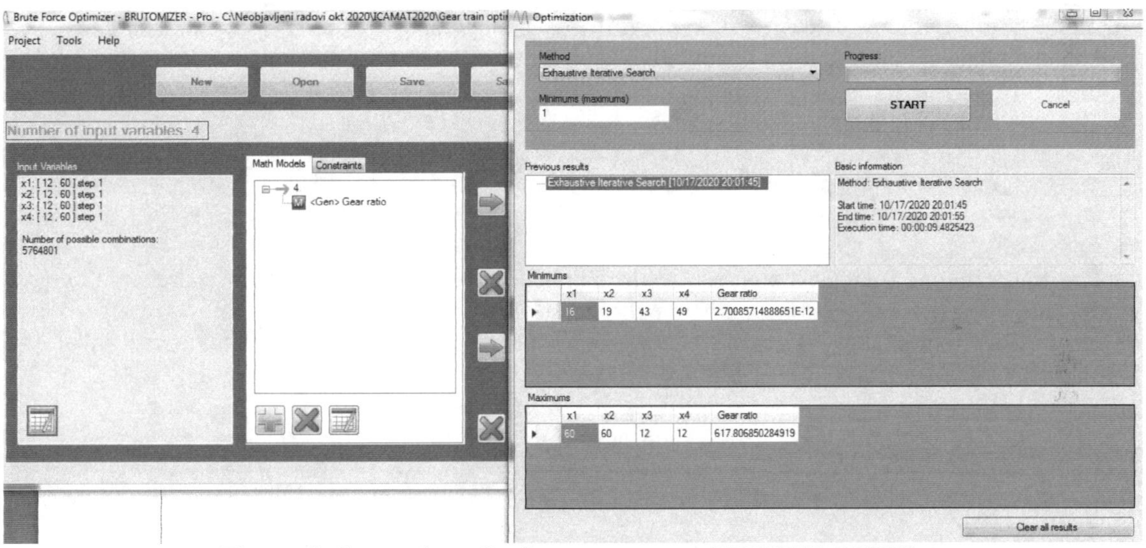

Figure 1. Screenshot of software prototype "BRUTOMIZER".

3. Engineering design optimization problems: case studies

To demonstrate the application and effectiveness of the developed software solution for solving mechanical engineering design optimization problems, four case studies, i.e. gear train design, tension/compression spring design and pressure vessel design were considered. For each case study, a comparative analysis of the solutions obtained by the software solution and the solutions obtained by the previous researchers using other optimization methods is given. In order to demonstrate the efficiency of the developed software prototype the computational time for each case study is presented. The developed software was run on Intel Core i7-2600 CPU@3.4GHz with 12 GB RAM.

3.1. Gear train design

Mathematically, gear train design problem is an unconstrained optimization problem. This design problem has four integer variables (T_d, T_b, T_a, T_f) and was introduced by Sandgren [14]. A gear train is to be designed to achieve a specific gear ratio between the driver and the driven shafts (figure 2).

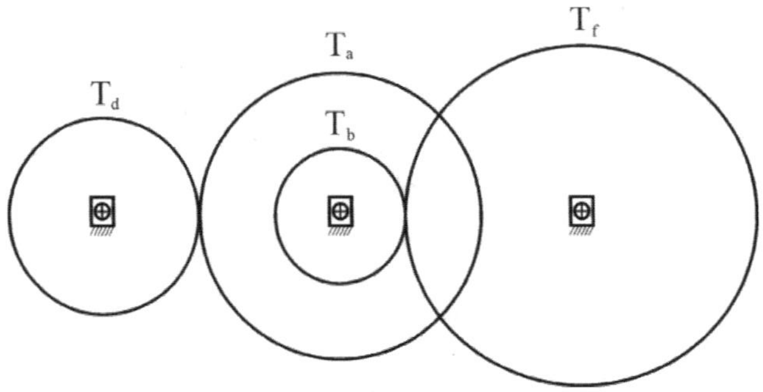

Figure 2. A gear train.

The objective of gear train design is to find the number of teeth in each of the four gears so as to minimize the error between the obtained gear ratio and the required gear ratio of 1/6.931 [6]. The gear ratio is defined as follows:

$$\text{Gear ratio} = \frac{T_d \cdot T_b}{T_a \cdot T_f} \tag{1}$$

where the number of teeth of each gearwheel is integer varying in the range 12–60.

The mathematical optimization formulation of gear train design optimization is as follows:

$$\text{Minimize} : f = \left(\frac{1}{6.931} - \frac{T_d \cdot T_b}{T_a \cdot T_f} \right)^2 \tag{2}$$
$$\text{Subject to} : 12 \leq T_d, T_b, T_a, T_f \leq 60$$

The optimization solution obtained by the application of the developed software solution and computational results obtained using different methods are given in table 1. As can be observed from this table, the optimization results are better than those reported by Sandgren [14] and Kannan and Kramer [15] and identical to those obtained by Deb and Goyal [6] and Gandomi et al. [5].

Table 1. Optimization solutions for the gear train design case study.

References	Sandgren [14]	Kannan and Kramer [15]	Deb and Goyal [6]	Gandomi et al. [5]	Present study
Design variables /Applied method	branch and bound method	Lagrange multiplier method	GA	CSA	software solution
T_d	18	13	19	19	19
T_b	22	15	16	16	16
T_a	45	33	49	43	43
T_f	60	41	43	49	49
Gear ratio	0.146667	0.144124	0.144281	0.144281	0.144281
f	2.387×10^{-3}	1.553×10^{-4}	1.643×10^{-6}	1.643×10^{-6}	1.643×10^{-6}

3.2. Pressure vessel design

The objective of this case study is to design a pressure vessel that has a given working pressure and volume so as to minimize the total cost, including the cost of the material, forming and welding. A cylindrical pressure vessel capped at both ends by hemispherical heads is presented in figure 3. In this case study there are four design variables: thickness of the shell (T_s), thickness of the head (T_h), inner radius (R) and length of the cylindrical section of the vessel, not including the head (L). The design variables L and R are continuous while T_s and T_h are discrete which are the available thicknesses of rolled steel plates (integer multiples of 0.0625 in).

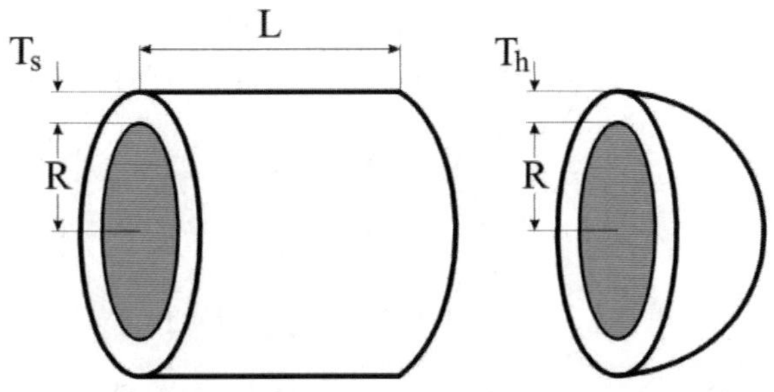

Figure 3. A pressure vessel.

After considering geometric constraints and other pressure vessel requirements, the final formulation of the pressure vessel design optimization problem is as follows:

$$\begin{aligned}
&\text{Minimize: } f = 0.6224 \cdot T_s \cdot R \cdot L + 1.7781 \cdot T_h \cdot R^2 + 3.1661 \cdot T_s^2 \cdot L + 19.84 \cdot T_h^2 \cdot L \\
&\text{Subject to: } g_1 = -T_s + 0.0193 \cdot R \le 0 \\
&g_2 = -T_h + 0.00954 \cdot R \le 0 \\
&g_3 = -\pi \cdot R^2 \cdot L - \frac{4}{3} \cdot \pi \cdot R^3 + (750 \cdot 1728) \le 0 \\
&0.0625 \le T_s, T_h \le 99 \cdot 0.0625 \\
&10 \le L, R \le 200
\end{aligned} \tag{3}$$

The optimization solution obtained by the application of the developed software solution and computational results obtained using different methods are given in table 2. It has been observed that determined optimization solution is significantly better than previously reported optimization solutions and at the same time satisfy all given constraints. Using the developed software solution one can easily check the optimization solutions reported from other authors. While solving this case study it was revealed that optimization solution reported by Sandgren [14] does not satisfy the third constraint g_3 and also a considerable difference in objective function values as reported by Cagnina et al. [4] and Gandomi et al. [5] was noticed.

Table 2. Optimization solutions for the pressure vessel design case study.

References	Sandgren [14]	Cagnina et al. [4]	Gandomi et al. [5]	Present study
Design variables /Applied method	branch and bound method	PSO	CSA	software solution
T_s	1.125	0.8125	0.8125	1.3125
T_h	0.625	0.4375	0.4375	0.625
R	48.97	42.098446	42.0984456	65
L	106.72	176.636596	176.6365958	11
f, Costs ($)	7980.894	6059.714335 *6179.1074834	6059.7143348 *6179.10741606	5424.62

* Corrected values

3.3. Tension/compression spring design

A very simple design problem in engineering is to design a minimum mass tension/compression spring to carry a given axial load subject to constraints of minimum deflection, shear stress, surge frequency and limits on design variables as shown in figure 4.

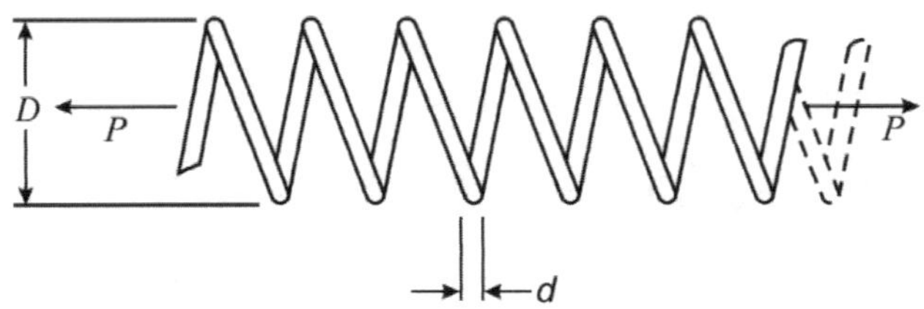

Figure 4. A tension/compression spring.

In this case study there are three design variables: the mean coil diameter (*D*), the wire diameter (*d*) and the number of active coils (*N*). The mathematical formulation of this problem is:

$$\begin{aligned}
&\text{Minimize: } f = (N+2)\cdot D\cdot d^2\\
&\text{Subject to: } g_1 = 1-\frac{D^3\cdot N}{71875\cdot d^4}\le 0\\
&g_2 = \frac{4\cdot D^2 - D\cdot d}{12566.4\cdot\left(D\cdot d^3 - d^4\right)}+\frac{1}{5108.3\cdot d^2}-1\le 0\\
&g_3 = 1-\frac{140.45\cdot d}{N\cdot D^2}\le 0\\
&g_4 = d + D - 1.5\le 0\\
&0.05\le d\le 2\\
&0.25\le D\le 1.3\\
&2\le \mathrm{N}\le 15
\end{aligned}\tag{4}$$

Table 3 presents the best optimization solution of this problem obtained using the developed software solution and compares the results with optimization solutions reported by other researchers. It is observed from table 3 that in some reported cases the number of coils (*N*) is not an integer at the optimum solution. Re-optimization results as reported by Arora [2] are given in brackets. This case study was solved considering number of coils as integer while other two design variables were given up to four decimal places resulting in around 716 millions of possible combinations and optimization solution determination for 45 seconds.

Table 3. Optimization solutions for the tension/compression spring design case study.

References	Coello [17]	Arora [2]	Cagnina et al. [4]	Present study
Design variables /Applied method	GA	numerical optimization technique	PSO	software solution
d	0.051989	0.0517 (0.0512)	0.051690	0.0508
D	0.363965	0.3565 (0.3454)	0.356750	0.3912
N	10.890522	11.31 (12)	11.287126	8
f	0.012681	0.01268	0.012665	0.010095

3.4. Eason and Fenton's gear train inertia function
Eason and Fenton's function [18] is a minimization problem for the inertia of a gear train. The mathematical formulation of this problem is:

$$\text{Minimize: } f = \frac{1}{10}\cdot\left(12 + x_1^2 + \frac{1+x_2^2}{x_1^2} + \frac{x_1^2\cdot x_2^2 + 100}{\left(x_1\cdot x_2\right)^4}\right) \tag{5}$$

$$\text{Subject to: } 0 \le x_1, x_2 \le 10$$

Table 4 presents the best optimization solution of this problem obtained using the developed software solution and compares the results with optimization solutions reported by other researchers.

Table 4. Optimization solutions for the tension/compression spring design case study.

References	Jaberipour and Khorram [7]	Lee and Geem [16]	Present study
Design variables /Applied method	Improved HSA	HSA	software solution
x_1	1.7435321	1.74338	1.743
x_2	2.0297445	2.03050	2.03
f	1.744152	1.74415	1.744152

While solving this case study design variables search space was given up to three decimal places. Thus, using the step of 0.001 for design variables x_1 and x_2, resulted in around 100 millions of possible combinations and the optimization solution was determined in 6 seconds. As in the case of other case studies, more precise determination of optimization solution can be achieved by successive runs of software solutions using smaller search intervals and smaller steps in the vicinity of the previous best solution.

3.5. Results and discussion
Many real world engineering design optimization problems involve mixed (involve continuous, binary, discrete and integer variables) design variables. The binary variables are usually involved in the formulation of the design problem to select alternative options. The discrete variables are used to represent standardization constraints such as the diameters of standard sized bolts or plate thicknesses. Integer variables are used to represent the number of gear teeth or number of coils [19]. However, as observed from literature and considered case studies, in an attempt to test new or improved optimization algorithms, majority of optimization studies neglect such facts in searching for optimal solution with a higher number of decimal places. This however, from engineering point of view is impractical since obtained solutions cannot be implemented in practice.

In general, engineering design optimization problems require different formulation and application of specific optimization approach and eventually optimization software to use. Since there is clearly no one universal algorithm that performs best in all categories of optimization problems, numerical experiments to compare new algorithms and software with existing ones is an important area complementing the theoretical analysis of optimization methods [20].

A number of researches and practitioners emphasize the speed of the algorithm to find a "good" solution, even though the solution found is not guaranteed to be the optimum. In this study, however, the presented software solution for solving engineering optimization design problems gives more attention to determination of practical solutions with guaranteed optimality.

The efficiency and practical usability of the developed software solution has also been recently confirmed while solving single and multi-objective manufacturing optimization problems (laser cutting and turning) with several process constraints [21, 22].

4. Conclusion

This paper proposes the use of deterministic approach for solving engineering design optimization problems using the developed software solution. The proposed approach was demonstrated while solving four referential engineering design optimization problems and performed well in terms of the quality of the optimization solutions found. The obtained optimization solutions were compared with solutions that had been determined by past researchers using meta-heuristic algorithms and mathematical programming techniques.

Although it uses brute force algorithm for the purpose of optimization and the amount of calculations when solving some case studies may be high, actual optimization time is realistic and acceptable. On the other hand the algorithm is parameter free, reliable and applicable for solving the given design optimization problem within a reasonable degree of accuracy, which can be specified by the user. Moreover, the developed software solution has high level of generality and can be applied to formulate and solve any type of engineering design optimization problem. One particular benefit of the developed software solution is its possibility to find user defined number of optimization solutions and to rank them. This is thought to be very practical in engineering design since design engineers are enabled to compare different designs which are, in terms of objective function values, very similar or identical. Finally, the application of the developed software solution is relatively easy to understand, even by less experienced users.

5. References

[1] Liao TW 2010 *Appl. Soft Comp.* **10** 1188
[2] Arora J 2004 *Introduction to Optimum Design* (Second Edition) (Academic Press) p 728
[3] Ravindran A, Reklaitis GV and Ragsdell KM 2006 *Engineering Optimization: Methods and Applications* (John Wiley & Sons) p 688
[4] Cagnina LC, Esquivel SC and Coello CAC 2008 *Inform.* **32** 319
[5] Gandomi AH, Yang XS and Alavi AH 2013 *Eng. Comp* **29** 17
[6] Deb K and Goyal M 1996 *Comp. Sci. Inform* **26** 30
[7] Jaberipour M and Khorram E 2010 *Commun. Nonlinear Sci. Numer. Simul.* **15** 3316
[8] Chung CJ and Reynolds RG 1996 A *testbed for solving optimization problems using cultural algorithms*. In Evolutionary programming (Academic Press) p 225
[9] Lin MH, Tsai JF and Wang PC 2012 *Appl. Math. Inform. Sci.* **6** 101
[10] Kovačević M, Madić M and Radovanović M 2013 *Expert Syst. Appl.* **40** 6985
[11] Madić M, Kovačević M, Marinković V and Radovanović M 2012 *Strojarstvo* **54** 161
[12] Himmelblau DM 1972 *Applied Nonlinear Programming* (McGraw-Hill) p 498
[13] http://www.virtuode.com
[14] Sandgren E 1990 *J. Mech. Des.* **112** 223
[15] Kannan BK and Kramer SN 1994 *J. Mech. Des.* **116** 405
[16] Lee KS and Geem ZW 2005 *Comp. Method. Appl. Mech. Eng.* **194** 3902
[17] Coello CAC and Montes EM 2002 *Adv. Eng. Inform.* **16** 193
[18] Eason ED and Fenton RG 1974 *J. Eng. Ind.* **96** 196
[19] Garg H 2014 *J. Ind. Manag. Optim.* **10** 777
[20] Khompatraporn C, Pintér JD and Zabinsky ZB 2005 *J. Glob. Optim.* **31** 613
[21] Madić M, Mladenović S, Gostimirović M, Radovanović M and Janković P 2020 *Proc. IMechE Part B: J. Eng. Manuf.* **234** 1323
[22] Trifunović M, Madić M and Radovanović M 2020 *Int. J. Adv. Manuf. Technol.* **110** 1893

Acknowledgments
This research was financially supported by the Ministry of Education, Science and Technological Development of the Republic of Serbia.

IOP Conf. Series: Materials Science and Engineering 1018(2021) 012026 doi:10.1088/1757-899X/1018/1/012026

Analysis and development of a key performance indicators model for the paper industry

A Tieber[1], D Manolache[2] and M Gheorghe[2]

[1] PhD Student, University POLITEHNICA of Bucharest
[2] University POLITEHNICA of Bucharest, Manufacturing Engineering Department, 313 Splaiul Independentei, 060042 Bucharest, Romania

E-mails: erfolg@alfredtieber.com[1], daniel.manolache@upb.ro , marian.gheorghe@upb.ro

Abstract. It is advisable to implement the use of key indicators by each company as a support for its continuous performant activity. Massive potential for optimisation can be found in the whole production industry. A scientific survey study has been designed and addressed to a number of companies from production industry in Europe. In the framework of this survey, the paper industry, as part of production industry, has been more profoundly analysed, in order to find the level of understanding and use of the key performance indicators and KPI systems, as an example. Based on the elements revealed by this analysis, a KPIs - Model for processes, production, financials, quality, supply chain, human resources and innovation is developed. This KPIs - Model can be a basis for the development of further KPIs – Models by any interested company, based on its specific conditions.

1. Introduction

In the past, the meaningfulness of key performance indicators, KPIs and key figure systems has been discussed intensively and extensively both in technical literature and in everyday work [1]. The potential for optimizing and improving company-specific business processes is often underestimated. In the course of day-to-day work, it is all too easy to neglect findings from key figure analysis or to take only limited or insufficient account of them [2]. The consequences are that defined goals and specifications are not achieved, adjustments of affected processes and organizational units are not carried out [3].

Current findings on the use and prevalence of various key performance indicator systems in the production industry are currently only available to a limited extent. This is a general problem of the production industry. A direct allocation to only one specific sector is to be excluded. Furthermore, all system partners are affected. Both manufacturers and suppliers show similar tendencies in the use of key performance indicators. The importance and the general benefit with regard to the use of KPIs is known to all concerned. Nevertheless, a widespread use of key performance indicators in the production industry does not exist [4]. The main reasons for this cited by the companies are the lack of know-how for the implementation of a KPI system, as well technical and IT problems during the implementation process. Another aspect mentioned is high implementation costs [5].

Ultimately, however, companies are facing ever greater challenges. Due to globalisation and a constant increase in the speed of communication, the use of key figures is essential for maintaining competitiveness. Large companies have a clear advantage over their suppliers here. SMEs in particular are overtaxed. They are confronted with constantly increasing requirements, which will be difficult to implement without appropriate control mechanisms.

Here it is necessary to take appropriate measures in good time and to install lean and cost-effective key performance indicator systems in the company organisation. Of course, SMEs already use KPIs. However, these focus mainly on the collection of financial ratios and to a lesser extent on production ratios. This is not enough for overall corporate management. Key figures for the HR management are insufficiently used [6].

In order to be able to operate successfully in international competition in the future, SMEs in the production industry must optimise their internal and external organisational processes in a targeted manner through the use of KPIs. This requires a strategic and operational objective in the companies. Taking into account the above elements, the development of a KPIs - Model for paper industry has been considered. The term *model* is used as a group of key indicators from the various modules [7], a meaningful collection of KPIs that can be assigned to one or more business processes/ business areas. This enables the user to select an optimized number of key performance indicators for his individual corporate management.

For development of an useful modelling system in industrial conditions, a scientific survey study has been designed and addressed to a number of companies from production industry in Europe [8]. This study is referring to Tier-1 and Tier-2 suppliers in paper industry, as part of a deployed survey study. It is to be noted that Tier-1 suppliers are the first communication partners for original equipment manufacturers (OEMs). Tier-1 companies have strong credibility, because they can generate reliable components on time and with strict adherence to standards procedures. Tier-2 suppliers are usually limited in what they can produce, smaller and with less technical advantages than Tier-1 companies.

2. Data analysis

2.1. Key facts

It is to be underlined that the content of this study is due to the fact that Tier-1 and Tier-2 suppliers and sub-suppliers of the paper industry are highly present in the considered survey study results. The share of paper industry enterprises in the whole survey was 22.8 %. In the course of study, the degree of SMEs uses of key performance indicators in the paper industry in the DACH region (DACH region - Germany, Austria and Switzerland/ German-speaking Europe) was analysed.

Fourteen Tier-1 and Tier-2 suppliers of the paper industry were basically used for this study. The distribution of Tier1 and Tier2 suppliers is approximately balanced (57 % to 43 %).

The selected individuals came from the managerial levels of reputable industrial companies and industrial suppliers of paper industry. The data were categorized appropriately. Statistical key figures were used to assess and analyse the data volume afterwards. A coding sheet or a scale rating matrix were used. The study results were verified on the basis of information and insights from specialist scientific works.

2.2. Method analysis

Let r and R be a result associated to the Tier-1 and Tier-2 suppliers paper industry, and the correspondent result associated at the level of production industry in deployed survey, respectively, i.e.

$$r = 100*s/t,\ R = 100*S/T, \text{ in } \% \tag{1}$$

where: s and S are the number of answers given by the companies to the considered opportunity, and to the topic/ question (all opportunities), respectively.

The results associated to the Tier-1 and Tier-2 suppliers from paper industry, type of r, have been compared with the correspondent results associated to whole production industry, type of R, in deployed survey, through the deviation (difference), v,

$$v = r - R \tag{2}$$

The values of the deviation $|v|$ were divided into the following categories:

- very strong, if predominantly were greater than 20%;
- strong, if predominantly were between 10% and 20%;
- weak, if predominantly were between 5% and 10%;
- none, if predominantly were less than 5%.

It is to be noted that the deviations are positive or negative. A positive/ negative deviation shows that the considered opportunity plays a more/ less important role in paper industry than at the level of production industry in the deployed survey.

2.3. Remarks

Suppliers to the paper industry rated hurdles related to the IT system and organizational challenges +12 % higher. Expenditure concerns was valued 10 % lower (-10%). Important differences were in the linking of key figures within the company/ divisions, where there have been changes of +21 % for the answer 'yes', -35 % for the answer 'no', and +13 % for the answer 'partly'.

There are strong differences in key figure communication in the companies. The Tier-1 supplier and Tier-2 supplier use communication by mail or email much more intensively (+12 %). Verbally communication is significantly weaker (-9 %).

The topic/ question about a customer requirement for using KPIs was answered as follows. There is a strong deviation, because plus 20 % answered the question with 'yes' and this is very high in shift to the whole survey. The suppliers to the paper industry further stated that only the quality assurance- and logistics KPIs are relative important (+12 %).

A shift between process improvement (+14%) and the fulfilment of own customer demands (-13%) happened. Process/ Manufacturing KPIs and innovation/engineering development KPIs grown up (+10 %).

Logistics and supply chain KPIs lost a bit (-5 %).

The personal statement about the use of KPI knowledge for everyday work, brought a strong deviation. A plus of 11% for answer 'agree completely' and - 13% happened for the answers 'do not agree' and 'do rather not agree' were given.

The further development of strategic partners and suppliers is promoted much more strongly, purposefully and sustainably as in other industry groups. The Tier-1 and Tier-2 suppliers in the paper industry are also much more convinced of the benefits of key figures than the industry as a whole. It can be stated that the Tier1- and Tier 2suppliers of the paper industry have recognized the meaningfulness of the use of key figures and actively integrate them into the work process on their own.

3. Development of KPIs – Model

Based on the results of the analysis on Tier-1 and Tier-2 suppliers in paper industry, partial above presented, a KPIs – Model has been developed for processes, production, financials, quality, supply chain, human resources, innovation, as follows.

The reference KPIs – Modules were graphically prepared in a general overview (figure 1) and show the variety of possible key figures available for the KPIs – Model development. It is to be underlined that each KPIs - Module is constituted of specific KPIs.

The KPIs – Model for paper industry has been developed as a mix of KPIs selected from KPIs – Modules associated to processes, production, financials, quality, supply chain, human resources, innovation key performance indicators as presented in figure 2.

Significant elements associated to some components of the KPIs - Model are as presented in table 1. In addition, the analytical representation of the considered KPIs – Model is proposed as the group of equations (3).

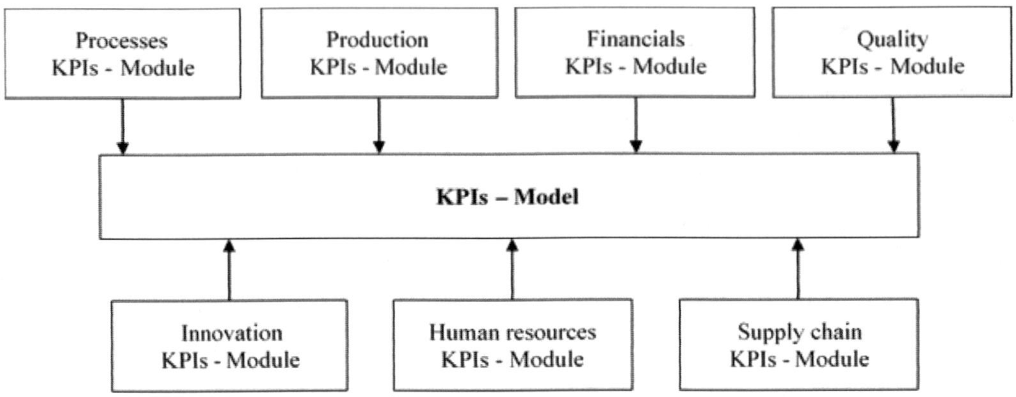

Figure 1. KPIs – Modules

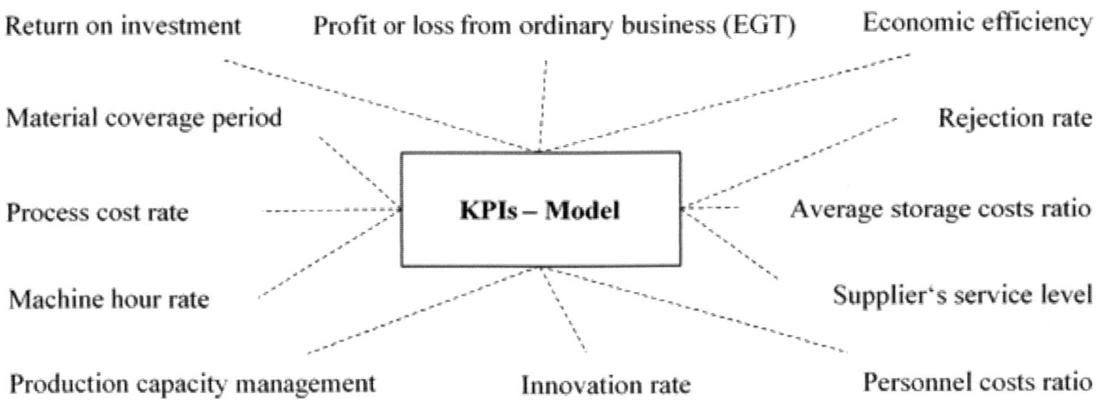

Figure 2. KPIs – Model

Table 1. Significant elements associated to some components of the KPIs – Model

	Defining elements	Unit of measurement
PCM	Production capacity management	%
EA	Effective availability	hours/week or month or ...
TOT	Total operation time	
MHR	Machine hour rate	%
RM	Related machine costs	euro/month or quarter ...
MR	Total machine runtime costs	
…	…	…
IR	Innovation rate	%
TSI	Turnover share of innovations	euro/quarter or quarter or ...
TNT	Total net turnover	

IOP Conf. Series: Materials Science and Engineering **1018** (2021) 012026 doi:10.1088/1757-899X/1018/1/012026

$$\mathrm{KPI}_1: PCM = 100 * EA/TOT,\ \mathrm{KPI}_2: MHR = 100 * RM/MR,\ \ldots,$$
$$\mathrm{KPI}_c: IR = 100 * TSI/TNT \quad (3)$$

The developed KPIs – Model can be applied in practice as a prototype for Tier- 1 and Tier-2 suppliers of the paper industry. Company-specific adjustments can be made at any time.

Company:	**Status report KPIs – Model**	**Date: DD.MM.YYYY**
Department: Prepared by: Approved by:	Review period: MM – MM	General status: ☐ On track ☐ Potential risks ☐ Roadblocks

No.	**Key performance indicator (KPI)**	**Status (☺ 😐 ☹)**	**Prospect (↑ ↔ ↓)**	**Notes**
1)	Production capacity management			
2)	Machine hour rate			
3)	Process cost rate			
4)	Material coverage period			
5)	Return on investments			
6)	Profit or loss from ordinary business (EGT)			
7)	Economic efficiency			
8)	Rejection rate			
9)	Average storage costs ratio			
10)	Supplier's service level			
11)	Personnel costs ratio			
12)	Innovation rate			

No.	**Reasons for deviations**
…	….

No.	**Further action**	**Due date**	**Responsible**
…	….		

Remarks		
Status report author:	Recipients:	Date of next status report: DD.MM.YYYY

Figure 3. Status report for KPIs – Model

A simple reporting document for information and continuous review, conventional or in digital system, is proposed as presented in figure 3. Some arguments to use this simple reporting document are as follows.

• With the status report, the development of the company can take place on the basis of key figure measurements during defined review periods.
• According to the survey results, these periodic reports are mostly made weekly or monthly for Tier-1 and Tier-2 suppliers in paper industry.
• A status report supports the continuous improvement and optimization of the entrepreneurial value creation and its process landscape.
• Measures and future activities can also be documented on the status report and provided with a completion date and a responsible person.

4. Conclusions

The comparative analysis of Tier-1 and Tier-2 suppliers in paper industry and the production industry in Europe, in the framework of the deployed survey, has revealed useful data.

A KPIs - Model for processes, production, financials, quality, supply chain, human resources and innovation is developed. This can be implemented quickly, cost-effectively and easily at suppliers and sub-suppliers in the paper industry. Further KPIs - Models can be developed ensuring a continuous corporate management by means of KPIs across all corporate divisions in the paper industry.

For success as a SME company in the paper industry, it is strongly advisable to intensify the use of key performance indicators. Transparency in analysis and daily work, as well as open and honest communication throughout all levels of the organization constitute key factors for success in this endeavor. Communication and continuous work on the relevant processes are imperative.

References

[1] Burkert M 2008 *Qualität von Kennzahlen und Erfolg von Managern: Direkte, indirekte und moderierende Effekte*, ISBN 978-3-8350-55619-2, e-book, Betriebswirtschaftlicher Verlag Dr. Th. Gabler, Wiesbaden

[2] Engroff B and Aulbach R 2005 *Erfahrungsbericht der AWF-Arbeitsgemeinschaft „Kennzahlen und Kennzahlensysteme zur Unternehmensführung und – steuerung“ – Praktischer Einsatz von Kennzahlen und Kennzahlensystemen in der Produktion*, volume 2, AWF Arbeitsgemeinschaften für Vitale Unternehmensentwicklung e.V. Groß-Gerau, www.awf.de

[3] Greiling D 2009 *Performance Measurement in Nonprofit-Organisationen*, ISBN 978-3-8349-9953-5, e-book, Gabler Verlag, Wiesbaden

[4] Becker J and Winkelmann A 2014 *Handelscontrolling: Optimale Informationsversorgung mit Kennzahlen*, ISBN 978-3-642-39530-7, e-book, volume 3 (Springer Berlin, Heidelberg)

[5] Günther W A and Dörnhöfer M 2014 *Einsatz von Kennzahlensystemen in der Automobillogistik Aktueller Entwicklungsstand und Handlungsbedarf*, Onlinerecherche, ISBN 978-3-941702-50-9, Technische Universität München, Lehrstuhl für Fördertechnik Materialfluss Logistik, http://mediatum.ub.tum.de/node?id=1236697, München (downloaded on 30.09.2019, 12:32)

[6] Pleier N 2008 *Performance-Measurement-Systeme und der Faktor Mensch: Leistungssteuerung effektiver gestalten*, ISBN 978-3-8349-8002-1, e-book, GWV Fachverlage GmbH, Wiesbaden

[7] Tieber A, Manolache D S and Gheorghe M 2019 Development of key performance indicators modules for small and medium-sized enterprises in production industry IOP Conf. Ser.: Mater. Sci. Eng. **682** 012020

[8] Tieber A, 2020, PhD Thesis – to be presented and sustained, University POLYTEHNICA of Bucharest, Bucharest

AUTHOR INDEX

Institute of Physics
Dirac House, Temple Back
Bristol BS1 6BE UK

ISSN: 1757-8981
ISBN 978-1-7138-3322-2

9 781713 833222

CURRAN ASSOCIATES INC.
proceedings.com